THE
NOISE
HANDBOOK

CONTRIBUTORS

W. A. AINSWORTH

D. R. DAVIES

P. S. EDELMAN

A. J. GENNA

R. GRIME

B. HAY

G. M. JACKSON

D. M. JONES

F. J. LANGDON

H. G. LEVENTHALL

K. A. MULHOLLAND

P. L. PELMEAR

W. TEMPEST

D. WILLIAMS

THE
NOISE
HANDBOOK

Edited by

W. TEMPEST

Department of Electronic and Electrical Engineering
University of Salford
Salford
England

1985

ACADEMIC PRESS

(Harcourt Brace Jovanovich, Publishers)

London   Orlando   San Diego   New York
Toronto   Montreal   Sydney   Tokyo

ACADEMIC PRESS INC. (LONDON) LTD.
24–28 Oval Road
LONDON NW1 7DX

United States Edition published by
ACADEMIC PRESS, INC.
Orlando, Florida 32887

British Library Cataloguing in Publication Data

The Noise handbook.
   1. Noise pollution.
   I. Tempest, W.
   363.7'4   TD892

Library of Congress Cataloging in Publication Data
Main entry under title:

The Noise handbook.

   Includes index.
   1. Noise—Hygienic aspects. 2. Noise—Physiological
effect. 3. Noise control. 4. Noise—Measurement.
5. Noise pollution. I. Tempest, W.
RA772.N7N66 1985   363.7'4   84-24519
ISBN 0–12–685460–2 (alk. paper)

PRINTED IN THE UNITED STATES OF AMERICA

85 86 87 88   9 8 7 6 5 4 3 2 1

Contents

PART I

1   NOISE MEASUREMENT

W. Tempest

PART II

2   NOISE AND HEALTH

P. L. Pelmear

v

3 NOISE AND HEARING
W. Tempest

4 NOISE AND COMMUNICATION
W. A. Ainsworth

5 NOISE AND EFFICIENCY
D. R. Davies and D. M. Jones

6 NOISE ANNOYANCE
F. J. Langdon

PART III

7 NOISE IN INDUSTRY
W. Tempest

8 NOISE ARISING FROM TRANSPORTATION
F. J. Langdon

9 NOISE IN TRANSPORTATION
D. Williams

10 NOISE IN THE HOME
G. M. Jackson and H. G. Leventhall

PART IV

11   NOISE CONTROL
K. A. Mulholland

12   NOISE AND THE LAW IN THE UNITED KINGDOM
R. Grime

13   NOISE AND THE LAW IN THE UNITED STATES
P. S. Edelman and A. J. Genna

14 EEC DIRECTIVES ON NOISE IN THE ENVIRONMENT
B. Hay

List of Contributors

Numbers in parentheses indicate the pages on which the authors' contributions begin.

W. A. AINSWORTH (69), Department of Communication and Neuroscience, University of Keele, Keele, Staffordshire ST5 5BG, England

D. R. DAVIES (87), Applied Psychology Division, University of Aston in Birmingham, Birmingham B4 7ET, England

PAUL S. EDELMAN (337), Kreindler and Kreindler, New York, New York 10017

A. J. GENNA* (337), Kreindler and Kreindler, New York, New York 10017

R. GRIME (303), Faculty of Law, University of Southampton, Southampton SO9 9NH, England

B. HAY (377), Department of Civil Engineering and Building, Coventry (Lanchester) Polytechnic, Coventry CV1 5FB, England

G. M. JACKSON (237), Atkins Research and Development, Woodcote Grove, Epsom, Surrey KT18 5BW, England

D. M. JONES (87), Department of Applied Psychology, University of Wales Institute of Science and Technology, Penylan, Cardiff CF3 7UX, Wales

F. J. LANGDON (143, 195), Hemel Hempstead, Hertfordshire HP1 3QH, England

H. G. LEVENTHALL (237), Atkins Research and Development, Woodcote Grove, Epsom, Surrey KT18 5BW, England

K. A. MULHOLLAND (281), Department of Construction and Environmental Health, University of Aston in Birmingham, Birmingham B4 7ET, England

P. L. PELMEAR (31), Ontario Ministry of Labour, Occupational Health Branch, Toronto, Ontario M7A 1T7, Canada

W. TEMPEST (3, 47, 179), Department of Electronic and Electrical Engineering, University of Salford, Salford M5 4WT, England

D. WILLIAMS (215), Department of Mechanical and Production Engineering, Stockport College of Technology, Stockport SK1 3UQ, England

* Present address: Victora & Genna, New York, New York 10150.

Preface

Noise has been described as sound undesired by the recipient. This definition is certainly valid, but needs to be extended to sound which is harmful or which interferes with normal activities, particularly with communication and efficiency.

The "Noise Handbook" aims to give a current picture of the effects of noise upon man, the incidence of noise in various environments and situations and the protection afforded by the law and by what is technically feasible in the way of noise control.

The book, it is hoped, will be of value to audiologists, architects, town planners, public health and factory inspectors, industrial hygiene staff, environmentalists, lawyers, sociologists, psychologists, communications engineers and machinery and aircraft designers, as well as teachers, students and research workers in these fields.

The book is divided into four parts:

Part I consists of one chapter, Noise Measurement. This covers the techniques of noise measurement which are most frequently encountered in connection with industrial and environmental noise, and goes on to consider noise units and noise evaluation procedures.

Part II comprises five chapters concerned with the effects of noise on the human organism. Chapter 2, Noise and Health, looks at the ways in which noise can cause temporary changes or permanent injury to the body (or mind). In view of its importance, the topic of noise-induced hearing loss is treated separately in Chapter 3, Noise and Hearing. The fourth chapter, Noise and Communication, deals with the effects of background noise on verbal communication. Noise and Efficiency, Chapter 5, examines the difficult area of the effects of noise on the performance of physical and mental tasks, referring mainly to the working environment. The final chapter in this part, Noise Annoyance, considers the problems of scaling and evaluating annoyance arising from noise.

Part III is made up of four chapters dealing with the major sources of noise exposure. Noise in Industry, Chapter 7, covers both the incidence of high noise levels and the use of hearing conservation programmes to minimise their effects. The two succeeding chapters are both concerned with transportation. Chapter 8, Noise Arising from Transportation, looks at the environmental aspects, while Noise in Transportation, Chapter 9, deals with noise levels and their effects on drivers and passengers. The last chapter in this part, Chapter 10, Noise in the Home, provides data on a wide range of noise-producing appliances within the home.

Part IV is concerned with remedies, and begins with Noise Control, Chapter 11, a necessarily brief account of what is technically feasible in this area. The following two chapters deal with Noise and the Law in the United Kingdom and the United States, respectively. Each chapter sets out the current situation in the area of compensation for occupational deafness and the legal regulation of noise in both the factory and the environment. The final chapter deals with the European Economic Community Directives on noise in the environment.

In total this book comprises the work of some fourteen contributors, and the editor would like to take this opportunity to thank them for their contributions, as well as the staff of Academic Press for their assistance in the production of the book and his wife Brenda for her help as assistant editor, proofreader, secretary and typist throughout the period of collecting and finalizing the manuscript.

St. Annes   W. TEMPEST

PART I

1

Noise Measurement

W. TEMPEST

Department of Electronic and Electrical Engineering
University of Salford
Salford
England

I. INTRODUCTION

The measurement and evaluation of noise is a two-stage process; first, the physical parameters of the noise must be measured, and then these physical data must be evaluated in some way to estimate how they will affect human observers.

A sound wave comprises a flow of energy through the air and involves the air in particle displacement, velocity and acceleration, together with corresponding fluctuations in pressure, density and temperature. All these quantities can be related mathematically (Kinsler *et al.,* 1982) and any one of them could be used to evaluate the sound intensity. In practice the great majority of measurements are of sound pressure level (SPL) and when necessary can be converted to sound intensity data or used with other data to calculate the total sound power emitted by a particular source.

Sound pressure levels are usually presented on the decibel scale and are calculated from the formula

$$\text{dB} = 20 \log_{10}(P_1/P_0)$$

where P_1 is the measured sound pressure and P_0 the reference sound pressure. This definition is not dependent on the choice of units, provided, of course, that P_1 and P_0 are defined in the same units. In practice, pressures are usually defined as the rms (root-mean-square) pressure in metric units. The unit in current use is the pascal (Pa), while in many older text books pressures are quoted either in newtons per square metre or in dynes per square centimetre:

$$1 \text{ Pa} = 1 \text{ N/m}^2$$

$$1 \text{ Pa} = 10 \text{ dyn/cm}^2$$

The reference-level P_0 is normally an rms pressure of 20 μPa (0.0002 dyn/cm^2) and corresponds approximately to the hearing threshold of a normal person at a frequency of 1000 Hz.

Sound intensity is defined as the sound energy flow per second across unit area perpendicular to the direction of propagation. Since intensity is proportional to the square of pressure, the decibel definition for intensity is

$$\text{dB} = 10 \log_{10}(I_1/I_0)$$

where I_0 is the reference intensity and I_1 the measured intensity. If the appropriate reference levels are chosen to be equivalent, then the sound pressure and intensity decibel levels will be identical for the same sound.

II. NOISE MEASUREMENTS

Noise measurements are, with very few exceptions, made by electro-acoustic methods in which the noise is converted to an electrical signal by a microphone and subsequently amplified electronically prior to some form of analysis. It is therefore relevant to briefly describe some of the microphones in current use and to comment on their characteristics.

Any device which converts mechanical to electrical energy could, in principle, be used as the basis of a microphone, as could some other devices in which the mechanical (or acoustical) input serves to control a current or voltage from an external source. In practice, most noise measurements are made by using one of three types of microphones: the capacitor, the electret and the piezo-electric.

A. The Capacitor Microphone

The capacitor (or electrostatic) microphone depends for its operation on the variation in capacitance between two electrodes, one fixed and one which moves in response to the incident sound pressure. The moving electrode is usually a thin, taut diaphragm mounted very close to the fixed electrode. Figure 1 shows an example of a high-quality measuring microphone of this type. In operation a sound wave incident on the diaphragm will move it alternately towards and away from the fixed backplate, thus causing a very small variation in the electrical capacitance between them. This does not, of itself, produce an electrical output; the microphone needs to be 'polarized' by means of an external electrical supply voltage fed to the fixed electrode via a high-value resistor. To obtain optimum performance from a capacitor microphone, a pre-amplifier mounted very close to the cartridge, usually within a few centimetres, is required.

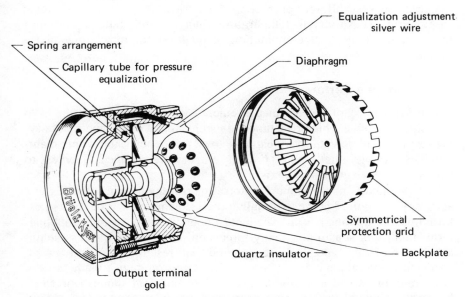

Fig. 1. Internal construction of a high-grade capacitor microphone. (Courtesy of Bruel and Kjaer, Ltd.)

B. The Electret Microphone

The electret microphone is a form of capacitor microphone in which an external polarizing potential is not needed. Instead the necessary polarizing field is provided by an 'electret' dielectric material which has a permanent polarization. Although the term 'electret' was coined by Oliver Heaviside in 1885, the practical development and application to microphone technology has been made only in the last two decades. While the electret microphone dispenses with the external polarization, it still needs a head pre-amplifier mounted close to, or even inside, the cartridge if optimum performance is to be obtained.

C. The Piezo-Electric Microphone

The piezo-electric effect is the property of certain materials to develop spontaneous electrical polarization under the influence of mechanical force. Numerous materials exhibit piezo-electric properties, notably quartz, Rochelle salt (sodium potassium tartrate) and a range of 'ferro-electric ceramic' materials. This last mentioned group are manufactured polycrystalline substances in which piezo-electric properties can be artificially induced by an electrostatic polarizing field. Two such materials are barium titanate and lead zirconate titanate. The various piezo-electric materials possess different properties of electrical output level, resistance to humidity, sensitivity to temperature changes, etc., but the ceramics have provided the best combination of qualities for measuring microphones.

D. Microphone Performance

The ideal microphone for measurement purposes would be cheap to manufacture, have a high resistance to mechanical damage and humidity, possess a completely uniform response to all frequencies and would maintain its sensitivity unaltered by time or temperature changes. Needless to say, such a device does not exist.

The 'standard' microphone for noise measurement when the highest accuracy is required is the externally polarized capacitor type. In its best form this microphone has an excellent frequency response and good long-term stability of sensitivity, many 'precision' sound-level meters use microphones of this type. Despite its good features, the capacitor microphone has several deficiencies; it is a high-precision device made to very close tolerances and is easily damaged by mechanical shock. Due to its high polarizing voltage, it needs very good insulation, and unless special precautions are taken, it cannot be operated under highly humid conditions. It needs a polarizing supply and a head amplifier, and these features, together with its precision construction, mean that it is relatively expensive.

The alternative types of microphone, the piezo-electric and the electret, both have advantages in terms of cost and robustness and are less susceptible to the effects of moisture. However, the piezo-electric types have found it difficult to compete with capacitor microphones for stability against time and temperature changes and uniformity of frequency response. The current situation is that piezo-electric microphones are used in many "type 2" (industrial) sound-level meters, while some manufacturers are now producing electret microphones for "type 1" (precision) meters.

III. SOUND-LEVEL METERS

The sound-level meter is the basic portable instrument used for the measurement of continuous noise. It consists in essence of a microphone to pick up the sound, an electronic amplifier, one or more frequency weighting networks and a meter to display the level. This simple description covers a wide range of instruments and a range of prices of about 30 : 1.

Sound-level meters are produced to meet various national and international standards, which define important aspects of their specifications. The principal standards involved are

International IEC 651   (IEC, 1979)
United States ANSI S 1.4 (ANSI, 1971)
United Kingdom BS 3489   (BSI, 1962)
United Kingdom BS 4197   (BSI, 1967b)

It is not practicable here to discuss all these standards in detail, and it would seem appropriate therefore to concentrate attention on the international standards, referring to the others only briefly.

International IEC 651 sets out requirements for four types of meter:

Type 0: The highest specification instrument, designed as a laboratory reference standard.

Type 1: For laboratory use, and for field use where the acoustical environment can be closely specified and/or controlled. This type corresponds closely to the precision-grade instrument (BS 4197, ANSI S 1.4 Type S1A) and is widely used, particularly in situations where the data obtained may be needed in connection with legal requirements.

Type 2: For general industrial and field applications. Similar to BS 3489 and ANSI S 1.4 Type S2A meter, industrial grade. Industrial-grade meters are adequate for the majority of noise survey work.

Type 3: Lowest grade of meter, intended for field noise survey applications to determine whether an established noise limit has been significantly violated.

The differences between the four types of meter are in the areas of frequency response limits, uniformity of frequency response, stability of performance, change of sensitivity with direction of incident sound, tolerance on level range controls and tolerance on the precision of the weighting networks incorporated.

In practice the type 0 specification has not been widely adopted, and certainly very few meters of this kind have been sold in the United Kingdom.

The type 1 specification is that chosen for a large proportion of the meters in current use. An example of a type 1 instrument, manufactured by GenRad, utilizes a 0.5-in. (12.7-mm) electret microphone and is capable of SPL measurements in the range 30–130 dB. This meter incorporates A, B and C weighting networks, together with 10-octave band filters covering the range 31.5–16 kHz.

A typical industrial sound-level meter (to the type 2 specification), the one illustrated was manufactured by Castle, is shown in Fig. 2. This instrument has a measurement range from 40 to 130 dB(A) and is much simpler, and therefore less expensive, than the precision instruments referred to earlier.

All meters provide a choice of response characteristics, 'fast', 'slow' and in some cases 'impulse' and 'peak'. The former two are simply different meter time constants applied to an rms reading instrument to cope with continuous sounds in the 'fast' position and sounds with short-term variability in the 'slow' position. In the latter, the meter provides a degree of averaging. In the 'impulse' position, the meter has a 'short rise time— long decay time' response, in accordance with the IEC 651 standard, and aims to provide some indication of the size of the impulse; however, this setting is of rather limited utility. In the 'peak' position, the rise time is much shorter (less than 50 μsec for a type 0 and 100 μsec for other types), and the instrument will effectively measure the peak of any transient within the frequency range of the microphone.

The provision of 'impulse' and 'absolute peak' detector characteristics draws attention to the difficulties which arise whenever impulsive, transient or highly variable sound levels need to be measured. In such instances, the sound-level meter may be useless or (worse still) may appear

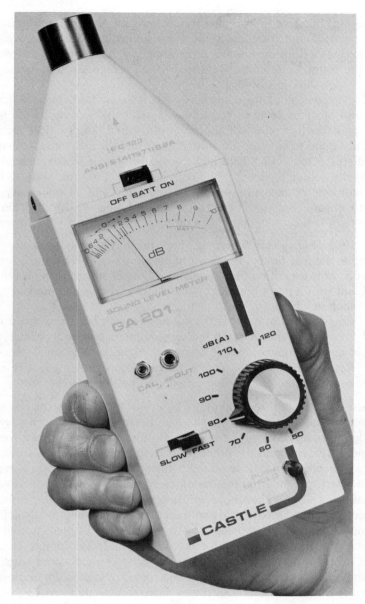

Fig. 2. Industrial-grade (IEC 651 type 2) sound-level meter. [Photograph courtesy of Castle Associates.]

to be useful but give incorrect readings. In general, the measurement of time-varying sounds requires some more sophisticated system, which *either* provides a full time history of the sound *or* determines some parameter of the sound (e.g., some measure of the average level) according to a precisely specified procedure.

In this discussion of noise measurement, it has so far been assumed that the sound-pressure level to be measured will be picked up by the microphone and that its value will appear on the meter. In practice, the situation is not quite so simple, since the presence of the measuring instrument itself disturbs the sound field and (particularly at high frequencies) can alter its magnitude by several decibels. It is therefore necessary to define the conditions of measurement more precisely and, if necessary, take steps to ensure that the correct quantity is being measured. There are three situations, each demanding a different definition of sound pressure.

(1) *Free field.* The 'free-field' pressure is the sound-pressure level existing in the free field at the point of measurement with the microphone removed. This condition implies plane waves of sound with no reflected sound and would pertain in the open air away from reflecting surfaces or in an anechoic chamber.

(2) *Pressure.* In this case the sound-pressure level is defined as the sound pressure at the microphone diaphragm, a condition which would apply to measurements made in a small volume, such as an artificial ear.

(3) *Random incidence.* This is the condition applying in a reverberant room with equal energy flow of sound in all directions.

To cope with these three situations different microphone cartridges and attachments can be used in order to avoid errors. There are two types of cartridge normally available for precision meters, one designed for free-field measurements and the other to be used for pressure measurements. For measurements in diffuse sound fields a 'random incidence corrector' can be fitted to a microphone to make it equally responsive to sounds from all directions. The choice of the size of cartridge depends on a number of factors. Most precision meters use a free-field (half-inch) unit as standard, and this is satisfactory for the majority of measurements. In some cases a larger microphone can be used, and its additional sensitivity will be useful when very low noise levels are to be measured, such as in assessing the background noise of a test room to be used for audiometry. However, larger-diameter cartridges have a less uniform response to sounds with varying angles of incidence, and this fact argues against their use unless other factors are more important. Very small cartridges (6.3 and 3.2 mm or 0.25 and 0.125 in., respectively) are used to measure extremely high noise levels or to detect airborne ultrasound, but they tend to have a high 'noise floor' and are not suitable for general use.

For indoor use either a hand-held sound-level meter or one mounted on a tripod without any additional equipment can be utilized. For outdoor use two problems arise, wind noise and damage to the instrument by rain or high humidity. Within certain limits wind noise can be reduced by fitting the microphone with a cap of polyurethane foam. This cap is transparent to sound but greatly reduces the generation of noise by the wind. For long-term all-weather use out-of-doors, microphones are available which can successfully resist damp conditions. One widely used type mounts the microphone on a dehumidifier, which ensures that the cartridge and pre-amplifier always operate under dry conditions.

IV. THE USE OF THE SOUND-LEVEL METER

The choice of the correct microphone cartridge and, if necessary, random incidence corrector has already been mentioned. Another requirement is to avoid reflections which can influence the readings; for instance, a position close to a wall will give an increased level while a point in a corner will show a further increase. Various standards set out procedures to avoid reflections; for instance, the U.K. standard procedure for traffic noise measurements (Anonymous, 1973a) specifies a microphone position 1 m in front of the most exposed window in the facade of the building.

When measuring the noise exposure of an individual in, for instance, the working environment, one must keep the microphone as close as possible to the subject's ear. In this case shadowing effects and reflections from fixed objects are not important, since it is the noise level at the ear, under actual work conditions, that needs to be measured.

The normal calibration of a sound-level meter is made with the microphone pointing towards the sound source, and the meter should, as far as possible, be used in this way. The presence of the meter user's own body may have some slight effect on readings, due to reflected sound, and this should be minimized. Obvious methods are the use of a meter tripod, the use of the meter at arm's length or the use of an extension piece between the meter and the microphone.

V. SOUND-LEVEL METER CALIBRATION

Sound-level meters are, by the standards of other electronic instruments, rather fragile and easily damaged. They need frequent calibration if the data they record are to be trusted.

Calibration is normally made by means of some stable noise source which provides a known sound-pressure level at the microphone. Three types are in common use; the pistonphone, the piezo-electric type and the falling ball calibrator. The pistonphone generates a known sound level by means of two motor-driven pistons working into a small volume tightly coupled to the microphone of the meter under test. Pistonphones usually operate at about 250 Hz and can calibrate to within ±0.2 dB. By comparison, the piezo-electric type is considerably cheaper and slightly less accurate. It consists of a battery-driven oscillator, which drives a piezo-electric transducer using a circuit designed to provide a maximum stability of output.

In general, calibrators are designed to fit one microphone type only and cannot be used with meters other than the one for which they were supplied. In normal use all sound-level meters should be calibrated before and after they are used on each occasion when measurements are made.

VI. THE MEASUREMENT OF TIME-VARYING NOISES

The sound-level meter is basically designed to measure continuous noise, which has no significant fluctuations. It can also cope with some types of variable level noise where the variation is rapid enough to result in a steady reading of the meter needle on the slow setting. There are two situations in which the sound-level meter is not satisfactory; these are the case of the noise where the variation causes the meter needle to swing significantly, and the noises where the peak levels are high enough to cause reading errors in the instrument. A number of special instruments are available to deal with these measurement problems. One of these is the L_{eq} meter, which computes the equivalent continuous sound-pressure level according to the equal energy principle:

$$L_{eq} = 10 \log_{10} \frac{1}{T} \int_0^T \frac{[P_a(t)]^2 \, dt}{P_0^2}$$

where L_{eq} is the equivalent continuous sound level, T the total period of time, $P_a(t)$ the instantaneous A-weighted sound pressure in μPa, P_0 the reference pressure (20 μPa) and t the time. This unit is widely used in the United Kingdom in evaluating the degree of hearing hazard produced by sounds of varying intensity (Department of Employment, 1972). The definition of L_{eq} simply defines the average noise level over the period of measurement. Two further definitions are sometimes used which specify the duration. The sound-exposure level (SEL or L_{AX} or L_{AE}) is the L_{eq} value referred to 1 sec. It is useful for comparing individual noise events of different durations, such as the pass-by noise of individual lorries at different speeds. The SEL is the level which, in 1 sec, will provide the same noise energy as the individual event measured.

The 'noise dose' is an L_{eq} measurement over a period of 8 hr and is used in the assessment of hazard to hearing. Often the noise dose is expressed as a percentage of maximum allowable daily exposure [usually 85 or 90 dB(A) for 8 hr]. Thus, in current U.K. practice, 90 dB(A) for 8 hr is described as a 100% dose, and 93 dB(A) L_{eq} measured over 8 hr would be 200%, since a 3-dB increase in level corresponds to a doubling of noise energy.

An example of an instrument to measure L_{eq} is the Bruel and Kjaer type 2230 illustrated in Fig. 3. This is a precision-grade meter (to IEC 651, type 1) and is normally fitted with a 0.5-in. diameter electret microphone. As a direct-reading sound-level meter it measures levels from 24 dB(A) to 130 dB(A) with the facilities of A-weighted, C-weighted or linear response, and the meter characteristics 'fast', 'slow', 'impulse' and 'peak'.

As an integrating meter it can determine L_{eq} over a measuring period up to 8 hr; the actual level is shown on a liquid-crystal display. The meter shown in the figure is fitted with an octave and one-third octave filter set for bandpass measurements.

The L_{eq} meter is a relatively large and heavy device, and it is expensive. The noise dosemeter is a much smaller and simpler device, designed to be worn by a worker for the full duration of a working day. It integrates the incident sound energy, and at the end of the day, the total can be read out on some form of display. The principle of operation is similar to that of the integrating meter.

While dosemeters used in the United Kingdom are calibrated to calculate the noise dose on the L_{eq} principle (that a 3-dB increase in level corresponds to a doubling of dose), instruments for use in the United States must be set up differently so as to equate a 5-dB increase in level to a doubling of dose.

Owing to their complexity, all types of integrating and averaging meters require frequent and careful calibration using the appropriate calibrator. The true integrating meter can be calibrated with a pistonphone or a similar device, making due allowance for the effect of A weighting on the reading [for example, a standard 250-Hz, 124-dB pistonphone should give 115.4 dB(A)]. The dosemeter needs a special calibrator that gives a tone defined in both intensity and time. This is usually achieved by providing a high level for a short period, for example, 117 dB(A) for 56.25 s is equivalent to 90 dB(A) for 8 hr.

For the evaluation of noise nuisance due, for example, to noise emission from industrial premises, it is often necessary to obtain measurements of noise parameters over long periods, in some cases as long as a

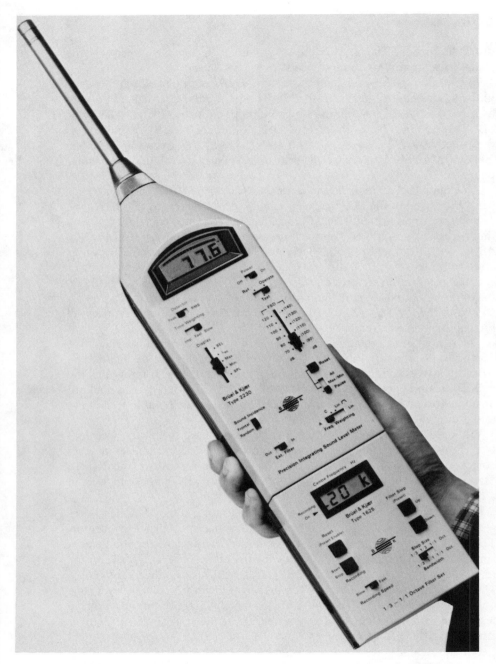

Fig. 3. Precision-grade (IEC 651 type 1) integrating sound-level meter fitted with one-third-octave and one-octave filter set. (Photograph courtesy of Bruel and Kjaer, Ltd.)

week. In addition to L_{eq}, already mentioned, other measures such as L_{10}, L_{50} and L_{90} are in use, with L_{10}, L_{50} and L_{90} defined as the decibels(A) level of the sound exceeded for 10, 50 and 90% of the time, respectively. The procedures laid down for these measurements often allow for computing at intervals as an alternative to continuous measurements. [See, for exam-

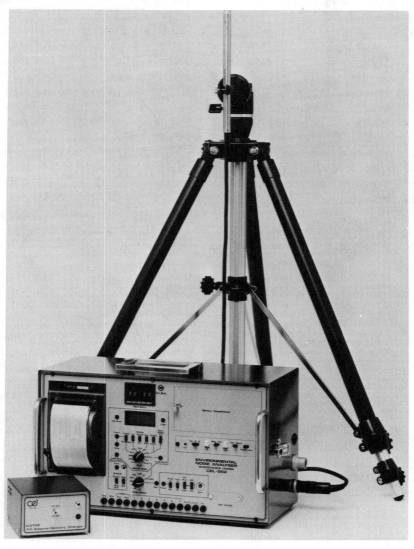

Fig. 4. Environmental noise analyser. [Photograph courtesy of CEL Instruments.]

ple, "Building and Buildings, The Noise Insulation Regulations" (Anonymous, 1973a).]

Even when sample periods of 15 min or less are permitted, the work involved in these procedures is considerable, and this has led to the development of various devices to simplify the task. An example of an environmental noise analyzer is shown in Fig. 4 (CEL Instruments). The instrument illustrated provides a wide range of facilities, which include the measurement and recording (by printout on a chart) of L_{eq} and up to seven programmed levels of L_N (e.g., L_1, L_{10}, L_{50}, etc.) The noise analyser comprises part of a measurement system which can be set up to operate unattended for up to a week powered by batteries or longer with an external power supply. It is microprocessor controlled and comprises a very powerful and flexible noise-measuring system.

An alternative approach to L_{eq} and L_N measurements is provided by instruments which can record sound levels in digital form on a standard cassette. By this means sample readings of noise levels can be obtained over a period of up to 7 days on a single cassette. The cassette can then be replayed into a data processor programmed to provide the required statistical analysis. In this way L_{eq}, L_{10}, etc., can be obtained by choice of the correct programme. This approach clearly has the advantage that it needs only changes in the data processor to take into account any new noise units which may come into use.

VII. THE MEASUREMENT OF INFRASOUND

Infrasound can be conveniently defined as the range of frequencies below 20 Hz. The lower limit is less easy to establish, since some studies of infrasound in the environment have been concerned with naturally arising noise energy down to 0.001 Hz. However, for purposes in which the measurement is concerned with the effect of sound on human observers, a lower limit of about 1 Hz is usually adequate.

This frequency range involves special problems at all points in the measuring chain—microphone, pre-amplifier, recording equipment and indicator. Various types of microphones have been used for measuring infrasound, but for many practical applications the capacitor type is the most suitable. The low-frequency response of a capacitor microphone is limited by the 'leakage' or air vent at the rear of the diaphragm. Such a leak is necessary, or changes in temperature and atmospheric pressure would affect the sensitivity unduly; but its presence means that at sufficiently low frequencies, it allows the sound pressure to reach both sides of the diaphragm simultaneously, and thus the microphone output is greatly reduced. The frequency at which the sensitivity falls off is, for general-purpose microphones, normally in the range 1–20 Hz, while special low-frequency units are available with a limit of 0.1 Hz or lower.

The pre-amplifier of a sound-level meter needs to have a flat response to frequencies down to 20 Hz. It may or may not extend usefully beyond this. Some sound-level meters have a response which extends below 2 Hz, but some recent models have deliberately limited the range to 10 Hz on the (quite reasonable) grounds that sensitivity at very low frequencies may sometimes give erroneous readings. It must therefore be emphasized that if any normal sound-level meter is to be used in the infrasonic range, it is necessary to first check its calibration in this range—a procedure which needs specialized laboratory facilities.

The most satisfactory way to measure infrasound is to use a special microphone and pre-amplifier designed for this purpose. This consists of a low-frequency capacitor microphone, used with a radio-frequency polarisation pre-amplifier, and such a system has a response to 0.1 Hz or lower if necessary.

When it is necessary to record infrasound, normal tape recorders are not suitable, and a frequency-modulation recorder is needed. Several recorders of this type (both laboratory and portable) are on the market, and some aspects of the design and use of such systems are discussed in the literature (Tempest and Bryan, 1967, 1972).

Due to its characteristics, it is often necessary to record infrasound for subsequent analysis in the laboratory, because there is no infrasonic equivalent of decibels(A) to give a single-figure evaluation of its level. Such analysis is frequently performed by feeding the recorded signal through selective filters (e.g., octave or one-third octave) and monitoring the output on a meter with a suitably slow response. Another approach to the analysis is to replay the tape recording at a higher speed, to complete the analysis by using some form of audiofrequency analyser and to refer the results back to the original frequencies.

VIII. IMPULSIVE AND IMPACT NOISE

An impulsive or impact noise is one in which the sound pressure rises rapidly to some maximum value and then decays, to be followed by a period of (relative) quiet. To define an impulsive sound, three parameters are needed. These are the peak amplitude A, the rise time t_r and the decay time t_d.

For a complete knowledge of the impulsive sound it is also necessary to know the shape of the waveform involved. Where the requirement is to determine the average sound level, an integrating or noise average meter will often be able to provide this information in the form of an L_{eq} value. This is normally sufficient for the assessment of hearing hazard. In certain situations it is necessary to determine the peak sound-pressure level, and many of the better sound-level meters will determine this parameter (sometimes called 'peak hold' or 'absolute peak'). However, peak measurements must be treated with caution, since some impulsive sc rces (for example, small-arms fire) show very short rise times of less than 50 μsec, and in these cases sound-level meters may not accurately determine the peak levels. When a full analysis of a very fast rising impulsive sound is to be made, it is usually necessary to use a microphone feeding an oscilloscope and to photograph the resulting trace or to record it on a digital event recorder, which can then be used to plot out the waveform on a chart recorder. Often the situation of a short rise time occurs with a very high peak pressure and necessitates the use of a small capacitor microphone or a special piezo-electric microphone. Such microphones have both a very short response time to high-speed transients and the ability to respond without distortion to high sound-pressure levels.

IX. NOISE AND LOUDNESS EVALUATION

A. Weighting Networks

Noise measurements are made because it is ultimately desirable to relate the noise to its effect on humans. It is therefore necessary to consider the methods by which noise levels can be evaluated in terms of their subjective effects. There are only a few cases, such as the effects of sonic booms on buildings or the fatiguing action of very intense noise on aircraft structures, in which the engineer might be concerned about the effect of noise on structures. These are not within the scope of this chapter.

In the 1920s, when the first sound-level meters were used, it became apparent that the simple decibel sound-pressure-level measurement had two deficiencies: it did not correctly relate the levels of sounds of different frequencies, and it did not yield a figure which related simply to the subjective magnitude (loudness) of the sound.

The failure of the sound-pressure-level decibel to cope with different frequencies arises from the shape of the threshold of hearing curve and the equal loudness contours. These are illustrated in Fig. 5, where the 0 (phon) curve shows the threshold of hearing as a function of frequency. The weakness of the decibel SPL is illustrated when one considers the audibility of a tone of 20 dB at a frequency of 2000 Hz (point A on Fig. 5). This will be some 20 dB above threshold, and clearly audible. At 40 Hz

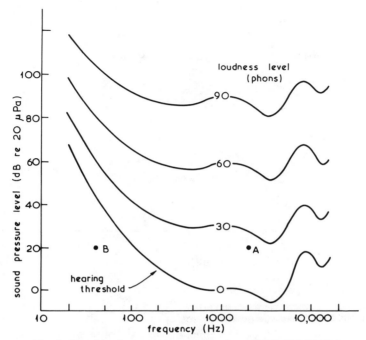

Fig. 5. Equal loudness contours. [Data from ISO R 226 (1961).]

(point B) it will be more than 20 dB below threshold and, consequently, inaudible.

The phon curves (labelled 0–90 on Fig. 5) are contours of tones of equal loudness, which link levels at different frequencies which are subjectively judged to be equally loud. Since the equal loudness contours rise steeply at low frequencies and less steeply at high frequencies, it would appear reasonable to build into the sound-level meter the same characteristic. This is the aim of the A weighting network (Fig. 6) which in shape is a simplified version of the 30-phon contour. The A weighting was an early attempt to produce a meter which would respond to noise signals with readings corresponding closely to subjective human response. In fact, despite its simple beginnings, the A weighting has been remarkably successful in achieving its intended purpose; it has not been superseded by any of the alternatives put forward since its inception. It should be noted that the aim of the A weighting is to allow the sound-level meter to be used to compare different noises; thus, to a good approximation, equal decibels(A) levels mean that two noises are subjectively about equally loud, and a greater decibels(A) level means a louder noise. Although the

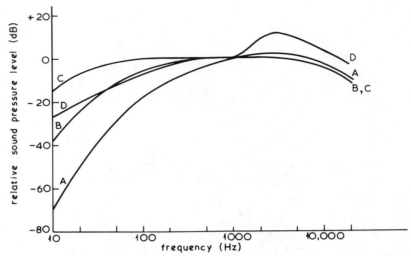

Fig. 6. A, B, C and D weighting networks. [Data from IEC 537 (1976) and IEC 651 (1979).]

A weighting allows different noises to be correctly rank ordered, the actual numerical decibels(A) values are not simply related to the subjective sensation of loudness. The relationship is a logarithmic one, an increase of 10 dB(A) corresponding roughly to a doubling of loudness.

At the time of its introduction, the A weighting was thought of as a scale to be used for fairly low-level noises in the 20- to 55-dB SPL range (Beranek, 1949). For higher levels, two other weightings were put forward: B, based on the 70-phon contour for levels from 50- to 85-dB SPL, and C for higher levels. In practice, B and C have been found (Scharf *et al.*, 1977) to correlate less well with subjective response than does A and are rarely used nowadays.

Other weightings are known as D, D1, D2, N and E. The most significant of these is the D weighting, formerly referred to as N or D1. This has been developed specifically for the measurement of aircraft noise and, as Fig. 6 indicates, has a maximum in response in the 3–4-kHz frequency range. The D weighting has been officially standardised (IEC, 1976).

The D2 (Kryter, 1970) and E (Stevens, 1972) weightings are still somewhat tentative; for further discussion and some data on their validity the reader is referred to Scharf *et al.* (1977).

The various weightings mentioned here simply modify the frequency response of the meter to something more like that of the ear and then rely on the meter (which normally has a root-mean-square response) to combine the various frequencies present into a single total decibel figure. This approach neglects some fundamental features of the human auditory system. Attempts to incorporate these features have led to the development of a number of procedures for 'loudness evaluation'.

B. Loudness Evaluation Procedures

Loudness evaluation procedures aim to permit physical measurements (normally sound-pressure levels measured in octave or one-third-octave bands) to be used to calculate the subjective loudness of a sound or in some cases a measure of its noisiness. Historically, this field owes a great deal to the work of AT&T Bell Telephone Laboratories in the United States, where Wegel (1922) determined the whole range of audibility from 20 Hz to 20 kHz and from the threshold of hearing to a level of 'maximum audibility'.

Once the range of hearing was established, the next stage was to describe sounds in terms of 'sensation units', the amount in decibels by which their sound-pressure levels exceeded the threshold at that frequency. The 'sensation unit' was put forward as a measure of the loudness of a sound, but it was soon realised (Fletcher and Munson, 1933) that this approach was too simple and that to produce an adequate procedure it would be necessary (a) to find a satisfactory way to combine components of different frequencies and (b) to relate the total to the magnitude of the subjective sensation.

The paper by Fletcher and Munson marked the beginning of a long period of study which in the post-war period was dominated by two workers, Stevens (1956, 1961, 1972) and Zwicker (1960, 1966). Both workers produced schemes for loudness evaluation which were basically similar but differed in detail. Stevens's procedures (Mark I to Mark VII) involved dividing the frequency spectrum into bands and measuring the SPL in each band. These pressure levels were then converted to sones, where the sone is a unit of loudness in which the numerical value is directly proportional to the magnitude of the subjective sensation. The contributions of the frequency bands are then added together according to a relatively simple formula which attaches a special weighting to the loudest band, and the resulting total gives a value to the overall loudness. Although the scheme was conceived in terms of sones, results are usually expressed in a logarithmic unit PLdB (perceived level decibel).

It is not feasible because of space limitations to discuss the various versions of the Stevens loudness evaluation procedures in detail here; however, it is perhaps useful to note some of the more important features and modifications. The first is the question of whether the loudness which these procedures aim to evaluate is in fact the same attribute of a sound as the 'noisiness' which can cause annoyance. Stevens (1972) himself has stated that 'the evidence from experiments in several laboratories suggests that loudness and noisiness may be considered essentially synonymous'. Many workers in the field would disagree with this statement, and there is a growing body of evidence that, in some circumstances, those attributes of a sound which might be called noisiness and therefore give rise to annoyance are not identical to loudness. A major development in this field is the use of perceived noise level in the evaluation of aircraft noise. The instantaneous perceived noise level (Federal Aviation Administration, 1976) is calculated from one-third-octave-band pressure levels in a manner which, in principle, closely follows Stevens's method. However, the detailed calculation uses noys instead of sones. The relationship between noys and one-third-octave-band sound-pressure level differs from that for sones in placing a greater emphasis on the frequencies above 1 kHz. A further feature of the perceived noise level calculation is the introduction of a correction for spectral irregularities (for example, discrete frequency components or tones).

A quite different example of the divergence between loudness and noisiness is in the evaluation of low-frequency noise annoyance (Tempest, 1976), for which it has been shown that conventional procedures, including PLdB and dB(A), underestimate the annoyance value of noises with spectra that fall rapidly with increasing frequency. It appears that sounds of this type, which can usually be described as having a rumbling sound, provoke adverse reactions in excess of those expected from their loudnesses.

Zwicker's (ISO, 1966) scheme is basically similar to Stevens's but considerably more complex in detail, since it takes into account the asymmetry of the masking pattern, whereby an intense band of noise inhibits the contribution to loudness of bands higher in frequency than itself to a much greater degree than of bands lower in the frequency scale. The Zwicker calculation can be performed graphically and involves the measurement of the area under a curve on which band pressures are plotted, with allowance for asymmetrical masking, on a non-linear frequency scale. Zwicker's method is somewhat less empirical than Stevens's and yields results which tend to be about 5 dB higher. However, in terms of success in correctly rank ordering different sounds, both procedures are about equally successful. The penultimate (Mark VI) version of Stevens's scheme (Stevens, 1961) has been modified by Robinson (1964), who put forward a modified procedure for summing the separate bands of noise. This modification effectively takes into account the asymmetry of masking and results in a scheme which is much closer to Zwicker's in effect, but is still simpler to apply.

The final Mark VII version of Stevens's scheme (Stevens, 1972) extends the procedure down to 1 Hz. This feature has been the subject of some criticism by Yeowart [see Tempest (1976)], who argues that it is not valid in the infrasonic range.

The situation can be summarised briefly by saying that Stevens's Mark VI, with the Robinson corrections, and Zwicker's procedure have both been proved valuable tools in loudness evaluation. Due, however, to their complexity, they are certainly not widely used in the United Kingdom in dealing with practical problems.

A different approach to the evaluation of noise levels, particularly in the internal environment of buildings, is the use of noise criterion (NC) and noise rating (NR) curves. The basis of these criteria is a family of curves of octave-band sound-pressure level curves versus frequency. These curves have a general resemblance to equal loudness contours, being concave upwards and rising at the low-frequency end more steeply at the lowest frequencies.

To use these curves the octave-band spectrum of the noise is overlaid on the family of curves and the value assigned to the spectrum corresponds to the highest curve which the spectrum touches. In practice, interpolation is usually needed to evaluate a noise spectrum and in most cases a single frequency level determines the rating.

The NC curves are clearly attributable to the work of Beranek *et al.* (1953), where they were introduced in a paper concerned with noise for ventilation systems. A slightly modified version was published by Schultz (1968), and this set of curves, known as "Beranek's noise criterion", has received considerable acceptance in such areas as office design, principally in the United States.

The particular value of the use of NC curves is that they show immediately which frequencies are responsible for the level of the noise and thus give an indication as to the acoustic measures required to reduce noise. However, Schultz (1972) has pointed out that if an environment is carefully designed to just fit a particular required NC curve, it may be rather unsatisfactory, since the low frequencies will sound 'rumbly' and the high frequencies 'hissy'.

X. THE EVALUATION OF TIME-VARYING NOISE

The procedures and weighting networks discussed so far are concerned with the evaluation of steady noise levels, or levels at which the variation can be averaged by the 'slow' setting of a sound-level meter. In order to deal with intermittent and variable level noises, a further range of procedures has been developed, and it must be admitted that at the current state of the art no dominant procedure or measurement technique has evolved.

One of the simplest measures of a variable noise is the equivalent continuous sound-pressure level L_{eq}. Defined thus, L_{eq} represents the average energy of the sound (usually A-weighted). It has gained wide acceptance in the United Kingdom and has become the accepted unit for the evaluation of noise as a hazard to hearing. Unless the sound has a very simple pattern of variation with time, L_{eq} must be measured by some form of averaging meter or derived from a statistical analysis of noise levels. As mentioned earlier in this chapter, L_{eq} has not been accepted in the United States as a measure of hearing hazard, although the U.S. method of evaluation, which regards a 5-dB increase in level as equal to a doubling of exposure time, can be built into an integrating dosemeter.

The equivalent continuous noise level L_{eq} has gained considerable acceptance as a measure of noise for determining noise's environmental effects. The International Standard Organisation (ISO, 1982) suggests its use as a means of evaluating the equivalent level of some intermittent and irregular noises.

Other units are very numerous and can only be discussed briefly. For the assessment of road traffic noise the TNI (traffic noise index) (Langdon and Scholes, 1968) was developed as

$$TNI = L_{90} + 4(L_{10} - L_{90}) - 30$$

where L_{10} and L_{90} are the decibels(A) levels exceeded for 10 and 90% of the time, respectively. The first term represents the background level, while the second takes into account the variability of the noise.

The traffic noise index was developed in the period 1968 to 1970 and represents a step forward from simpler measures in that it took considerable account of the fact that the variability of a noise was an important factor in its annoyance.

In the United Kingdom the 18-hr average of L_{10} in decibels(A), known as 18-hr L_{10} dB(A), has been adopted as a measure of road traffic noise.

In order to obtain 18-hr L_{10} dB(A) it is necessary to measure L_{10} dB(A) each hour from 6 A.M. to midnight, and the resulting 18 levels are then arithmetically averaged; see "Calculation of Road Traffic Noise" (Anonymous, 1973b). More recent work (Rossall, 1978) has demonstrated that, in fact, quite good estimates of 18-hr L_{10} can be obtained from samples as short as 15 min measured during the working day.

In the field of aircraft noise evaluation for assessment of community reaction, many units have been proposed, as might be expected from the scale of the problem and the large-scale economics involved. One of the earlier units was the noise and number index (NNI), defined in the Wilson report (Wilson, 1963):

$$\text{NNI} = (\text{average peak noise level}) + 15 \log_{10} N - 80$$

where N is the number of aircraft heard in one period of time (one day or one night) and the average peak noise level is the average of the perceived noise level (PNL) values calculated according to Kryter's procedure (Kryter, 1959). Perceived noise level is calculated for the highest sound-pressure level attained in each octave as the plane flies over and is not corrected for duration or for the presence of pure tones.

Perceived noise level has already been mentioned and has been used particularly to evaluate aircraft noise. In the NNI, the duration of the noise is not considered; however, the U.S. unit effective perceived noise level (EPNL) takes a different view (Bishop and Horonjeff, 1967). Here EPNL is defined by the equation

$$\text{EPNL} = \text{PNL} + D + F$$

where PNL is maximum calculated perceived noise level at an instant during the flyover; $D = 10 \log_{10}(t/15)$, where t is the time in seconds for which the noise level is within 10 dB of the maximum PNL; and F is a correction for the presence of discrete frequency components. The purpose of EPNL is to attach a numerical value to the noise produced at a point on the ground by one overflight.

Derived from EPNL is the noise exposure forecast (NEF), which is a means of summarising the EPNL values for many overflights into a single figure. It is essentially an energy summation, but it involves different weightings for day- and night-time flights. Values of NEF can be projected onto a map in the form of contours. These contours can then be related to the possibility of complaints.

The diversity of units mentioned so far includes only a fraction of those which have been put forward at various times. A much more detailed account is to be found in Schultz (1972).

One self-evident fact is that the approaches to different noise sources have been quite independent and largely unrelated. This fact led Robinson (1969) to look at the possibility of designing a unified system of noise assessment. He considered NNI, TNI and other data and put forward the basic definition of a noise pollution level L_{np} defined by

$$L_{np} = L_{eq} + k\sigma$$

where L_{np} is the noise pollution level, σ the standard deviation of the instantaneous level, k a constant, possibly assigned the value 2.56, and L_{eq} the 'energy mean' of the noise level over a specified period. It should be noted that L_{eq} is here the 'mean energy' of the noise level as measured on an appropriate scale, such as decibels(A), decibels(D) or perceived noise decibel.

In fact, such high correlations exist between these units (Parkin, 1964; Botsford, 1969) that the difference between them is not likely to influence the validity of the L_{np} concept one way or the other.

Robinson puts forward detailed arguments to demonstrate that as a unit, L_{np} is at least comparable with such concepts as TNI and NNI.

A somewhat different approach to noise evaluation is to be found in the British Standard, BS 4142 (BSI, 1967a), which is concerned with the assessment of environmental noise and the possibility of its giving rise to annoyance. The rating procedures comprise three separate tasks: the measurement of the intrusive noise, the measurement or estimation of the background noise and a comparison of the intensity of the noise with the background.

The measurement of noise is made in decibels(A) with the meter in the slow setting, and if the level is steady or fluctuates less than 10 dB, it is averaged by eye. The standard provides for corrections to take into account impulsive or tonal noises and, in the case of intermittent noises, has different correction procedures for day-time and night-time. The corrected noise level (or noise levels in some cases, since a steady level with peaks superimposed is treated as two separate noises for evaluation purposes) is compared with the measured ambient noise of the area in the absence of the intrusive noise.

In cases in which the ambient cannot be obtained by measurement, a procedure is set up by which it can be estimated from a 'basic criterion' plus corrections. The comparison of corrected noise level and ambient is used to estimate the possibility of complaints on the basis that an excess of 5 dB over ambient is of marginal significance, but an excess of 10 dB or more will lead to complaints.

REFERENCES

Anonymous (1973a). Statutory Instrument 1363/1973, "Building and Buildings, the Noise Insulation Regulations." Her Majesty's Stationery Office, London.
Anonymous (1973b). "Calculation of Road Traffic Noise." Her Majesty's Stationery Office, London.
ANSI (1971). ANSI 1.4, "American National Standard Specification for Sound Level Meters." American National Standards Institute, Boulder, Colorado.
Beranek, L. L. (1949). "Acoustic Measurements." Wiley, New York.
Beranek, L. L., Reynolds, J. L., and Wilson, K. E. (1953). Apparatus and procedures for predicting ventilation system noise, *J. Acoust. Soc. Am.* **25,** 313–321.
Bishop, D. E., and Horonjeff, R. D. (1967). "Procedures for Developing Noise Exposure Forecast Areas for Aircraft Flight Operations." Bolt, Beranek and Newman Inc. Washington, D.C., Report No. 1151; FAA DS-67-10.

Botsford, J. H. (1969). Using sound levels to gauge human response to noise, *J. Sound Vib.* **3**, 16–28.

BSI (1962). BS 3489, "Specification for Sound Level Meters—Industrial Grade." British Standards Institution, London.

BSI (1967a). BS 4142 (with amendments 1661, 1975; 2956, 1980; and 4036, 1982), "Method of Rating Industrial Noise Affecting Mixed Residential and Industrial Areas." British Standards Institution, London.

BSI (1967b). BS 4197, "Specification for a 'Precision' Sound Level Meter." British Standards Institution, London.

Department of Employment (1972). "Code of Practice for Reducing the Exposure of Employed Persons to Noise." Her Majesty's Stationery Office, London.

FAA (1976). "Code of Federal Regulations," Vol. 14, Aeronautics and Space, Parts 1–59; Part 36, Appendix B. Federal Aviation Administration, Washington, D.C.

Fletcher, H., and Munson, W. A. (1933). Loudness, its definition, measurement and calculation, *J. Acoust. Soc. Am.* **5**, 82–108.

IEC (1976). Publication 537, "Frequency Weighting for the Measurement of Aircraft Noise ('D'-Weighting)." International Electrotechnical Commission, Geneva, Switzerland.

IEC (1979). Publication 651, "Sound Level Meters." International Electrotechnical Commission, Geneva, Switzerland.

ISO (1961). ISO R 226. "Normal Equal-Loudness Contours for Pure Tones and Normal Threshold of Hearing under Free Field Listening Conditions." International Organisation for Standardization, Geneva, Switzerland.

ISO (1966). ISO R 532, "Method of Calculating Loudness Level." International Organisation for Standardization, Geneva, Switzerland.

ISO (1982). ISO 1996/1, "Acoustics—Description and Measurement of Environmental Noise, Part 1: Basic Quantities and Procedures." International Organisation for Standardization, Geneva, Switzerland.

Kinsler, L. E., Frey, A. R., Coppens, A. B., and Sanders, J. V. (1982). "Fundamentals of Acoustics." Wiley, New York.

Kryter, K. D. (1959). Scaling human reactions to the sound from aircraft, *J. Acoust. Soc. Am.* **31**, 1415–1429.

Kryter, K. D. (1970). "The Effects of Noise on Man." Academic Press, New York.

Langdon, F. J., and Scholes, W. E. (1968). The traffic noise index: A method of controlling noise nuisance, *Arch. Journ.* **147**, 813–820.

Parkin, P. H. (1964). On the accuracy of simple weighting networks for loudness estimates of some urban noises, *J. Sound Vib.* **2**, 86–88.

Robinson, D. W. (1964). A note on the subjective evaluation of noise, *J. Sound Vib.* **1**, 468–473.

Robinson, D. W. (1969). "The Concept of Noise Pollution Level." NPL Aero Report Ac 38, National Physical Laboratory, Teddington, London.

Rossall, A. W. (1978). The Measurement and Analysis of Road Traffic Noise. M.Sc. thesis, Salford University, Salford, England.

Scharf, B., Hellman, R. P., and Bauer, J. (1977). "Comparisons of Various Methods for Predicting the Loudness and Acceptability of Noise." EPA Report 550/9-77-101, Washington, D.C.

Schultz, T. J. (1968). Noise-criterion curves for use with the U.S.A. S.I. preferred frequencies, *J. Acoust. Soc. Am.* **43**, 637–638.

Schultz, T. J. (1972). "Community Noise Ratings." Applied Science Publ., London.

Stevens, S. S. (1956). Calculation of the loudness of complex noise, *J. Acoust. Soc. Am.* **28**, 807–832.

Stevens, S. S. (1961). Procedure for calculating loudness: Mark VI, *J. Acoust. Soc. Am.* **33**, 1577–1585.

Stevens, S. S. (1972). Perceived noise level by Mark VII and decibels(E), *J. Acoust. Soc. Am.* **51**, 575–601.

Tempest, W. (ed.). (1976). "Infrasound and Low Frequency Vibration." Academic Press, London.

Tempest, W., and Bryan, M. E. (1967). A simple frequency modulation tape recorder system, *Electron. Eng.* **39**, 87–89.

Tempest, W., and Bryan, M. E. (1972). Low frequency sound measurement in vehicles, *Appl. Acoust.* **5**, 133–139.

Wegel, R. L. (1922). The physical characteristics of audition, *Bell Systems Tech. J.* **1**, 56–68.

Wilson, A. (1963). "Noise—Final Report of the Committee on the Problem of Noise." Command 2056. Her Majesty's Stationery Office, London.

Zwicker, E. (1960). Ein Verfahren zur Berechnungder Lautstarke, *Acustica* **10**, 304–308.

Zwicker, E. (1966). Lautstarkeberechnungsverfahren im Vergleich, *Acustica* **17**, 278–284.

PART II

2

Noise and Health

P. L. PELMEAR

Ontario Ministry of Labour
Occupational Health Branch
Toronto, Ontario
Canada

I. INTRODUCTION

The World Health Organization defines health as "the state of complete physical, mental and social well-being, and not merely an absence of disease and infirmity". Noise, or "unwanted sound", obviously diminishes well-being, so in this sense health is adversely affected, and it is generally appreciated that noise can physically damage the inner ear. But there are other less well reported effects which may be profoundly disturbing. The adverse reactions of annoyance, communication interference, work interference and hearing loss from noise are dealt with in the other chapters, so we are concerned here only with the remaining effects on mental and physical health.

II. SLEEP

Interference with rest or sleep is a most widespread and disturbing manifestation, and the majority of people will have experienced difficulty at some time to a lesser or greater extent. The consequences may be serious if the body is not able to adapt to persistent noise for a prolonged period, because adequate periods of rest and sleep are physiologically necessary. Chronic loss of sleep may impair performance and cause psychological distress. In fact, severe disturbances of sleep precede and accompany most acute psychiatric illnesses. To achieve sleep in the presence of noise and thereby avoid the serious consequences from repeated interference, the body, with remarkable powers of adaptation, may suppress the appreciation of noise and hence the interference. Consequently some people can become as accustomed to noise as to other environmental factors and disregard the disturbance which initially may have been very troublesome. Others are less fortunate. Some households may sleep through traffic noise or chiming clocks which wake the unaccustomed guest, while other households or individuals always have difficulty in adapting to noise and may never succeed.

It is apparent that noise is more likely to prevent sleep if it is unusual and intermittent, particularly if there is resentment against the noise. The overall sound level is also very important, and loud high-frequency noise is more disturbing. A social survey carried out by the Central Office of Information on the effects of noise in the vicinity of London Airport (Wilson, 1963) revealed that 22% of people living near the airport were sometimes kept from going to sleep by the noise of aircraft and that the proportion rose to more than 50% with very high levels of noise. A still higher proportion, also increasing with noise intensity, complained that they were sometimes awakened by the noise.

At the University of North Carolina a team of investigators measured how much the sounds of jet aircraft would disturb sleep and impair alertness and efficiency next morning (LeVere et al., 1972). While volunteers slept in a laboratory, sounds of jets flying overhead lasting 20 sec were played at irregular intervals for a total of 9 hr per night at a loudness of 80 dB. Continuous monitoring of the brain rhythms showed that, compared with control periods, the rhythms after the noises were faster and remained so for over 5 min indicating that sleep was lighter.

In the morning the volunteers were required to perform a task in which they had to press one of four buttons to extinguish one of four lights that kept flashing on and in which they had to remember which button corresponded to which light in a relationship that changed at intervals. After nights of aircraft noise, the performance of this task was much poorer than after an ordinary night's sleep. Moreover, while the task was being carried out, the electrical brain rhythms contained many more slow-wave components, suggesting that the brain was still tired and sleepy.

The auditory thresholds of awakening during sleep are functions of several variables. These include stimulus intensity, stage of sleep, subject differences, accumulated sleep time, time of night, amount of prior sleep deprivation and the subject's past experience with the stimuli. Sleep varies in depth in the same person at different times, and during deep sleep periods awakening by noise is less likely. During light sleep, awakening is easy and much fainter sounds will arouse. Young men tend to be heavy sleepers and it is older people—especially women—who sleep badly (McGhie and Russell, 1962).

In addition to the intensity and nature of the sound, the significance has a powerful influence on its awakening effect. Sleep is disturbed to a greater degree by sounds that have some personal significance or that have been thought about as important just before sleep (Oswald, 1970). Familiar sounds which do not normally require a response are ignored and so are less likely to awaken. Examples include ticking clocks and air-conditioning sounds. While sounds having special significance for a particular person may awaken easily even if faint; for example, a baby's coughing or cry will arouse an anxious mother because it is highly meaningful and may demand action.

That disturbed sleep will ultimately cause fatigue with impairment in performance was confirmed by Stanbridge (1951). Flying personnel in the 1948 Berlin Airlift became jeopardised by aircrew fatigue until resolute action was taken to provide them with quiet surroundings for rest and undisturbed sleep.

Hospital junior doctors form another group who traditionally may not get enough sleep. In a New York study 14 interns were checked for their accuracy and speed in detecting arrhythmias in electro-cardiograms and were offered prizes for good performance (Friedman *et al.*, 1971). After nights when they had averaged under 2 hr of sleep, they made twice as many errors and were much slower than after a night of 7 hr, as well as feeling more depressed, irritable and lacking in confidence.

The psychological and social consequences of sleep-disturbing stimuli are greater for middle-aged and older persons, for day-time sleepers and for the physically and mentally ill than for the young male volunteers usually studied in sleep experiments. Investigations need to be extended to these more sensitive groups. Meanwhile, with such subjective and varied responses to noise during rest or sleep, it is difficult to recommend threshold levels to avoid disturbance. It is even more difficult with intermittent noise, because its presence or absence may be equally disturbing.

Lukas (1978) in a review of noise and sleep notes the difficulty in experimental work of defining what may be the most significant response to noise during sleep. Is it behavioural awakening, movement in bed, cortical desynchronisation as determined by computer analysis or electroencephalographic response? The latter appears to be the most common denominator used.

Sleep is typically classified into five stages. Stage 1 is the brief period between awake and stage 2. Stage 2 is the most prevalent, occupying 50% of the usual sleep period. Stage delta (a combination of stages 3 and 4) may be thought of as deep sleep, while the rapid eye movement (REM) stage is usually associated with dreaming. These stages occur periodically and sequentially throughout the night, but REM tends to occur more frequently and with longer duration in the last half of the night's sleep.

Experimental work has demonstrated that the older the individual, the more likely he is to be awakened or to change sleep stage from exposure to noise; that the sleep arousal thresholds are lower in women than in men; and that although subjects respond differently to noise during the sleep stages, the specific distribution of responses is apparently a function of the age-group. However, because of the cyclical occurrence of the sleep stages throughout the night and because stage delta occurs infrequently during the later hours of sleep, relative responsiveness to noise during the different stages is confounded with the time of night and number of hours of accumulated sleep. With this proviso it may be said that an increase in stimulus intensity generally results in increased frequencies of behavioural awakening and arousal and reductions in the frequency of no change in the EEG.

One would expect that disturbance of sleep would result in measurable behavioural effects following sleep disturbance, but the demonstration of such effects has proved to be most difficult following brief disturbance of sleep. Poor performance, dependent on the type of task, has been demonstrated, however, following sleep disturbance for long periods of time and following shorter periods subsequently.

For environmental control, research data would seem to suggest two alternatives, i.e., to limit sleep disturbance to less than a change in sleep stage or to limit the frequency of arousal and behavioural awakening. A more reasonable policy as suggested by Lukas (1978) would seem to be to limit noise levels indoors so that responses to noise are limited to EEG changes of no more than one sleep stage.

III. PHYSICAL ILLNESS

There has been little evidence to date that noise has been the cause of permanent physical illness, apart from hearing loss. In an attempt to demonstrate physical changes due to noise, heart rate, blood pressure, muscular activity, metabolic rate and other responses have been studied. Davies (1967), Kryter (1970, 1972) and Glorig (1971) have discussed the general physiological responses to noise and referred to the work of many investigators. Anticaglia and Cohen (1970) in reviewing the literature for potentially harmful extra-auditory effects of noise on man present evidence to illustrate how noise acting as a non-specific physiological stressor can alter endocrine, cardiovascular and neurological functions and cause biochemical changes with possible significance to health.

Loud sounds, intense light, immobilisation, anxiety, forced exercise, surgery, cold and many other stressful agents increase the secretion of cortico-trophin (ACTH) from the pituitary gland. It appears that noise activates cortico-trophin secretion via a hypothalamic–hypophyseal mechanism. The resulting evaluation in plasma concentration of ACTH causes an increase in the secretion of adrenocortical as well as gonadal and thyrotropic hormones (Bugard, 1964; Sackler et al., 1959). The additional corticoid secreted is that characteristic of the particular species under stress (Bush, 1953). Thus, loud sounds raise plasma concentrations of corticosterone in the rat and 17 hydrocorticosterone in man and monkeys.

Related to noise effects on the adrenals are associated electrolytic imbalances (potassium, sodium, calcium, magnesium) and changes in the blood glucose level (Grognot, 1961; Cox et al., 1973). Anitesco and Contulesco (1972) have reported an increased secretion of catecholamines (urinary epinephrine and norepinephrine) in noise-exposed subjects followed by an increased urinary excretion of vanillyl mandelic acid.

Reactions provoked by noise on glands influencing sexual and reproductive functions are as yet unclear (Zondek and Tamari, 1960), since noise has been observed to induce both inhibition and stimulation of gonadotrophin and ovarian hormones. In experiments with rats Tamari (1970) has demonstrated that noise does not affect adult males, as indicated by normal spermatogenesis, but in adult females although the sex organs are stimulated, the reproductive functions—fertility, productivity and maintenance of pregnancy—are inhibited. In the present state of knowledge it would be premature to conclude that there are any serious adverse or beneficial effects from noise on sexual functions in humans.

Similarly, the evidence concerning the action of noise on thyroid activity is quite variable and inconclusive. There appears to be an inverse relationship between the rates of secretion of ACTH and of thyroid stimulating hormone (TSH) by laboratory animals subjected to stresses. No successful demonstration, however, has been made of stress-induced change in the thyroid function of man (Shipley and MacIntyre, 1954).

As regards the cardiovascular system, noise has been reported to cause vasoconstriction and fluctuations in arterial blood pressure (Lehmann and Tamm, 1956; Singh *et al.*, 1982). Vasoconstriction of the small arterioles of the extremities occurs with acute noise exposures of moderate level (about 70 dB) and can become progressively stronger with increasing noise intensity. The raised arterial blood pressure may reflect compensating heart action to overcome the vasoconstrictive effects of noise.

Mosskov and Ettema (1977) in a series of studies on 12 young, healthy males investigated auditory effects of short-term (15-min) and long-term (3-hr) exposure to aircraft noise [84–91 dB(A)], traffic noise [L_{eq} 83.5 dB(A)] and textile factory noise [98 dB(A)]. All produced similar effects, namely, an increase in diastolic blood pressure and respiratory rate and decrease in systolic pressure, pulse pressure, sinus arrhythmia and mental capacity. They concluded that long-term exposure to noise is a risk factor for cardiovascular disease in daily living and working conditions. Ettema and Zielhuis (1977) in reviewing the results of these studies and others by Knipschild (1977) hypothesize that sensitive people who tend to have hypertension with advancing age will develop it earlier (about 5 years) when exposed to noise for long periods.

Jonsson and Hansson (1977) observed that the systolic and diastolic blood pressure was significantly higher in 44 male industrial workers with a noise-induced auditory impairment (65 dB at 3000, 4000 or 6000 Hz) than in 74 males of the same age with normal hearing. Moreover, significantly more individuals with hypertension (resting recumbent blood pressure 160/100 mm Hg) were found in the group with noise-induced loss of hearing. They suggested that repeated and prolonged exposure to a stressful stimulus (industrial noise severe and prolonged enough to cause a permanent loss of hearing at the relevant frequencies) may be a contributing factor to the rise in blood pressure through a mechanism involving structural adaptation of blood vessels in response to repeated peaks of raised blood pressure.

Jansen (1961) reported a higher incidence of circulatory irregularities among steel workers in noisy jobs as compared with other worker groups in less noisy plant areas, and Shetalov *et al.* (1962) also found workers in noisy ball-bearing and steel plants to have a relatively greater prevalence of cardiovascular irregularities such as bradycardia.

There have been reports, however, of a decrease in the systolic and an increase in the diastolic blood pressure in workers exposed to industrial noise during the course of their working day (Ponomarenko, 1966). Others have reported precordial distress and discomfort from an apparent increase in heart action and breathlessness (Shetalov *et al.*, 1962; Corporale and DePalma, 1963). Miyazaki (1971) observed an increase in cerebral blood flow in 10 subjects exposed to a noise of 100 phon. Headache and discomfort were a complaint from all and disturbance of sleep from two. The fact that all persons exposed to noise do not show cardiovascular disorders is consistent with the likelihood that noise affects the health of susceptable individuals when combined with other stresses, such as work pressure and population density.

Noise, particularly of sudden onset, can cause reductions in salivary and gastric secretions and a general slowing of digestive functions. These changes together with other effects on respiratory dynamics seem to be part of a generalised stress reaction to noise.

In general it can be said that in response to unfamiliar noise there may be a vasoconstriction of the peripheral blood vessels, slow deep breathing, an increase in gastrointestinal motility (particularly with higher intensities of sound) and an increase in blood glucose and urinary 17-ketosteroids. The levels revert to normal if the noise stops or if it continues and is accepted. These several responses cannot necessarily be due to fear or anxiety because some of them can be associated with emotion-arousing and some with emotion-suppressing activities of the autonomic nervous system. The rationale of the changes is as yet not clearly understood, and there is a need for much more extensive and deeper research. There are no valid data to show that the responses carry over to produce permanent effects, but some investigators have postulated that these temporary effects may become chronic if they re-occur frequently over long periods of time. Further studies in the comparison of non-noise-exposed groups and noise-exposed groups with respect to the several phenomena are essential before valid conclusions can be drawn.

IV. EXTRA-AUDITORY EFFECTS ON THE SPECIAL SENSES

A. Vision

The first observation or record of the effect of noise on vision has been credited to the Copenhagen anatomist Thomas Bartholinus, who in 1669 noted that those who were partially deaf could hear better in the light than in the dark. Since then the effect of noise on vision has been noted by many authors (Kravkov, 1936; Serrat and Karwoski, 1936; Benko, 1959; Letourneau and Zeidel, 1971), and most experimenters now agree that the visual effects from noise are probably caused by centrally located mediating processes (Letourneau, 1972). Sensitivity to white and green light increases under the influence of sound, while sensitivity to red remains unchanged or diminishes. Peripheral vision is usually heightened. Although the laboratory effects from high noise levels are temporary in nature (Grognot and Perdriel, 1959), permanent concentric narrowing of visual fields in subjects exposed to intense noise for many years at work has been reported (Benko, 1962).

Loeb *et al.* (1976) evaluated the effect of continuous and impact noise on non-auditory sense functions and on the performance of tasks dependent on those functions under controlled laboratory conditions. The continuous noise ranged from 105 to 110 dB(A) and the impact sounds were 136 dB at peak level. The results indicated that the noise conditions impaired tracking performance but not peripheral light monitoring.

The effect of noise on the critical flicker frequency phenomenon in which alternating dark and light visual fields will become blurred (cease to flicker) at some frequency of alteration has been studied by numerous researchers. The results have been inconsistent (Kryter, 1970).

An interesting correlation exists between auditory stimulation and the condition of the iris. The concentration of melanin in the stria vascularis of the cochlea is directly proportional to that of the iris (Bonaccorsi, 1965). The stria vascularis is almost de-pigmented when the iris is blue, but has high concentrations of pigment when the pigmentation of the iris is dark brown.

Tota and Bocci (1967) studied the relationship between the colour of the iris and the temporary threshold shifts for hearing that are caused by exposure to sound. One hundred healthy, young subjects were subdivided according to the colour pigmentation of their irises and were exposed to a sound-pressure level of 100 dB centred at 1000 Hz for 3 min. The authors reported the greatest threshold shifts and the longest recovery time in subjects with lightly pigmented irises, while the opposite was noted in those individuals with darkly pigmented irises. The authors interpreted these results as support for the hypothesis that there is a relation between iris melanin and resistance to acoustic trauma. They concluded that the resistance to auditory fatigue is directly proportional to the pigment concentration in the iris.

B. Vestibular Function

Complaints of nystagmus and vertigo have been reported under noise conditions in the laboratory as well as in field situations. The levels needed to cause such effects are quite high, typically 130 dB or more (Ades *et al.,* 1957). But less intense noise may upset one's balance, particularly if the noise stimulation is unequal at the two ears. This was most clearly shown in a laboratory study in which subjects were required to balance themselves on rails of different widths (Nixon *et al.,* 1966). All these effects are believed to be due to noise directly stimulating the vestibular organ of the inner ear (McCabe and Lawrence, 1958).

C. Skin

Exposure to high-intensity sound can evoke changes in the electrical potential and resistance between two points of the skin. Plutchik (1962) exposed 18 subjects to brief periods of intermittent sound in the 100–120-dB range. There was no effect on skin temperature, but galvanic skin response showed a linear decrease in voltage with an increase in the intensity of sound.

Atherley *et al.* (1970) showed that the subjective importance of the noise was a material factor affecting changes in electrical resistance of the skin. Noises of high subjective importance—e.g., aircraft and type-writers—showed measurable physiological changes, whereas those of low subjective importance (white noise) showed no significant change compared with control levels. The authors suggest that "moderate" noise does not appear to act as a "conventional" stresser, but postulate that it may result in a characteristic syndrome which is comparable with a mild form of anxiety and depression.

D. Audio-Analgesia

Gardner and Licklider (1959) have described an audio-analgesic device which suppressed pain in dental patients. The patients listened to music or white noise through earphones but operated a switch button to control the sound intensity, increasing the levels as needed to suppress pain. In a series of 1000 patients Gardner *et al.* (1960) reported that audio-analgesia was complete or sufficiently effective in 9 out of 10 patients, so that no ordinary anaesthetic was required. Others have corroborated the clinical efficacy of audio-analgesia in paediatric patients, in patients undergoing minor and major dental operations and in obstetric patients (Sidney, 1962; Monsey, 1960; Burt and Korn, 1964).

In a review of the literature on the extra-auditory physiological effects of sound on the special senses, Anticaglia (1970) concludes that sound has important effects on the special senses but that further work is required to determine the prevalence of its effects and to evaluate their pathological significance.

V. STRESS AND MENTAL ILLNESS

Noise can produce a number of reflex reactions known as orienting, startle and defensive responses (Landis and Hunt, 1939; Booker *et al.*, 1965; Westman and Walters 1981). The orienting response involves rotation of the head and eyes towards the source of low or moderately intense sound in order to receive and respond to the sound stimulus sensation. The startle response usually caused by the occurrence of loud, unexpected sounds includes flexion of the arms, arching of the trunk, opening of the mouth and blinking of the eyes. The startle itself is of no health significance, but may cause injury if the subject is in a risk situation.

Although usually an extension of the orienting and startle responses, the defensive response may occur independently of them. It does not require sound of high intensity, but the sound must be perceived as threatening. The response includes alerting of the cerebral cortex, emotional arousal and preparation of the body for action—pupillary dilatation, skeletal muscle tension, increased pulse rate, reduction in salivary and gastric secretions, etc.

The level of mental arousal is influenced whether or not a sound stimulus is consciously perceived as a stressor. During the stage of early sleep sound can produce orienting and defensive responses and alter the quality of sleep without awakening. At the other extreme, an anxious individual can experience heightened sensitivity to a sound stimulus.

Sound will interact with other sensory stimuli to produce significant effects. Related visual stimuli will enhance the effect of sound, while sound may have an analgesic effect as demonstrated by its use in dentistry.

The ongoing motor activity of an individual is an important factor in sound appreciation. Higher levels of arousal by sound stimuli exist while complex tasks are being performed than with routine, monotonous activities.

The meaning of a sound is one of the most important factors that determine an organism's response, as is its predictability. Unpredictable noise results in lower tolerance for frustration and greater impairment of performance than does predictable noise. Attitude towards the noise source will also determine its perceived annoyance. The final factor is individual sensitivity.

An epileptic seizure is an untoward effect of auditory stimulation occurring in a small select group of patients who by their own predisposition are harmed by auditory stimulation which for others is entirely neutral (e.g., telephone bells or even pleasurable music). The auditory stimulus evokes a seizure because the patients are afflicted with sensory-evoked or reflex epilepsy (Forster, 1970).

Reflex epilepsy is not a large problem in the sense of numbers of patients. While epilepsy affects close to 1% of the population, it is estimated by Symonds (1959) that only 6.5% of the epileptic patients have their seizures evoked by sensory stimulation. The usual stimulus is visual, but auditory stimulation is by no means rare. It may be as relatively simple and non-specific as unexpected noise or as highly sophisticated and indeed difficult to determine as it is in musicogenic epilepsy.

The most common of the auditory-induced seizures is the so-called startle or acoustico-motor epilepsy. In this form the patient when presented with an unexpected loud noise will have a short minor seizure. There is a second group of startle patients, usually with brain damage and hemiplegia, in whom the presentation of noise evokes clonic movements on the paralysed side. In rare instances this type of sensory stimulus evokes a major seizure or convulsion. The stimulus itself can be highly specific, e.g., the clicking of billiard balls. The most complex types of auditory-evoked seizures are those elicited by voice or musical characteristics.

Continuous and intermittent noise may cause annoyance, and it might be expected that in many cases mental or nervous illness would result if the noise became very irritating. To date there is little published evidence to support this view. Abey-Wickrama *et al.* (1969) in a retrospective study covering two years of admissions to a psychiatric hospital compared the admissions from a maximum noise area caused by aircraft movement at London Airport, or by aircraft approaching to land, with those from outside this area. The authors admit that several factors can govern whether a patient is admitted to the hospital or treated in the community and do not suggest that aircraft noise itself can cause mental illness. But the results clearly showed that the admission rates were significantly higher from the maximum noise area than from outside it, for both the total person admissions and first admissions. Furthermore, the type of person most affected is the older woman who is not living with her husband and who suffers from neurotic or organic mental illness.

Knipschild (1977) from a series of epidemiological studies into the medical effects of aircraft noise on communities in the vicinity of Schipol Airport (Amsterdam) found that in areas with more aircraft noise, more people were under medical treatment for heart trouble and hypertension. Furthermore, the use of sedatives and hypnotics, and for female patients anti-hypertensive drugs, was higher. Drug consumption decreased when the number of flights was diminished. He concluded that aircraft noise, as prevalent around many airports, constitutes a very serious threat to public health in all its aspects: affection of well-being, mental disorders, somatic symptoms and disease (especially cardiovascular disease).

Noise-related annoyance in itself probably is not a cause of mental illness, although psychiatric patients do constitute a group vulnerable to the adverse effects of noise. Determining whether noise is an important external factor in precipitating mental illness and whether noise aggravates established mental states requires further study.

VI. ULTRASOUND

Acoustic energy at frequencies above 20,000 Hz is inaudible to man and is therefore "ultrasonic". Common sources are jet engines, high-speed dental drills, spinning machines, ultrasonic cleaners and mixers. In addition to ultrasound all may emit audible sound at sound-pressure levels exceeding safe exposure limits, so the effects of ultrasound on exposed subjects are not easily discernable. The symptoms of fatigue, dizziness, headache, nausea, tinnitus and fullness of the ears often reported by subjects exposed to these sources (Mohr *et al.*, 1965; Acton and Carson, 1967) may not be due to ultrasound, but to the audible frequencies. Energy around 16,000 Hz seems particularly critical in this regard. The physiological basis for these reactions is still not known, but workers having a hearing loss for these high-frequency sounds do not seem to display the same ill effects as those who have normal hearing (Acton, 1968). No information is available as regards the effects of long-term exposure and possible adaptation.

Low-intensity ultrasound is widely used as a therapeutic and diagnostic aid in medical practice. High-intensity focussed ultrasound can produce focal destructive lesions and may also be used to treat selected neurological diseases (Fry *et al.*, 1970a,b). The pathogenesis of this biological effect has not been completely elucidated, but both heating and cavitation have been identified as mechanisms, also a mechanical effect. Although nervous tissue has been shown to be susceptible to ultrasonic treatment, cerebral vessels have been considered immune at threshold values for mammalian nervous tissue (Fry *et al.*, 1970a,b). However, in future studies of the effect of ultrasound on tissues, the possibility of associated vascular damage should be taken into account because Dunn and Fry (1971) have confirmed that at high intensities, i.e., above 2000 W/cm^2, disruption of small intracerebral blood vessels may occur. Others (Dyson *et al.*, 1971) have observed stasis in the microcirculation during exposure.

Bell (1957), using focussed ultrasound of 1 MHz on mouse livers, showed that lesions were produced selectively at the surface of the irradiated lobe opposite to the portal of entry due to a combination of thermal and physical effects. Curtis (1965), also using a frequency of 1 MHz on mouse livers, showed that an intensity of less than 10 W/cm^2 was insufficient to cause infarction of the liver (as occurred at higher intensities), but produced evidence of liver damage of a characteristic distribution. The damage was especially noted around the central vein and spared the periphery of the lobule. At intensities of 105 W/cm^2 at frequencies in the 5–6-MHz range, macromolecules in malignant tissue will break down.

It is now well established that the effects of ultrasound vary with the frequency, the intensity and the duration of exposure; with the form (focussed or planar) of the beams; and with the form (pulsed or continuous) of the radiation. Taylor and Connolly (1969) from research on rat livers drew attention to two other important variables, namely, the absolute length of the pulse and the method by which energy is delivered to the tissues. Although ultrasonic irradiation can cause some aberrations in the chromosomes of human lymphocytes under experimental conditions (Kunze-Muhl and Golob, 1972), most researchers are of the opinion that ultrasound is harmless until the sound-pressure level approaches 140 dB.

VII. INFRASOUND

Only in recent years has it been suggested that infrasound may have an adverse effect on human beings. Infrasound is usually taken to mean sound below the audible range—those frequencies below 20 Hz. This is not strictly true because at high intensities low-frequency notes may be heard, but the threshold of hearing rises rapidly as the frequency drops, i.e., from 65 dB at 32 Hz to 95 dB at 16 Hz, 100 dB at 10 Hz, 120 dB at 3 Hz and 140 dB at 1 Hz. These thresholds may be exceeded in many situations, e.g., in compressor rooms, ship's engine rooms and cars travelling at speed with the windows open (Hood and Leventhall, 1971).

Although infrasound may be audible, there is little evidence that short-term exposure is injurious (Stephens, 1969; Pimonow, 1971), and there is no evidence to date that the unpleasant symptoms which have been experienced have led to pathological changes. Because of the impedance mismatch between airborne acoustic energy and the body, acoustic energy has little or no effect upon the body other than the ear until the levels become quite intense.

To date observations concerning infrasonic noises have been restricted to short-term, high-level exposures such as those encountered in launching rocket boosters or in certain phases of space vehicle acceleration (Von Gierke, 1965). It has been noted that noise levels which produced symptoms in the laboratory were greatly in excess of those recorded in the field survey (Mohr et al., 1965).

In addition to stimulation of the vestibular system causing nystagmus, giddiness and earache, sound at 10 to 75 Hz may cause resonant vibration of the abdomen, chest and throat. The abdominal vibrations may cause distress and sickness, whilst excessive thoracic vibrations may interfere with normal respiratory movement. In animal experiments ventilatory movements have stopped on exposure to high-intensity infrasound because the air movement associated with low-frequency pressure changes was sufficient to ventilate the lungs (Johnson, 1973). American work in connection with the space programme has indicated that the maximum permitted short-term exposure should be in the region of 140 to 150 dB. Beyond this the chest walls of test subjects vibrate, with a sensation of gagging and blurring of vision. The resonant frequency of the eyeball is near 5 Hz. Speech becomes modulated. At somewhat lower sound-pressure levels, chest vibrations are a welcome effect at discotheques and pop concerts, when the music is felt as well as heard—the total experience. Tempest (1973) from investigations of several noise annoyance complaints noted the emergence, in cases where the energy was at 32 Hz or lower, of a type of adverse reaction to sound which was quite different from loudness and seems best described as disturbance or unease. This description has been used by subjects in industrial work situations, too; and where a level of 99 dB at 34 Hz was recorded, the additional symptoms were occipital headaches and giddiness.

There is no real evidence that infrasound below the threshold of hearing is detected by the body through any other receptor; however, prolonged infrasound exposure with elevated acoustic energies at frequencies up to 1000 Hz will result in the development of well-recognized ill-health syndromes. If the vibration exposure is whole body, the syndrome of vibration sickness occurs with nausea, general malaise, weight loss and fatigue. The condition is reversible if vibration exposure is discontinued.

When vibration exposure is localized or point distributed to the hands, it can cause Raynaud's phenomenon (blanching of the fingers on exposure to cold). Vibration white finger, as the phenomenon is better known when so induced, may or may not be reversible when vibration exposure is discontinued.

VIII. CONCLUSION

It will be apparent to the reader that our present knowledge of the effects of noise on health is inadequate and fragmentary. We need not only more knowledge of cause and effect, but a better understanding of the physiological reactions. There is much scope for research but with the subjective impression being such an important factor in the evaluation of the effects of noise, the quality of the research must be high if false assumptions and conclusions are to be avoided.

REFERENCES

Abey-Wickrama, I., A'Brook, M. F., Gattoni, F. E. G., and Herridge, D. F. (1969). *Lancet* **2,** 1275–1277.

Acton, W. I. (1968). *Ann. Occup. Hyg.* **11**, 227–234.

Acton, W. I., and Carson, M. G. (1967). *Brit. J. Ind. Med.* **24**, 297–304.

Ades, H. W., Graybiel, A., Morrill, S. N., Tolhurst, G. C., and Niven, J. I. (1957). Research Project NM 130199, Subtask 2, Report 6, U.S. Navy Research Laboratory, Pensacola, Florida.

Anitesco, C., and Contulesco, A. (1972). *Arch. Mal. Prof.* **33**, 365–371.

Anticaglia, J. R. (1970). *In* "Physiological Effects of Noise" (B. L. Welch and A. S. Welch, eds.), pp. 143–150. Plenum, New York, London.

Anticaglia, J. R., and Cohen, A. (1970). *Amer. Industr. Hyg. Ass. J.* **31**, 277–281.

Atherley, G. R. C., Gibbons, S. L., and Powell, J. A. (1970). *Ergonomics* **13**, 536–545.

Bell, E. (1957). *J. Cell. Comp. Physiol.* **50**, 83–103.

Benko, E. (1959). *Opthalmologica* **138**, 449–456.

Benko, E. (1962). *Opthalmologica* **140**, 76–80.

Bonaccorsi, P. (1965). *Annali di Laringologia, Otologia, Rinologia, Faringologia (Torino)* **64**, 725.

Booker, H. E., Forster, F. M., and Klove, H. (1965). *Neurology* **15**, 1095–1103.

Bugard, P. (1964). *Folia Med. (Naples) Anno* **47**(8), 717–729.

Burt, R., and Korn, G. (1964). *Amer. J. Obst. Gynecol.* **88**, 361–365.

Bush, I. E. (1953). *J. Endocrin.* **9**, 95–100.

Corporale, R., and DePalma, M. (1963). *Riv. Med. Aeronaut. Spaz.* **26**, 273–290.

Cox, T., Simpson, G. C., and Rothschild, D. (1973). IRCS Environmental Physiology, Effects of Stress (73-9) 28-12-2.

Curtis, J. C. (1965). *In* "Ultrasonic Energy" (Elizabeth Kelly, ed.), pp. 85–116. Univ. of Illinois Press, Urbana, Illinois.

Davies, D. R. (1967). In *Proc. Symp. Psychological Effects of Noise* (W. Taylor, ed.), Soc. Occup. Med., London.

Dunn, F., and Fry, F. J. (1971). *IEEE Trans. Biomed. Eng.* **18**, 253–256.

Dyson, M., Woodward, B., and Pond, J. B. (1971). *Nature* **232**, 572–573.

Ettema, J. H., and Zielhuis, R. L. (1977). *Int. Arch. Occup. Environ. Health* **40**, 205–207.

Forster, F. M. (1970). *In* "Physiological Effects of Noise" (B. L. Welch and A. S. Welch, eds.), pp. 151–158. Plenum, New York, London.

Friedman, R. C., Bigger, J. T., and Kornfield, D. S. (1971). *New Engl. J. Med.* **285**, 201–203.

Fry, F. J., Kossoff, G., and Eggleton, R. C. (1970a). *J. Acoust. Soc. Am.* **48**, 1413–1417.

Fry, F. J., Heimburger, R. F., and Gibbons, L. V. (1970b). *IEEE J. Sonics Ultrasonics* **17**, 165–169.

Gardner, W., and Licklider, J. (1959). *J. Am. Dent. Assoc.* **59**, 1144–1149.

Gardner, W., Licklider, J., and Weisz, A. (1960). *Science* **132**, 32.

Glorig, A. (1971). *Sound Vib.* **5**(5), 28–29.

Grognot, P. (1961). *Presse Thermale Climat* **98**, 201–203.

Grognot, P., and Perdriel, G. (1959). *Med. Aeronaut.* **14**(1), 25–30.

Hood, R. A., and Leventhall, H. G. (1971). *Acustica* **25**, 10–13.

Jansen, G. (1961). *Adverse Eff. Noise Iron Steelworkers* **81**, 217–220 (in German).

Johnson, D. L. (1973). *J. Acoust. Soc. Am.* **53**, 293.

Jonsson, A., and Hansson, L. (1977). *Lancet* **1**, 86–87.

Knipschild, P. (1977). *Int. Arch. Occup. Environ. Health* **40**, 185–204.

Kravkov, S. V. (1936). *Acta Opthalmologica* **14**, 348–360.

Kryter, K. D. (1970). "The Effects of Noise on Man." Academic Press, New York.

Kryter, K. D. (1972). *Am. J. Publ. Health* **62**, 389–398.

Kunze-Muhl, E., and Golob, E. (1972). *Humangenetik* **14**, 237–246.

Landis, C., and Hunt, W. A. (1939). "The Startle Pattern." Farrar, New York.

Lehmann, G., and Tamm, J. (1956). *Intern. Z. Angew. Physiol.* **16**, 217–227.

Letourneau, J. E. (1972). *Eye, Ear, Nose Throat Monthly* **51**, 441–444.

Letourneau, J. E., and Zeidel, N. S. (1971). *Am. J. Optometry Arch. Am. Acad. Optometry* **48**, 133–137.

LeVere, T. E., Bartus, R. T., and Hard, F. D. (1972). *Aerospace Med.* **43**, 384.

Loeb, M., Jones, P. D., and Cohen, A. (1976). "Effects of Noise on Non-Auditory Sensory Functions and Performance." NIOSH Report, HEW No. 76-176, Washington, D.C.

Lukas, J. S. (1978). Noise and sleep: A literature review and a proposed criterion for assessing effect. *In* "Handbook of Noise Assessment" (D. N. May, ed.), pp. 313–334. Van Nostrand Reinhold, New York.

McCabe, B. F., and Lawrence, M. (1958). *Acta Oto-Laryngol.* **49**, 147–157.

McGhie, A., and Russell, S. M. (1962). *J. Mental Sci.* **108**, 642.

Miyazaki, M. (1971). *Jpn. Circulaj. J.* **35**, 931–936.

Mohr, G. C., Cole, J. N., Guild, E., and Von Gierke, H. E. (1965). *Aerospace Med.* **36**, 817.

Monsey, H. (1960). *J. Calif. Dental Assoc.* **36**, 432–437.

Mosskov, J. I., and Ettema, J. H. (1977). *Int. Arch. Occup. Environ. Health* **40**, 165–184.

Nixon, C. W., Harris, C., and Von Gierke, H. E. (1966). Report AMRL-TR-66-85. Wright Patterson Air Force Base, Ohio.

Oswald, I. (1970). "Sleep," 2nd ed. Penguin, Harmondsworth.

Pimonow, L. (1971). *Med. Hyg. (Geneve)* **29**, 1072–1075.

Plutchik, R. (1962). Contract No. NONR-2252 (01), Office of Naval Research, Hofstra College, Hempstead, New York.

Ponomarenko, I. (1966). *Gigiena i Sanit.* **31**, 188–193.

Sackler, A. M., Weltman, A., Bradshaw, M., and Justshuk, P. (1959). *Acta Endocrinol.* **31**, 405–418.

Serrat, W. D., and Karwoski, T. (1936). *J. Exper. Psychol.* **19**, 604–611.

Shetalov, N. N., Saitanov, A., Bradshaw, M., and Glotova, K. (1962). *Labour Hyg. Occup. Disease* **6**(7), 10–14.

Shipley, R. A., and MacIntyre, F. H. (1954). *J. Clin. Endocrinol. Metabl.* **14**, 309–317.

Sidney, B. (1962). *J. Am. Pediatr. Assoc.* **7**, 503–504.

Singh, A. P., Rai, R. M., Bhatia, M. R., and Nayor, H. S. (1982). *Int. Arch. Occup. Environ. Health* **50**, 169–174.

Stanbridge, R. H. (1951). *Lancet* **2**, 1–3.

Stephens, R. W. B. (1969). *Ultrasonics* **7**, 30.

Symonds, C. (1959). *Brain* **82**, 133–146.

Tamari, I. (1970). *In* "Physiological Effects of Noise" (B. L. Welch and A. S. Welch, eds.), pp. 117–130. Plenum, New York.

Taylor, K. J. W., and Connolly, C. C. (1969). *J. Pathol.* **98**, 291–293.

Tempest, W. (1973). *Acustica* **29**, 205–209.

Tota, G., and Bocci, G. (1967). *Rivista Oto-Neuro-Oftalmologica (Bolonga)* **42**, 183–192.

Von Gierke, H. E. (1965). *Arch. Environ. Health* **11**, 327–339.

Westman, J. C., and Walters, J. R. (1981). *Environ. Health Persp.* **41**, 291–309.

Wilson, A. (1963). "Noise—Final Report." Cmnd. 2056, Her Majesty's Stationery Office, London.

Zondek, B., and Tamari, I. (1960). *Am. J. Obstet. Gynecol.* **80**, 1041–1408.

3

Noise and Hearing

W. TEMPEST

Department of Electronic and Electrical Engineering
University of Salford
Salford
England

I. INTRODUCTION

Intense noise of any kind can cause temporary and/or permanent damage to the human hearing process. This chapter is primarily concerned with the phenomenon of permanent noise-induced hearing loss (NIHL), although temporary threshold shift (TTS) will also be briefly discussed.

Permanent hearing loss can be defined for our purposes as a permanent shift in the hearing threshold, and for steady-state noise exposure it does not seem to occur for exposure levels below 80 dB(A); however, it is significant at 85 dB(A) and becomes a major hazard to hearing once a level of 90 dB(A) is exceeded. The actual damage to the hearing mechanism takes place in the inner ear in the form of selective destruction of the hair cells which convert acoustic energy into electrical impulses to be fed, via the nervous system, to the brain. Since the hair cells are incapable of regeneration, the process is irreversible, and the pattern of the damage is such that the hearing threshold rises. The process is in some ways similar

47

to the natural deterioration of hearing with age in that it raises the threshold but leaves the perception of loud sounds unimpaired. The view has been put forward (Evans, 1975) that the essential lesion in cochlear deafness is damage to the second filter mechanism of the cochlea. Such damage leads not only to loudness recruitment, but also to a deterioration in the ability of the ear to discriminate between differing frequencies. Such deterioration impairs the ability of the ear to distinguish between vowel sounds, leading to a reduction in the intelligibility of speech. This cannot be successfully corrected by hearing aids, which are therefore of only limited value to the sufferer from NIHL.

While damage due to steady-state noise is usually confined to the inner ear in the form already mentioned, other types of damage can occur at very high intensities. The most frequent of these is ear-drum rupture due to 'impulsive' noise, usually blast from an explosion or gunfire. While a ruptured ear drum will in many cases heal with little or no permanent loss of hearing, blast can also injure either the middle or inner ear. In the former case, either dislocation or fracture of the ossicles may occur, situations which can often be alleviated surgically. In the latter case of direct inner ear blast damage, it is possible that rupture of the inner ear membranes may occur with a consequent serious or total permanent deafness.

The type of blast noise damage just mentioned can usually be associated with a single event; however, there is some evidence from otologists of a third, possibly intermediate situation in which a severe loss of hearing occurs after exposure to very intense noise over a period of weeks or months, but in which there is no evidence of a single blast-type exposure. These cases involve exposure to a steady-state noise of a very high level, often with an impulsive content, or to a repeated impulsive noise, and result in a severe hearing loss of the type associated with blast rather than steady-state noise exposure. It is suspected that they are due to some type of catastrophic inner ear damage; but due to the relative infrequency of such cases, and the difficulty in quantifying both exposure and the nature of the damage, the evidence for these cases must be regarded as somewhat speculative.

In order to discuss the effect of noise on hearing in more detail it is convenient therefore to treat the two cases of steady-state and impulsive/impact noise separately.

II. HEARING LOSS DUE TO STEADY-STATE NOISE

The existence of noise-induced hearing loss has been known for at least two centuries (Ramazzini, 1713 [1964]), but the precise quantification of the relationship between exposure and the resultant loss of hearing has only taken place in the last two decades, and indeed, there are still areas of controversy in this field.

It was accepted by about 1960 [see Bryan and Tempest in Robinson (1971) for a discussion of the relevant literature] that long-term exposure to noise levels of 85 dB(A) and over causes a permanent loss of hearing; however, at that time there were no adequate data available to permit reliable predictions of hearing loss from known noise exposure. A serious attempt to relate exposure to loss had been made in the report Z24-X-2 "The Relations of Hearing Loss to Noise Exposure" (ANSI, 1954). This report concentrated its attention on the development of hearing loss over a period of time in relation to exposure to particular octave bands of noise. It successfully demonstrated the way in which losses developed with time and also showed examples of the types of audiograms produced by noise exposure. Although the report showed that octave band levels in excess of 90 dB led to substantial hearing losses, it did not go on to propose any criteria for the limits of safe exposure.

The first two studies which succeeded in bringing some sort of order to the problem appeared quite close together in 1968. The first was the study by Passchier-Vermeer (1968) in which collected data from various sources on 4600 people were brought together. The second report (Robinson, 1968; Burns and Robinson, 1970) was based on a study of about 1000 persons. The reports of Passchier-Vermeer and Burns and Robinson represent quite different approaches to the problem, and, although in most areas they are in broad agreement, their methods of data handling and presentation have little in common. In view of their importance it would seem appropriate to consider them individually before commenting on their relative importance.

A. Hearing Loss and Noise Exposure: Passchier-Vermeer

Passchier-Vermeer's report is concerned with the effects of steady-state broad band noise on the hearing levels of people exposed to noise for 8 hr per day, at least 5 days per week. It is based on data for 20 groups of workers, derived in turn from the publications of 8 authors. Exposure times range from 10 to 40 years, and noise levels are basically defined in terms of noise rating (NR) numbers from 500 to 2000 Hz. The actual range of levels considered, NR 75 to 98, corresponds approximately to 79 to 102 dB(A).

Passchier-Vermeer's data are presented in the form of both median data for subjects of normal sensitivity and audiograms showing the 25 and 75% levels of the distribution.

Figure 1 shows median audiograms obtained after 10 years for noise-

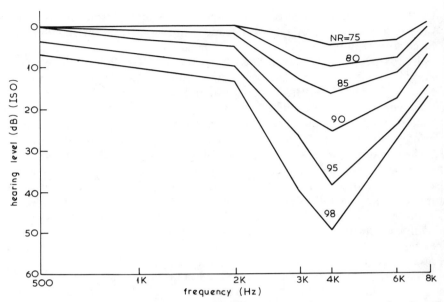

Fig. 1. Median noise-induced hearing loss after 10 years of exposure as a function of noise-level NR (500–2000 Hz). [Data from Passchier-Vermeer (1968).]

exposure levels of NR 75 to 98 [79 to 102 dB(A)]. Figure 2 gives similar data for 40 years of exposure.

These curves illustrate the way in which the damage to hearing increases with increasing noise level for a constant exposure time. The general pattern is that loss is always greatest at 4000 Hz and that the magnitude of the loss increases rapidly with increasing noise level.

Figure 3 shows, to some extent, the range of individual variation in sensitivity to noise damage by presenting audiograms showing the extent of hearing loss experienced by 25, 50 and 75% of the population for exposure to NR 90 for 10 years.

The manner in which hearing loss grows with time at a constant level of noise exposure is illustrated by Fig. 4, which is drawn from the data of Taylor *et al.* (1965) and Pearson (1977). These data (Taylor's were included in Passchier-Vermeer's larger study) show that the loss begins in the form of a 'notch' at 4000 Hz, which first develops and, at a later stage, widens to include 3000 and 2000 Hz.

Within its limits (noise levels from NR 75 to 98, exposures of 10 and 40 years) Passchier-Vermeer's data provide an excellent basis for the evaluation of expected hearing loss. For intermediate periods of exposure be-

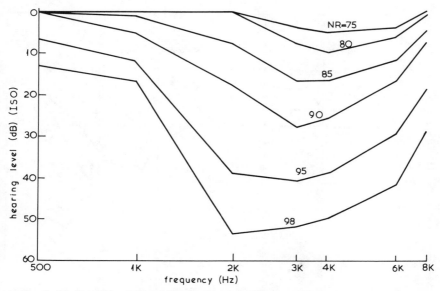

Fig. 2. Median noise-induced hearing loss after 40 years of exposure as a function of noise-level NR (500–2000 Hz). [Data from Passchier-Vermeer (1968).]

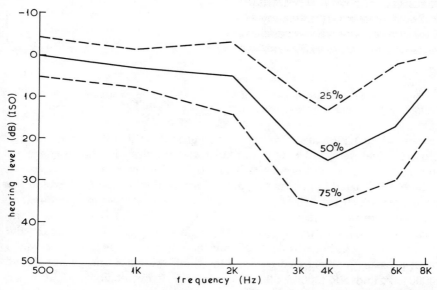

Fig. 3. Noise-induced hearing loss after 10 years of exposure at NR 90 (500–2000 Hz), showing 25, 50 and 75% levels of the distribution. [Data from Passchier-Vermeer (1968).]

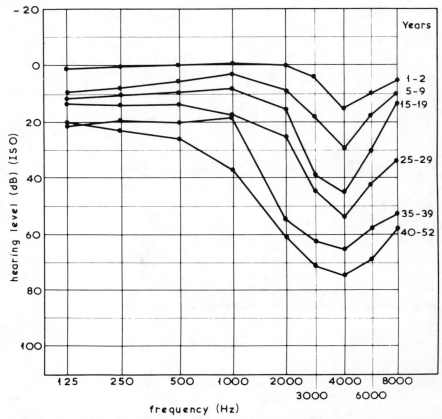

Fig. 4. Development of noise-induced hearing loss with time for an exposure level of 101 dB(A), including presbycusis. [Data from Taylor *et al.* (1965) and Pearson (1977).]

tween 10 and 40 years, a simple linear growth of loss with time is proposed. The data have two limitations. The first is the absence of information on levels above NR 98 and the second is the particular way in which NR is used in this report, in the form of the highest NR level in the range 500–2000 Hz. The restriction of the definition to this range means that the approximate relationship dB(A) = NR + 4 only holds if the highest levels lie in the 500–2000-Hz region. In practice this covers the great majority of industrial noises, but there are cases (for example, small high-pressure compressed air jets) in which the main energy lies above 4000 Hz, and NR (500–2000 Hz) could differ from decibels(A) by much more than four units.

The most important quality of Passchier-Vermeer's work is its basis in the data of many other workers, making it unlikely that the whole range of results could be seriously affected by any minor inaccuracies in the published work on which it is based.

B. Hearing Loss and Noise Exposure: Burns and Robinson

The extensive study of the relationship between noise exposure and hearing loss reported by Burns and Robinson (Burns and Robinson, 1970; Robinson, 1971) was based on a survey of hearing and noise exposure in industry in Britain, commissioned in 1961 by the Ministry of Pensions and National Insurance.

In this study the final analysis was based on the hearing and noise exposure of 759 noise-exposed persons, plus 97 non-noise-exposed controls. Burns and Robinson then attempted a sophisticated analysis, aiming to bring together in a unified form:

(1)　the effect of noise levels for 75 to 120 dB(A),
(2)　the influence of duration from 1 month to 50 years,
(3)　the distribution of hearing loss from the least to the most sensitive members of the population,
(4)　the affect of presbycusis and
(5)　the difference in hearing loss between the sexes.

The results of this analysis are expressed in the form of an equation which provides the decibel hearing level at frequencies from 500 Hz to 6 kHz for any given noise level, exposure period, susceptibility to hearing loss and age. The details of the calculations are somewhat complex and the interested reader is referred to the paper by Robinson (1971) for a clear account in which a series of tables are provided to simplify the procedure.

One of the most important features of the Burns and Robinson study was the development of the concept of noise immission level (NIL), the total A-weighted noise energy received by the ear. They showed that this parameter, which combines noise level and duration into a single figure, could provide the basis of their calculation of hearing loss:

$$\text{NIL} = L_A + 10 \log_{10} T$$

where L_A is the level of noise exposure in decibels(A) over the normal working period (approximately 2000 hr per year) and T the exposure time in years. In the book "Hearing and Noise in Industry", Burns and Robinson (1970) provide a useful nomogram which permits the expected hearing

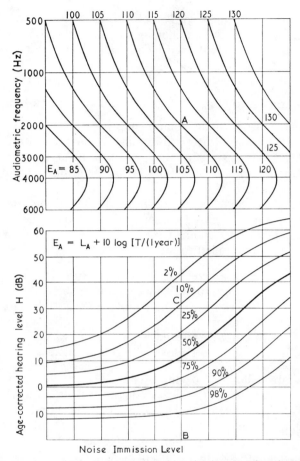

Fig. 5. Nomogram used to calculate expected hearing loss as a function of noise immission level and susceptibility. [From Burns and Robinson (1970).]

loss to be obtained directly from the NIL. The nomogram (Fig. 5) is used as follows:

(1)   Calculate the noise immission level (e.g., 110).
(2)   Enter the appropriate curve in the upper nomogram and follow it to the appropriate audiometric frequency (e.g., 2000 Hz at point *A*).
(3)   Descend vertically from this intersection to the lower nomogram (line *AB*).
(4)   Select the appropriate centile of the distribution (e.g., 10%, point

TABLE I. Constant Used to Calculate Presbycusis Loss[a]

Frequency (Hz)	500	1000	2000	3000	4000	6000
C_i	0.004	0.0043	0.006	0.008	0.012	0.014

[a] Based on Burns and Robinson (1970).

C). (Note: Burns and Robinson describe high susceptability by a *low* centile; Passchier-Vermeer does the opposite.)

(5) Read out the expected hearing level on the vertical axis (e.g., at point C, 32 dB).

Having used Fig. 5 to obtain the extent of the hearing loss due to noise, Burns and Robinson provide a term to take into account the effects of presbycusis. This takes the form

$$H_P = C_i(N - 20)^2$$

where N is age in years for workers over 20. Here N is taken as 20 for ages less than 20 years and C_i is a constant obtained from Table I. The total expected hearing loss is then found by adding the noise induced and presbycusis term

$$H_T = H_N + H_P$$

This procedure thus provides an estimate of expected hearing loss in any case of exposure to continuous noise up to 120 dB(A) maximum level.

Both these sets of data (those of Passchier-Vermeer and Burns and Robinson) can provide valuable estimates of expected hearing loss in individual cases. A comparison of the two studies shows that in most situations they yield fairly similar data. In the case of exposure to levels of around 100 dB(A) over 20 to 40 years, Burns and Robinson's procedure yields smaller hearing loss figures than have been found in some other studies. These differences were noted by Robinson (1971), who discussed them but did not resolve them. A further analysis by Kell (1975) discusses the discrepancy in detail and suggests that it may be due to some aspects of the subject selection procedure used by Burns and Robinson, and also to the small numbers in their own sample who were exposed for very long periods of time.

III. HEARING LOSS DUE TO IMPULSIVE AND IMPACT NOISE

An impulsive noise is normally defined as a short-duration sound, particularly characterised by a shock-front pressure waveform. Such a noise has a very short rise time and is usually generated by an explosion or by the blast from a gun. Impact noise is normally generated by non-impulsive means, such as metal-to-metal impacts in industrial situations. In these instances there may be a less steep shock front or no shock front, but the impact is followed by a substantial amount of reverberant sound. The duration of the impact is usually defined as the time taken for the sound to fall by 20 dB from its initial peak, and this is known as the B duration. The B duration is typically in the range of 0.1 to 2 s, depending mainly on the acoustics of the enclosure in which the impact takes place.

As can be seen from the previous sections, in the case of steady-state noise the problem of evaluating the noise exposure does not present serious difficulties. Decibels(A) or NR can be used (sometimes in slightly modified forms), but the results are not overly sensitive to the units chosen.

In the case of short-duration sounds of high intensity, the situation appears to be more complex. If the peak instantaneous SPL is sufficiently great, then clearly the ear may suffer immediate, serious damage (e.g., from an explosive blast). The first requirement, therefore, is to establish a maximum SPL to which the unprotected ear may be exposed and to fix this at such a level that there is no significant risk of permanent damage from a single event. Above this level it is therefore to be assumed that *no* exposure is safe or permissible.

Below this peak level there will clearly be some quantitative relationship between noise exposure and expected hearing loss. To date we have very few reliable data as to the peak level at which immediate hearing damage can occur or to the noise exposure versus hearing loss relationship at lower levels at which the noise is of an impulsive type.

Historically, damage risk criteria for impulsive noise have been based primarily on the assumption that the degree of TTS (temporary threshold shift) produced (at, say, the end of a day) is related to the risk of NIHL. This approach led to the development of the CHABA criterion (Ward, 1968), which was defined in terms of the maximum allowable instantaneous peak pressure, the duration of the impulse and the number of impulses per day. This criterion permitted a hundred exposures per day to a peak pressure of 162 dB, with a total duration of 10 ms, rising to an upper limit of 174 dB for one exposure per day of not more than 25 μsec. The CHABA criterion was based on a maximum TTS of 10 dB at 100 Hz and below, 15 dB at 2000 Hz or 20 dB at 3000 Hz and above at the end of an exposure period, normally a day. The TTS is that existing 2 min after cessation of exposure to the noise and the criteria are designed to protect 95% of the exposed personnel. Despite its empirical basis, the CHABA criteria gained wide acceptance as the best available for dealing with high-intensity, short-duration impulses.

However, more recent developments are tending to suggest that the complexity of the CHABA criterion may well be replaced by an extension of the L_{eq} (equivalent continuous noise level) concept already accepted for continuous noise. The continuous noise studies (Passchier-Vermeer, 1968; Burns and Robinson, 1970) both excluded impulsive and impact noise from their results. Since then several workers (Atherley and Martin, 1971; Martin and Atherley, 1973; Kuzniarz *et al.*, 1976) have examined in detail the application of the L_{eq} concept to repeated impact noise, such as that found in the drop-forge industry. The work of both groups concluded that L_{eq} was an appropriate measure of the ability of a noise exposure to damage hearing. They found that the relationship between noise immission and expected hearing loss was the same as that for continuous noise. The extension of the use of L_{eq} [and hence noise immission level (NIL)] to impact noise has raised the question of whether it could be extended further to cover the range of short-duration, high-pressure impulses covered by the CHABA criterion. This was considered by Rice and Martin (1973), who reviewed the impulse–impact damage risk criteria and suggested that the unification of these might be possible through using the immission principle of A-weighted noise energy as for continuous noise. They compare the L_{eq} criteria [with a maximum exposure equivalent to 90 dB(A) over a working day] with the more complex CHABA standard and show that for all but the briefest of exposures the L_{eq} approach leads to a slightly *lower* permitted peak level than does CHABA.

The current state of knowledge was discussed in the report of the workshop on impulse noise and auditory hazard held at Southampton University (von Gierke *et al.*, 1982). The main conclusions of this workshop can be summarized as follows.

(1) The risk of hearing impairment should ideally be based on the relationship between permanent threshold shift and noise exposure, but in lieu of such data information on TTS as well as recovery curves could be accepted.

(2) There is no reason not to accept decibels(A) L_{eq} over a year's time to evaluate *all* types of noise (with a frequency range of 10 to 20000 Hz and an unweighted peak instantaneous sound-pressure level not exceeding 145 dB), using ISO DIS/1999: 1982 to evaluate the hazard to hearing. [The ISO draft standard ISO/DIS 1999 (ISO, 1982a) has been prepared to ultimately replace the current standard ISO—R 1999 (ISO, 1971) and is based on a number of studies in Europe and the United States. At the time

of writing (May 1984) it has not been finally adopted, and it should be noted that the draft version referred to earlier appears to contain some typographical errors.]

(3)   There is no clear evidence to separate impulse from non-impulse noise, or their effects, but in the interests of caution, hearing conservation programmes should be initiated at a 5-dB-lower level for impulse noise if, on the basis of available studies, a hazard in excess of that for continuous noise is suspected.

(4)   For sound-pressure levels above 145 dB there is no universally accepted relationship between exposure and predicted hearing loss, but, provided that hearing protection is worn, such levels are very rarely encountered in industry.

IV. PRESBYCUSIS

Presbycusis is the deterioration in clinically normal ears which takes place with advancing years in the absence of any injury or disease. Although it is not primarily due to noise, knowledge of its effects and magnitude are essential to any evaluation of noise-induced hearing loss.

Perhaps the most widely used data are those of Hinchcliffe (1959), who measured the hearing threshold from 125 Hz to 12 kHz of a sample of 400 subjects drawn from a rural area in southwest Scotland. Hinchcliffe found that between 2 and 8 kHz, the male ears are consistently inferior to the female, but that the difference is not significant at 12 kHz. He concludes that the greater presbycusis shown by the male is not due to some basic differences in the ageing process, but to other adventitious factors, such as acoustic trauma. A further comment on Hinchcliffe's data is noted by Pearson (1977), who points out that while in 1959 it was assumed that there would be no exposure to noise in a rural population, it was subsequently discovered that many of the males had been exposed to gunfire.

In addition to Hinchcliffe's data there have been numerous other presbycusis studies, and the paper by Pearson (1977) has endeavoured to bring some of these [including Hinchcliffe's (1959)] together in the form of a single set of curves. Pearson's study used data from five populations [see Kell et al., (1970), Taylor et al., (1967), Hinchcliffe (1959), and Glorig and Nixon (1962)] and from these derived a series of curves relating threshold to age and frequency. The data from the various sources were combined by referring them all to the hearing at age 25 as a base and then fitting a parabolic function to the result. Pearson's curves are shown in Fig. 6. Pearson comments that 'since the estimates are based on the results from a number of widely differing populations, it is suggested that

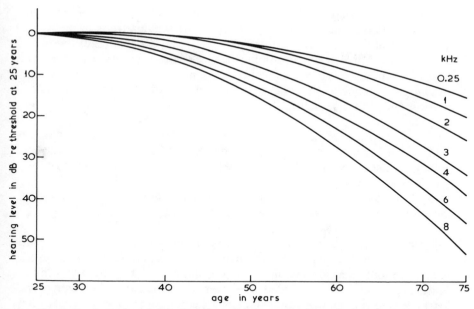

Fig. 6. Presbycusis as a function of age and frequency. [Data from Pearson (1977).]

they may be considered to represent general values for presbycusis applicable in a wide range of situations, in particular industrial audiometry'.

The view that a single set of curves can suffice to define presbycusis for
both males and females does not appear to have the support of the International Organization for Standardization, since their discussion draft on
presbycusis (ISO, 1982b) puts forward separate curves for the sexes. At
the time of writing (May 1984) it is understood that this standard has not
yet been adopted.

V. NOISE-INDUCED HEARING LOSS AND PRESBYCUSIS

The largest single cause of significant hearing loss in the population (at
least in Western civilizations) is presbycusis, the natural deterioration of
hearing with age. Noise-induced hearing loss must therefore be viewed
against this background.

Noise-induced loss and presbycusis are basically similar in that both
occur in the inner ear, both have their initial effect on hearing at the higher
frequencies and both develop slowly in the individual. However, the two

effects differ in that NIHL shows a maximum at around 4 kHz, while presbycusis continues to increase towards the highest frequency tested. The practical problem which must be faced in any evaluation of the hearing of a person with a noise-induced loss is the separation of the two effects whenever it is believed that both are present to a significant extent.

One can speculate that the existence of presbycusis could influence noise-induced loss in three different ways:

(1)   The effects of noise and age are independent and can be added.

(2)   The presence of one loss of this nature (either noise or age induced) reduces the sensitivity of the ear to noise and hence to damage and provides a protection against further loss.

(3)   An ear damaged by age or noise is less robust than normal and is therefore particularly vulnerable to further damage.

Mollica (1969) made a study of some 200 industrial workers, who had spent from 2 to 40 years in a noisy environment; ages ranged from 25 to 65 years. He looked at groups of workers of different ages, but with similar noise exposures. He concluded that after the subtraction of presbycusis losses [using the data of Johansen (1943)], the groups showed that "the subject showed a hearing deficit almost on the same level, independent of age". No statistical evidence is offered to support the conclusion. Mollica therefore concludes that the effects of noise damage and presbycusis are additive without influencing one another.

Mollica's work was concerned with the effect of noise on persons of different ages; by contrast Macrae (1971) has looked at the converse problem, the effects of ageing on subjects who suffered noise damage at an earlier period in their lives. Macrae began his research by selecting from the audiological records of male war veterans (presumably Australians) 350 subjects who satisfied the following criteria:

(1)   No conductive component in the ear (or ears) to be included in the investigation,

(2)   a hearing level at 4000 Hz, 20 dB or more greater than predicted by Glorig and Nixon's (1962) presbycusis curves,

(3)   a hearing level at 1000 Hz not more than 10 dB greater than predicted by the presbycusis curves,

(4)   on the basis of the presbycusis data of Spoor (1967) a change of at least 10 dB in the 4000 Hz hearing level predicted due to the passage of time since the initial audiogram and

(5)   no sign of hereditary or non-organic hearing loss presented by the subject.

From the 350 records selected and followed up, 240 subjects agreed to undergo an audiometric re-test. These 240 were examined otoscopically and questioned about possible causes of hearing loss, i.e., middle ear infection, non-organic loss, noise exposure since military service, use of hearing aids, illness and ototoxic drugs, blows to the head, Menieres disorder and acoustic neuroma.

After the rejection of all cases in which there was evidence of one or more of the preceding possible causes of additional hearing loss, the sample was reduced to a total of 285 ears on which audiometric tests were carried out. The resulting audiometric data obtained at 1000 Hz and 4000 Hz were analysed in several ways, but showed that at 4000 Hz, all age groups showed fairly close agreement with the expected changes in hearing level due to presbycusis. At 1000 Hz the agreement was again fairly close, but in this instance 15 out of 16 groups showed larger increases than those predicted from the presbycusis data. Macrae concludes that the effects of presbycusis are independent and additive at 4000 Hz and that at 1000 Hz, the deterioration in hearing has occurred at approximately the rate predicted by Spoor's (1967) equation.

A rather different approach to the combined effect of noise and age has been made by Robinson (1968) in his large survey of hearing and noise in industry. Robinson found that the variances of 19 out of his 20 'exposure-duration cells' were reduced when he subtracted a presbycusis correction based on Hinchcliffe's (1959) data.

In conclusion, it can be said that available data support the hypothesis that presbycusis and noise-induced hearing loss occur independently and are additive. However, this conclusion is based on limited experimentation in an area where the accuracy required for a definitive study would be extremely difficult to achieve. It would therefore seem wise to regard the hypothesis of additivity as a useful practical approximation to be used until something better is produced. A further important question, which remains entirely unanswered, is that of interindividual differences in susceptibility to hearing loss. Although it is well established that there are wide differences among individuals in both presbycusis and noise damage sensitivity, there is no information as to whether these susceptibilities are correlated.

VI. INDIVIDUAL SUSCEPTIBILITY TO NOISE DAMAGE

Both Passchier-Vermeer and Burns and Robinson, and numerous other workers, have shown that there are large variations in individual susceptibility to noise damage. Similarly, it is the experience of all who undertake industrial audiometry that, for any group of workers with a similar noise exposure, some will have suffered a negligible loss of hearing, while others will be more severely impaired.

The cause of these differences is clearly some substantial natural variability in resistance to noise damage, but it has so far not proved possible to predict in advance who will be susceptible and who will not. A report by Burns (1971) shows that some positive correlation exists between susceptibility to NIHL and to TTS. Unfortunately, these correlations are too weak to be of value in predicting which individuals will be susceptible. The lack of any available technique by which individual susceptibility can be predicted is a handicap to the optimum implementation of any hearing conservation programme, since it implies that all schemes must be designed to protect the most susceptible and cannot be applied selectively to those who most need them.

VII. NOISE-LEVEL CRITERIA

The current U.K. criterion (Department of Employment, 1972) is to be found in the "Code of Practice for Reducing the Exposure of Employed Persons to Noise" and proposes a maximum exposure level of 90 dB(A) over an 8-hr working day. This criterion is defined in terms of the equivalent continuous sound level L_{eq} and therefore permits higher noise levels for shorter periods on the 'equal energy' basis of allowing an increase of 3 dB for each halving of the duration, up to an absolute maximum of 135 dB(A). Table II sets out the corresponding levels and permitted maximum durations. A further limit is applied to impulsive noise at a maximum instantaneous pressure of 150 dB.

In the United States the OSHA standard (Department of Labor, 1974) is also set at a maximum of 90 dB(A) for 8 hr per day, but the rule governing shorter exposures is different, requiring a halving of permitted duration for each 5 dB increase in level (see Table III). It will be seen from the tables that the maximum permitted level in this case is 115 dB(A), however short the exposure. A further stipulation in the OSHA standard is that no impulsive noise should exceed a 140-dB sound-pressure level.

An examination of Passchier-Vermeer's and Burns and Robinson's data shows that, at noise levels up to about 80 dB(A), even a working lifetime's exposure will cause very little damage to a person of average susceptibility and no serious damage to the more susceptible members of the population. At around 90 dB(A), however, the situation is quite different, and Fig. 7 shows the expected hearing levels, including the effects of

TABLE II. Maximum Permissible Noise
Exposure Level Versus Daily
Exposure Time[a]

Sound level [dB(A)]	Maximum daily exposure
88	12 hr
90	8 hr
93	4 hr
96	2 hr
99	1 hr
102	30 min
105	15 min
108	7.5 min
111	3.75 min
114	1.9 min
115	1 min
118	30 sec
121	15 sec
124	7.5 sec
127	3.8 sec
130	1.9 sec
133	1 sec
135	0.6 sec

[a] From Department of Employment (1972).

presbycusis, for a worker aged 60, who has worked for 40 years in a 90-dB(A) noise level, calculated according to the Burn and Robinson (1970) procedure. It demonstrates that for the more sensitive individuals (90% level) this level of noise exposure can have serious consequences.

TABLE III. Maximum Permissible Noise
Exposure Level Versus Daily
Exposure Time[a]

Duration per day (hours)	Sound level [dB(A)]
8	90
6	92
4	95
3	97
2	100
1.5	102
1	105
0.5	110
0.25 or less	115

[a] From Department of Labor (1974).

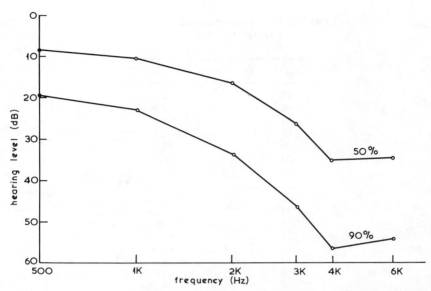

Fig. 7. Expected hearing loss, including presbycusis, for a 60-year-old worker who has spent 40 years working in 90 dB(A); 50% curve is median data, while 90% is for a highly susceptible individual. [Data from Burns and Robinson (1970).]

Another point of view on the 90 dB(A) criterion is to use available data on noise and hearing loss to estimate what percentage of the workforce will experience a significant loss of hearing after a lifetime's exposure at the maximum permitted level. Data on this are available from several sources, the international standard (ISO—R 1999: 1971) tabulated the percentage of the labour force at risk after 40 years of service as 21% due to noise alone. This standard defines a significant loss of hearing as an average hearing level of 25 dB or more at the three frequencies 500, 1000 and 2000 Hz.

The British standard (BS 5330, 1976) uses a different definition of impairment—an average loss of 30 dB at 1000, 2000 and 3000 Hz—and finds that 17% of those exposed to noise are at risk after 40 years when the effects of noise and presbycusis are taken into account.

The differing data bases and definitions of significant impairment leave scope for detailed discussion at great length as to the exact degree of risk involved in the 90 dB(A) criterion, but all available data agree that there is *some* degree of risk, and they clearly indicate that a lower criterion is desirable. The ISO—R 1999 recommendation suggests that an 85 dB(A) criterion would roughly halve the risk and that an 80 dB(A) criterion would eliminate it.

VIII. TEMPORARY THRESHOLD SHIFT

Temporary threshold shift is the short-term elevation of the hearing threshold after an exposure to a high noise level. It is often noticed by visitors to a noisy industrial environment, who after a period (of say an hour) in a noisy workplace find their cars sound unusually quiet on the return journey.

The extent to which the hearing threshold is raised depends on the intensity and duration of noise exposure and is at a maximum just after the exposure is terminated. Recovery occurs over a number of hours, and, according to the data of Ward *et al.* (1958), is logarithmic; i.e., the graph of log (recovery time) versus temporary threshold shift in decibels is a straight line. The frequency of maximum threshold shift is related to the frequency of the noise stimulus and, for high levels of noise, is about half an octave above the stimulus. Some indication of the way in which TTS develops and recovers can be obtained from Fig. 8. The data show that even at a level of 88 dB(A) more than 5 dB of TTS can be acquired in less than 2 hr, while at 106-dB exposure level, the TTS reaches 40 dB. The data in Fig. 8 suggest (by extrapolation) that the 30 dB of TTS caused by 112 min exposure at 100 dB recovers in about 1000 min, or just over 16 hr. Ward (1960) has shown that in the case of higher-level exposures giving rise to TTS in the region above 40 dB, the recovery process is no longer logarithmic with time, but occurs at a steady rate of about 0.012 dB/min.

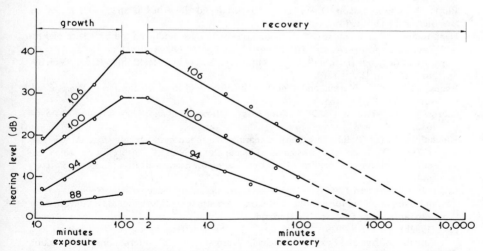

Fig. 8. Composite graph showing growth of, and recovery from, temporary threshold shift. [Data from Ward *et al.* (1958).]

The fact that recovery from TTS takes 16 hr or longer has raised questions as to the nature of the process involved. Such long periods would suggest that more than the normal metabolic restorative processes are involved and that recovery from tissue alteration is occurring.

The problem of recovery from TTS is of great practical importance whenever it is necessary to conduct audiometric tests on noise-exposed workers. As the above data show, even quite moderate exposures, such as 90 dB(A) for 2 hr, can affect the hearing threshold. Normal practice in industrial audiometry is to allow at least 15–16 hr between noise exposure and testing. This is usually achieved by carrying out the tests at the start of the shift or working day. This should be satisfactory in all but the noisiest of working environments, and nowadays there is an increasing likelihood that protection will be worn if levels are very high. The alternative to 15–16 hr without exposure is to make sure that each worker to be tested wears adequate hearing protection for the part of the day or shift before the hearing test. This condition is often difficult to enforce.

REFERENCES

ANSI (1954). Report Z24-x-2, "The Relations of Hearing Loss to Noise Exposure." American National Standards Institute, Boulder, Colorado.
Atherley, G. R. C., and Martin, A. M. (1971). Equivalent continuous noise level as a measure of injury from impact and impulse noise, *Ann. Occup. Hyg.* **14,** 11–28.
BSI (1976). BS 5330, "Method of Test for Estimating the Risk of Hearing Handicap due to Noise Exposure." British Standards Institution, London.
Burns, W. (1971). *In* "Occupational Hearing Loss" (D. W. Robinson, ed.), pp. 63–70. Academic Press, London.
Burns, W., and Robinson, D. W. (1970). "Hearing and Noise in Industry." Her Majesty's Stationery Office, London.
Department of Employment (1972). "Code of Practice for Reducing the Exposure of Employed Persons to Noise." Her Majesty's Stationery Office, London.
Department of Labor (U.S.) (1974). "Occupational Safety and Health Standards," Vol. 39, No. 125, Part II.
Evans, E. F. (1975). Normal and abnormal functioning of the cochlear nerve, *Symp. Zool. Soc. Lond.* **37,** 133–165.
Glorig, A., and Nixon, J. (1962). Hearing loss as a function of age, *Laryngoscope* **72,** 1596–1610.
Hinchcliffe, R. (1959). The threshold of hearing as a function of age, *Acustica* **9,** 303–308.
ISO (1971). ISO—R 1999, "Assessment of Occupational Noise Exposure for Hearing Conservation Purposes." International Organization for Standardization, Geneva, Switzerland.
ISO (1982a). ISO/DIS 1999, "Acoustics—Determination of Occupational Noise Exposure and Estimation of Noise-Induced Hearing Impairment." International Organization for Standardization, Geneva, Switzerland.
ISO (1982b). ISO/DIS 7029, "Acoustics—Threshold of Hearing by Air Conduction as a Function of Age and Sex for Otologically Normal Persons." International Organization for Standardization, Geneva, Switzerland.

Johansen, H. (1943). "Den Altersbedingete Tunghored." Munksgaard, Copenhagen.

Kell, R. L. (1975). Hearing loss in female jute weavers, *Ann. Occup. Hyg.* **18,** 97–108.

Kell, R. L., Pearson, J. C. G., and Taylor, W. (1970). Hearing thresholds of an island population in north Scotland, *Audiology* **9,** 334–349.

Kuzniarz, J. J., Swierozynski, Z., and Lipowczan, A. (1976). Impulse noise-induced hearing loss in industry and the energy concept: A field study, *in* "Disorders of Auditory Function II" (S. D. G. Stephens, ed.), pp. 63–68. Academic Press, London.

Macrae, J. H. (1971). Noise induced hearing loss and presbycusis, *Audiology* **10,** 323–333.

Martin, A. M., and Atherley, G. R. C. (1973). A method for the assessment of impact noise with respect to injury to hearing, *Ann. Occup. Hyg.* **16,** 19–26.

Mollica, V. (1969). Acoustic trauma and presbycusis, *Audiology* **8,** 305–311.

Passchier-Vermeer, W. (1968). "Hearing Loss due to Exposure to Steady-State Broadband Noise." IG-TNO Report 35, Research Institute for Public Health Engineering, TNO, Delft, the Netherlands.

Pearson, J. C. G. (1977). Prediction of presbycusis, *J. Soc. Occup. Med.* **27,** 125–133.

Ramazzini, B. (1964). "Disorders of Workers." Hafner, New York and London. [First published in Latin in 1713.]

Rice, C. G. (1974). Damage risk criteria for impulse and impact noise. Paper read at the 1974 Noise Shock and Vibration Conference, Monash University, Melbourne, Australia.

Rice, C. G., and Martin, A. M. (1973). Impulse noise damage risk criteria, *J. Sound Vib.* **28,** 359–367.

Robinson, D. W. (1968). "The Relationship between Hearing Loss and Noise exposure." NPL Aero Report Ac 32. Teddington, England.

Robinson, D. W. (ed.) (1971). "Occupational Hearing Loss." Academic Press, London.

Spoor, A. (1967). Presbycusis values in relation to noise-induced hearing loss, *Int. Audiol.* **6,** 48–57.

Taylor, W., Pearson, J. C. G., Mair, A., and Burns, W. (1965). A study of noise and hearing in jute weaving, *J. Acoust. Soc. Am.* **38,** 113–120.

Taylor, W., Pearson, J. C. G., and Mair. A. (1967). Hearing thresholds of a non-noise-exposed population in Dundee, *Brit. J. Ind. Med.* **24,** 114–122.

von Gierke, H. E., Robinson, D. W., and Karmy, S. J. (1982). Results of a workshop on impulse noise and auditory hazard, *J. Sound Vib.* **83,** 579–584.

Ward, W. D. (1960). Recovery from high values of TTS, *J. Acoust. Soc. Am.* **32,** 497–500.

Ward, W. D. (ed.) (1968). "Proposed Damage-Risk Criterion for Impulse Noise (Gunfire)." NAS-NRC Committee on Hearing, Bioacoustics and Biomechanics, Working Group 57, Report. Washington D.C.

Ward, W. D., Glorig, A., and Sklar, D. L. (1958). Dependence of TTS at four kc on intensity and time, *J. Acoust. Soc. Am.* **30,** 944–954.

4

Noise and Communication

W. A. AINSWORTH

Department of Communication and Neuroscience
University of Keele
Keele, Staffordshire
England

I. INTRODUCTION

A communication system consists of three parts: a transmitter, a channel and a receiver. In speech communication the transmitter is the talker's vocal apparatus controlled by his musculature and brain, the channel is the air in which he lives and the receiver is the auditory system and brain of the listener. For long-distance communication the channel may also include a device transducing the acoustic vibrations into an electrical signal for transmission either directly, as in telephone systems, or via electromagnetic radiation, as in radio systems. In these cases the channel will include devices for reconstituting the acoustic signal for the listener to hear.

Noise may enter the system at any point. The talker himself may generate the noise by stuttering, by mispronunciation or because of some speech impediment. Noise may be introduced by the auditory system of the listener (tinnitus). In normal speech communication, however, noise is introduced mainly in the channel. The signal is usually contaminated by noise as a result of unwanted sound sources located in the vicinity of the talker or listener.

The acoustic structure of speech signals is very complex. Consequently different types of noise have different effects on speech communication. The loudness of the noise is just one of many parameters which must be considered. In order to understand the effects of noise on speech communication, it is useful to know something of the structure of speech signals.

II. STRUCTURE OF SPEECH SIGNALS

In order to produce speech sounds air is drawn into the lungs by lowering the diaphragm and expanding the rib cage. The diaphragm is then raised and the rib cage contracted, increasing the pressure of the air in the lungs and providing the energy for the sound.

Air flows up the windpipe until it encounters the larynx, a bony structure covered by a skin containing a slit called the glottis. This structure is known as the vocal cords. The pressure of the air opens the slit, but the resulting flow causes the pressure to drop and the tension in the cords closes it. The pressure then builds up again and the process repeats. Thus pulses of air enter the vocal tract. The period and intensity of these pulses are controlled by the pressure of air in the lungs and the tension in the vocal cords.

These pulses of air excite the resonances of the vocal tract. The frequencies of these resonances depend upon the positions of the articulators, the tongue, the lips, the jaw and the velum. This is the mechanism which produces the 'voiced' speech sounds.

Alternatively, the vocal cords are held open and the articulators are maneuvred to produce a constriction in the vocal tract. This constriction produces turbulent airflow which has a noisy, aperiodic waveform. Speech sounds produced in this manner are said to be 'unvoiced'.

The sounds of speech are best represented by phonetic symbols. These are shown, together with their phonetic classes, in Table I.

The vowel sounds are produced by voiced excitation, the different vowels being formed by the position of the tongue hump and the amount of opening of tract caused by lowering the jaw. Vowel sounds are classed as 'front' if the tongue hump is at the front of the mouth or 'back' if it is at the back of the mouth and 'close' if the vocal tract is narrow or 'open' if the tract is wide.

As the positions of the articulators determine the resonances of the

TABLE I. The Speech Sounds of English

Vowels		Voiceless fricatives	
i	team	s	six
ɪ	him	ʃ	ship
ε	head	f	fog
æ	pad	θ	thought
ɑ	card	h	house
ɒ	rod		
ɔ	ford	**Voiced fricatives**	
ʊ	good		
u	true	z	zoo
ʌ	mud	ʒ	azure
з	bird	v	van
ə	the	ð	that
Glides		**Voiceless plosives**	
w	won	p	peep
r	roll	t	tap
l	luck	k	kick
j	yacht		
Nasals		**Voiced plosives**	
		b	big
m	mad	d	dog
n	night	g	get
ŋ	sang		
		Affricates	
		tʃ	church
		dʒ	jack

vocal tract, the differences between the speech sounds are most clearly seen by analysing the signals into their component frequencies. An instrument for performing this task is the sound spectrograph. This produces a display of the short-term spectrum of an acoustic signal as a function of time. This display is known as a spectrogram. Time is displayed along the x-axis, frequency along the y-axis and intensity as the blackness of the trace.

Spectrograms of the vowels /i, u, a/ are shown in Fig. 1a. It will be seen that three or four horizontal bars are clearly visible. These are the major resonances of these vowels and show the frequencies at which the energy is concentrated. These concentrations of energy are known as formants.

The glides are produced in a manner similar to the vowels except that instead of the articulators being held in the appropriate position for a significant time, they only attain that instantaneously. Thus in order to produce /w/ the mouth is made narrow and the lips are rounded, as in /u/, but as soon as the vocal cords begin vibrating the positions of the articula-

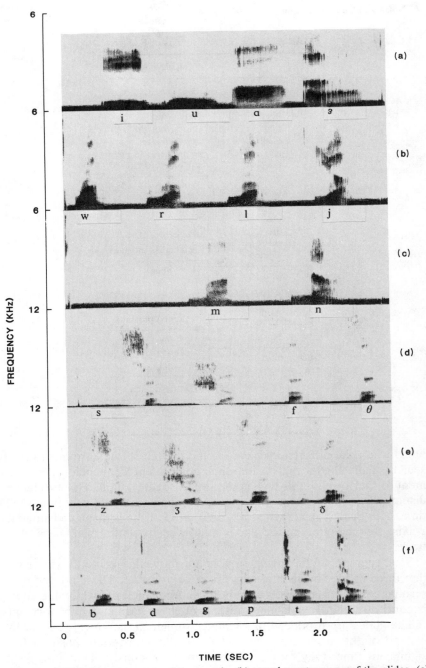

Fig. 1. (a) Sound spectrograms of the vowels, (b) sound spectrograms of the glides, (c) sound spectrograms of the nasals, (d) sound spectrograms of the voiceless fricatives, (e) sound spectrograms of the voiced fricatives and (f) sound spectrograms of the plosives.

tors are moved to those appropriate for the following vowel. This results in changing resonances which are shown as formant transitions on spectrograms (Fig. 1b).

Nasals are also produced similarly, but in this case the oral tract is closed at some point and the velum is opened so that the air flows down the nasal tract and out of the nostrils instead of the mouth. This longer path results in a low-frequency formant which is characteristic of nasals (see Fig. 1c). If the oral tract is closed at the lips, an /m/ results; but if it is closed by placing the tongue against the roof of the mouth, an /n/ is produced. When the velum is closed and the oral tract is opened, there is an abrupt change in the formant structure of the sound.

The fricatives are produced by noise excitation generated at a constriction. Constrictions at different positions of the vocal tract lead to aperiodic signals with energy spreading over different regions of the spectrum. There are two types of fricatives, 'unvoiced' fricatives in which the vocal cords are held open and excitation is random (Fig. 1d) and 'voiced' fricatives in which the vocal cords are allowed to vibrate. In this latter case periodic and random signals excite the resonances of the vocal tract simultaneously (Fig. 1e).

Finally there are the plosive consonants. These are produced by interrupting the flow of air for a short period and then suddenly releasing the pressure that has built up. If the vocal cords are allowed to vibrate at the moment of release, a voiced plosive /b, d, g/ is produced; but if there is a significant delay before the cords begin vibrating, an 'unvoiced' plosive /p, t, k/ is generated (Fig. 1f). An affricate consists of a plosive followed immediately by a fricative.

A spoken utterance can be represented by a string of the symbols shown in Table I. For a complete description, however, an intonation contour, the frequency of vibration of the vocal cords at each instant, must be added.

The result of a spoken utterance is thus a complex waveform whose vibrations represent the positions and motions of the articulators of the talker. In order to understand the message which has been spoken, the listener must analyse and interpret these vibrations. Any additional signals which distort or mask these vibrations will interfere with this analysis. Thus noises of different structures will affect the communication of the various speech sounds in diverse ways.

A. Relative Intensities of Speech Sounds

Because of the different processes involved in producing speech sounds, their intensities differ considerably. Less intense phonemes natu-

TABLE II. Relative Intensity of Each Phoneme Compared
with the Intensity of the Weakest Phoneme[a]

Phoneme	Relative intensity (dB)	Phoneme	Relative intensity (dB)
ɔ	28.3	n	15.6
ɑ	27.8	dʒ	13.6
ʌ	27.1	ʒ	13.0
æ	26.9	z	12.0
ʊ	26.6	s	12.0
ɛ	25.4	t	11.8
u	24.9	g	11.8
ɪ	24.1	k	11.1
i	23.4	v	10.8
r	23.2	ð	10.4
l	20.0	b	8.5
ʃ	19.0	d	8.5
ŋ	18.6	p	7.8
m	17.2	f	7.0
tʃ	16.2	θ	0.0

[a] Data from Fletcher (1953).

rally sound less loud and are thus most easily masked by the presence of
noise.

One useful measure of the intensity of a phoneme is the maximum value
of the energy radiated per 10 msec. whilst the phoneme is being spoken.
The relative intensity of each phoneme is shown in Table II. The weakest
phoneme is /θ/, and the intensities of the other phonemes are shown
relative to this.

In general the vowels are more intense than the consonants, so most of
the energy of the speech wave of a typical utterance is accounted for by
the energy of the vowels. Conversely a good deal of the information about
the message is conveyed by the consonants. Hence it is quite possible to
be able to hear speech without being able to understand its meaning.

III. SPEECH INTELLIGIBILITY IN NOISE

The standard method of assessing a speech communication system is to
have a talker read a list of words or sentences and a group of listeners
write down what they hear. The responses are scored by calculating the
percentage of words correctly heard. The figure obtained is known as the
articulation score, or intelligibility, of the system.

One method of measuring the intensity of speech is by subjectively comparing its loudness with that of a continuous broadband signal of known characteristics. This is then known as the speech level. For example, speech level is defined as being of equal loudness to that of a signal obtained by passing a random noise through a filter with A-weighting characteristics (Rothauser, 1969). Similarly, noise level is the A-weighted sound level of the noise. The signal-to-noise ratio in decibels is then the difference between the speech level and the noise level.

Articulation scores depend upon the intensity and clarity of the voice of the talker, the hearing systems of the listeners and the speech material employed, as well as on the properties of the system. The most-left-hand curve in Fig. 2 shows the articulation score as a function of speech intensity at the ear of the listener for a system containing no noise and employing consonant–vowel–consonant (CVC) nonsense syllables. A practiced announcer read the syllables and the listeners were young adults with no hearing deficits.

It will be seen that an intensity of about 35 dB is necessary for 50% of the syllables to be heard correctly. The maximum intelligibility is achieved at a level of about 80 dB; above this level the intelligibility begins to fall.

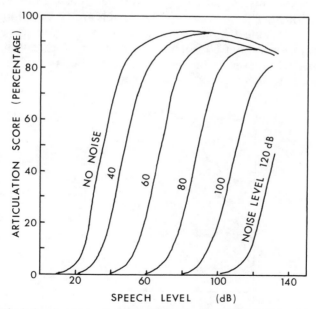

Fig. 2. Articulation score as a function of intensity for CVC syllables with no noise and with added random noise. [After Fletcher (1953).]

When white noise (noise having equal intensities at all frequencies) is added to the speech, the relationship of intelligibility with intensity is similar, but is shifted to the right (Fig. 2). Over the range of intensities normally encountered in speech (40–100 dB) the level of the speech has to be increased by an amount equal to the intensity of the added noise in order to maintain the same level of intelligibility.

A. Effects of Context

The intelligibility of speech in noise depends upon the type of speech material employed. This effect has been demonstrated by Miller *et al.* (1951). They measured the relative intelligibility of the digits, words in sentences and nonsense syllables for various signal-to-noise ratios. The level of the speech remained constant at about 90 dB.

The results obtained are shown in Fig. 3. It will be seen that the signal-to-noise ratios for 50% correct responses are approximately −14 dB for digits, −4 dB for words in sentences and +3 dB for nonsense syllables.

At a *S/N* value of −10 dB practically none of the nonsense syllables were heard correctly, and yet nearly all of the digits were correctly identified. A possible explanation is that all of the digits except 'five' and 'nine' contain different vowels and can therefore be recorded correctly if the vowel is identified. The intensities of the vowels are 10–20 dB higher than the consonants (Table II), so most digits can be identified even though some of their constituent consonants are not perceived. With the nonsense syllables, however, both consonants and vowels must be identified for the syllable to be recorded correctly.

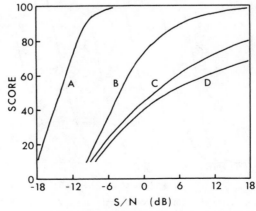

Fig. 3. Intelligibility of spoken digits (A), words in sentences (B), isolated words (C), and nonsense syllables (D) as a function of signal-to-noise ratio. [Data from Miller *et al.* (1951).]

TABLE III. Articulation Score (%) with Various Sizes of Vocabulary as a Function of Signal-to-Noise Ratio[a]

S/N (dB)	Size of vocabulary						
	2	4	8	16	32	256	Monosyllables
−18	51	27	17	13	5		
−15	67	52	32	19	20	2	
−12	87	69	57	51	39	14	3
−9	98	92	89	85	61	28	
−6		94	95	82	81	39	17
0				97	95	70	37
+6						76	53
+12						90	70
+18							82

[a] Data from Miller *et al.* (1951).

Miller *et al.* (1951) also investigated the effect of size of vocabulary on the ability to identify spoken words at various signal-to-noise ratios. They constructed vocabularies from phonetically balanced monosyllables of 2, 4, 8, 16, 32 and 256 words. For the 2-, 4- and 8-word vocabularies the listener ticked the appropriate word on an answer sheet, whereas for the 16-, 32- and 256-word vocabularies he selected the word he heard from a list that he had been given.

The results are shown in Table III, together with those obtained with unrestricted nonsense syllables. It can be seen that as the range of alternatives increases a larger signal-to-noise ratio is required for a given level of intelligibility to be maintained.

In normal conversations, the word which occurs at any instance is restricted by the meaning of the sentence and the grammar of the language. It would thus be expected that words in sentences would be easier to understand than words spoken in isolation. Miller *et al.* (1951) tested this hypothesis by measuring the intelligibility of words in meaningful sentences and the same words, scrambled, and spoken in isolation at various signal-to-noise ratios. The results are shown in Fig. 3. At 6 dB about 90% of the words in the sentences were heard correctly (a high enough level of intelligibility for conversation to take place), and yet only 60% of the same words were identified in isolation.

B. Speech Interrupted by Noise

In everyday situations, such as in the office or factory, the background noise is often not continuous, but occurs in bursts. It is therefore interesting to examine the effects of the frequency and level of noise bursts on the intelligibility of speech.

The relevant experiments were performed by Miller and Licklider (1950). They measured word articulation scores for speech which had random noise added to it 50% of the time. They varied the signal-to-noise ratio and the frequency at which the noise was turned on and off. The speech level was held constant at 90 dB.

The results are shown in Fig. 4. For the intense noise bursts ($S/N = -18$ dB) at low frequencies (5 sec of speech followed by 5 sec of speech plus noise), the noise masks half of the words, so the score is approximately 50%. At noise interruption frequencies of greater than 100 Hz, the noise masks the speech completely. Between these two extremes (1–100 Hz) parts of the words are masked but sufficient of the rest of the word is heard for the word to be identified. The articulation score rises with a maximum at about 10 Hz. At higher signal-to-noise ratios the pattern is similar except that the less intense noise has less masking effect on the speech so the articulation scores are higher. At a signal-to-noise ratio of +9 dB the interrupting noise has little effect on intelligibility.

Miller and Licklider repeated these experiments with irregularly spaced

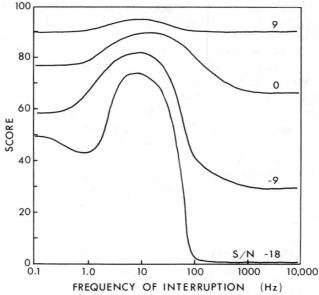

Fig. 4. Intelligibility of speech interrupted by noise 50% of the time, as a function of interruption rate. [After Miller and Licklider (1950).]

bursts of noise instead of periodically interrupted noise. The results they obtained were similar.

C. Effects of Other Sounds

Noise is sometimes defined as unwanted sound. According to this definition there is no requirement for it to possess a random waveform or a flat spectrum. Signals such as pure tones and square waves can be considered to be noise if they interfere with speech communication. Whistles, sirens and fog horns fall into this category.

The effects of sine waves, square waves and pulses on speech communication was studied by Stevens *et al*. (1946). The technique they employed was to listen to a high-fidelity recording of Adam Smith's "The Wealth of Nations" (chosen for its uniform level of difficulty and interest) with the interfering signal added to it. The intensity of the speech signal was adjusted until the meaning could just be followed.

The results for added sine waves showed that the level to which the speech had to be set for various intensities of tone was a function of the frequency of the tone. The maximum interference occurred at about 500 Hz for the weak tones (60 dB) and at about 300 Hz for the more intense tones (110 dB). This might be predicted from experiments with pure tones from which it is known that intense low-frequency tones mask higher tones but not the other way around. This phenomenon is known as upward spread of masking.

Similar results were obtained with square waves and pulses of 10 μsec duration, except that in these cases the maximum interference was caused when the repetition frequency of the signal was about 200 Hz and this peak was not dependent upon intensity. This lack of intensity dependence was probably due to the fact that square waves and pulses have harmonics throughout the spectrum.

D. Consonant Confusions

So far the effects of noise on speech communication have been considered only in terms of the articulation score. This is a useful measure for summarising the effectiveness of a communication system, but it has little diagnostic value. It is more helpful to know which speech sounds are confused. A good deal is now known about the structure of speech sounds and about what features of the signals are relevant for speech perception (Ainsworth, 1976). If the confusions which occur are known, it is sometimes possible to redesign the communication system so that the relevant features are not masked by noise.

Miller and Nicely (1955) have examined two confusions which occur among some of the consonants when random noise is added to spoken syllables. They chose the consonants /p, t, k, f, θ, s, ʃ, b, d, g, v, ð, z, ʒ, m, n/ combined with the vowel /ɑ/. It will be seen from Table II that apart from the affricates /tʃ/ and /dʒ/, which are compounds, and the nasal /ŋ/, which does not occur in pre-vowel position in English, these 16 are the least intense of the phonemes and so are the ones most likely to be confused when noise is added.

Five American female subjects served as talkers and listeners in the experiments. They took turns so that one read the list of syllables and the other four recorded what they heard. The tests were repeated at signal-to-noise ratios of −18, −12, −6, 0, +6 and +12 dB with a frequency bandwidth of 200 to 6500 Hz.

The average scores are shown in Table IV. Miller and Nicely give confusion matrices for the 16 consonants for each of the S/N ratios, but the most interesting is at −6 dB, where the average score was near to 50%. The confusion matrix for this level is given in Table V. The symbols in the left-hand column show the consonant spoken and the symbols in the top row show the consonant recorded by the listener. Numbers in the cells show the number of times that the spoken consonant was heard as each consonant.

It will be seen that at S/N of −6 dB the main confusions are between the consonants within each class (the voiceless plosives /p, t, k/, for example, are mostly confused with voiceless plosives). The main confusions among classes are between voiceless plosives and voiceless fricatives and between voiced plosives and voiced fricatives. Nasals are mainly confused with each other.

At higher signal-to-noise levels there are fewer inter-class confusions, but the intra-class confusions remain. At +6 dB the voiceless plosives are

TABLE IV. Analysis of Correct Responses to Consonants in Noise in Terms of Linguistic Features[a,b]

S/N (dB)	Percentage correct	Voicing	Nasality	Affriction	Duration	Place
−18	7.8	57.2	80.0	50.8	65.0	40.0
−12	27.0	89.7	93.6	62.3	77.4	46.1
−6	46.5	96.9	98.0	80.1	85.7	59.7
0	71.6	99.4	99.8	92.7	94.5	77.5
6	83.4	99.6	99.9	96.5	97.8	86.8
12	90.9	99.6	99.9	97.9	99.5	92.3

[a] Data from Miller and Nicely (1955).
[b] The percentage of times each feature was correctly recognised is shown for each signal-to-noise ratio.

TABLE V. Confusion Matrix for Consonants at a Signal-to-Noise Ratio of −6 dB[a]

									Heard									
	p	t	k	f	θ	s	ʃ	b	d	g	v	ð	z	ʒ	m	n
p	80	43	64	17	14	6	2	1	1	0	1	1	0	0	2	0
t	71	84	55	5	9	3	8	1	0	0	0	1	2	0	2	3
k	66	76	107	12	8	9	4	0	0	0	0	1	0	0	1	0
f	18	12	9	175	48	11	1	7	2	1	2	2	0	0	0	0
θ	19	17	16	104	64	32	7	5	4	5	6	4	5	0	0	0
s	8	5	4	23	39	107	45	4	2	3	1	1	3	2	0	1
ʃ	1	6	3	4	6	29	195	0	3	0	0	0	0	0	0	1
b	1	0	0	5	4	4	0	136	10	9	47	16	6	1	5	4
d	0	0	0	0	0	0	8	5	80	45	11	20	20	26	1	0
g	0	0	0	0	2	0	0	3	63	66	3	19	37	56	0	3
v	0	0	0	2	0	2	0	48	5	5	145	45	12	0	4	0
ð	0	0	0	0	6	0	0	31	6	17	86	58	21	5	6	4
z	0	0	0	0	1	1	1	7	20	27	16	28	94	44	0	1
ʒ	0	0	0	0	0	0	0	1	26	18	3	8	45	129	0	2
m	1	0	0	0	4	0	0	4	0	0	4	1	3	0	177	46
n	0	0	0	0	0	0	0	1	5	2	0	7	1	6	47	163

(Spoken, rows at left)

[a] After Miller and Nicely (1955).

little confused with the voiceless fricatives, but they are still confused with each other.

In order to summarise their results, Miller and Nicely classify the phonemes in terms of the features of the articulatory process used to generate the sounds. The linguistic features adopted were as follows.

(1) *Voicing:* In /b, d, g, v, ð, z, ʒ, m, n/ the vocal words vibrate, whereas in /p, t, k, f, θ, s, ʃ/ they do not. The first set have a periodic acoustic structure, but the second set have a noisy structure.

(2) *Nasality:* The phonemes /m, n/ contain a nasal resonance, whereas the others do not.

(3) *Affriction:* In /f, θ, s, ʃ, v, ð, z, ʒ/ there is a constriction through which there is a turbulent flow of air, whilst in /b, d, g, p, t, k, m, n/ the oral tract is closed.

(4) *Duration:* The phonemes /s, ʃ, z, ʒ/ have a longer duration than the others. They also contain intense, high-frequency noise.

(5) *Place:* This feature describes the position of the major constriction of the vocal tract. The phonemes /p, b, f, v, m/ are classed as front, /t, d, θ, s, ð, z, n/ as middle and /k, g, ʃ, ʒ/ as back.

The results are given in Table IV, which shows the percentage of times each feature was most correctly recognised. It will be seen that place was most affected by added noise. Affriction and duration were less affected than place, but voicing and nasality were least affected.

It is interesting to note that place, which is most affected by added noise, is the feature which is most easily seen by lip reading, whereas voicing and nasality are invisible.

E. Speech Babble

So far only the interfering effects of mathematically well-defined signals on speech perception have been considered. These are useful for predicting the effects of random sounds (such as traffic noise) and periodic sounds (such as that generated by machinery) on speech communication, but there is another common signal which interferes with speech comprehension: speech from other talkers.

Cherry (1953) has reported that when two passages spoken by the same talker are played simultaneously, the listener has great difficulty in separating them. If the mixed signal is repeated many times, however, the listener manages eventually to understand each passage. On the other hand, if the two passages are presented to different ears, the listener has no difficulty in following the passage in one ear to the almost total exclusion of the one in the other ear.

Carhart *et al.* (1975) have investigated the effects of adding different numbers of other voices on the intelligibility of speech. They used isolated words (spondees) as their test material and compared the masking of combinations of talkers with the masking obtained with modulated noise.

They found that combinations of talkers (speech babble) had a greater interfering effect than did the noise. Adding one talker affected the intelligibility score by the same amount as increasing the noise by 6.2 dB. Two talkers had the same effect as increasing the noise by 7.2 dB, and three talkers were equivalent to a noise increase of 9.8 dB. Thereafter increasing the number of talkers had less effect, with 64 talkers being equivalent to a noise increase of only 3 dB.

IV. SPEECH AUDIOMETRY

The standard method of assessing hearing is by pure-tone audiometry. The intensity of a tone is adjusted until it is just audible. This process is either repeated at a few discrete frequencies or the tone is swept through the spectrum and some tracking method is used to determine the threshold.

Many cases have been found in which the audiogram suggests that sounds can be heard throughout a sufficient range of frequencies, and yet the person has great difficulty in comprehending speech. In order to detect such cases a technique known as speech audiometry has been developed.

In this technique, spoken words are presented, and the subject is asked to repeat what he hears. Suitable lists of words are given by Fry (1961). The number of items (usually expressed as a percentage) correctly identified at a particular sound-pressure level is known as the discrimination score. The curve drawn through a number of scores obtained over a range of sound-pressure levels is the speech audiogram. The results obtained are similar to the curves shown in Fig. 2, but if the subject has defective hearing, the curves are displaced to the right and the articulation score may reach a peak at a lower value (Cooper and Cutts, 1971). These tests are difficult to administer since many lists have to be used to achieve statistical significance, and the intelligence and linguistic experience of the subject may distort the results.

In addition, the results of these tests are difficult to interpret. In speech perception it is necessary not only to be able to hear a tone (as tested by pure-tone audiometry), but also to be able to hear one tone in the presence of another (tone discrimination). As well as this the auditory impression must be compared in some way with a stored version of the word obtained from previous experience. Speech audiometry fails to distinguish between these latter two mechanisms.

Speech audiometry, however, is employed for two main purposes: to help diagnose the cause of hearing disorders and to test the effectiveness of hearing aids. Sometimes these objectives are combined, for example, in selecting a suitable hearing aid for alleviating a particular hearing disorder.

There are two main techniques employed in speech audiometry. In one, the talker reads a list of words or sentences and the listener repeats them or writes them down. In the other, recorded material is used. Advocates of the first technique insist that speech communication takes place in a live, face-to-face situation, and using recorded material is unnatural to such an extent that it invalidates any results obtained. Some authors have argued that even reading lists of sentences is unnatural, and a better method is for the talker to read a passage phrase by phrase, and the listener repeat it back to him. If any errors are made, the last phrase is repeated until it has been heard correctly. A score is obtained in terms of the time taken to complete the passage. This technique, of course, is no good for diagnosis, but it could be used to compare the effectiveness of two hearing aids.

The main virtue of using recorded material is that of standardisation. The same test may be administered to many subjects on different occasions. However, depending upon the memory of the listener and the interval between tests, it is unwise to use the same list with the same subject on more than one occasion. For this reason, a number of sets of lists have been developed containing equal numbers of familiar and unfamiliar words. One of the earliest of these were the Harvard phonetically balanced lists (PB lists). These consist of 20 lists of 50 monosyllabic words, each containing the proportions of frequent and infrequent words and also approximately the same number of phonemes of each class. A good deal of testing was carried out to ensure that the articulation score was the same in a variety of listening conditions (Egan, 1948).

Lists such as these can be used for diagnostic purposes since a phoneme confusion matrix can be constructed from an analysis of the results. (A patient suffering from a high-frequency hearing loss, for example, might be expected to confuse voiceless fricatives.) This analysis, however, is time-consuming and tedious, and other tests have been developed to obtain the same results more efficiently. One technique is to restrict the listeners' choice of response, for example, as in the six-alternative rhyme test of Fairbanks (1958).

One modern test is the two-alternative forced-choice test developed by Grose and Pick (1979). In this test the subject is given a response sheet listing pairs of words which differ in only a single phoneme (e.g., 'cop' and 'top') and is required to circle the one he hears. A similar test called the FAAF (four-alternative auditory feature test) has been developed by Foster and Haggard (1979). As the name implies, the listener is able to choose one from four alternative responses. Both of these tests are somewhat artificial, but they are very efficient means of determining whether certain features of speech can be distinguished.

One attempt to make speech audiometry more natural was made by Kalikow *et al.* (1977). They suggest that instead of employing isolated words, it is more natural to use sentences. The complications involved in scoring the comprehension of sentences are avoided by asking the listener to repeat (or write down) just the last word of the sentence.

Using sentences enables the predictability of the test word to be controlled. For example, in the sentence 'The boat sailed across the bay' the predictability of the word 'bay' is high, whereas in the sentence 'He was talking about the bay' the predictability of 'bay' is low. Kalikow *et al.* found that with a signal-to-noise ratio of -2 dB (the noise consisting of speech babble from six talkers), the intelligibility of high-predictability words in sentences was 79%, whereas that of low-predictability words was 42%.

It is expected that the scores obtained with these tests will enable the ability of a hearing-impaired listener to perform in everyday communication situations to be predicted with much more reliability than has been possible hitherto.

V. CONCLUSIONS

There are many factors which influence speech communication in noise. The variety of intensities and structures of speech sounds make it inevitable that some sounds will be masked by lower levels of noise than others. Yet the inherent redundancy of natural language enables the listener to decode the meaning of a sentence when many of the words are obscured by noise.

Different kinds of noise (random, periodic, interrupted) have different effects on speech communication, and the intensities of the frequencies present have a crucial influence on whether the message is received. It is interesting to note, however, that the kind of noise which causes the greatest disruption to speech communication is speech itself.

REFERENCES

Ainsworth, W. A. (1976). "Mechanisms of Speech Recognition." Pergamon, Oxford.
Carhart, R., Johnson, C., and Goodman, J. (1975). Perceptual masking of spondees by a combination of talkers, *J. Acoust. Soc. Am.* **58,** 535(A).

Cherry, E. C. (1953). Some experiments on the recognition of speech, with one or two ears, *J. Acoust. Soc. Am.* **25,** 975–979.

Cooper, J. C., and Cutts, B. P. (1971). Speech discrimination in noise, *J. Speech Hear. Res.* **14,** 332–337.

Egan, J. (1948). Articulation testing methods, *Laryngoscope* **58,** 955–991.

Fairbanks, G. (1958). Test of phonemic differentiation: The rhyme test, *J. Acoust. Soc. Am.* **30,** 596–600.

Fletcher, H. (1953). "Speech and Hearing in Communication." Van Nostrand, New York.

Foster, J. R., and Haggard, M. P. (1979). FAAF—An efficient analytical test of speech perception, *Proc. Inst. Acoust.* (Autumn), 9–12.

Fry, D. B. (1961). Word and sentence tests for use in speech audiometry, *Lancet* **2**(July 22), 197–199.

Grose, J. H., and Pick, G. F. (1979). The calibration and validation of a two-alternative forced-choice test for hearing loss of cochlear origin, *Proc. Inst. Acoust.* (Autumn), 13–16.

Kalikow, D. N., Stevens, K. N., and Elliott, L. L. (1977). Test of speech intelligibility in noise, *J. Acoust. Soc. Am.* **61,** 1337–1351.

Miller, G. A., and Licklider, J. C. R. (1950). The intelligibility of interrupted speech, *J. Acoust. Soc. Am.* **22,** 167–173.

Miller, G. A., and Nicely, P. E. (1955). An analysis of perceptual confusions among some English consonants, *J. Acoust. Soc. Am.* **27,** 338–352.

Miller, G. A., Heise, G. A., and Lichten, W. (1951). The intelligibility of speech as a function of the context of speech materials, *J. Exper. Psych.* **41,** 329–335.

Rothauser, E. H. (1969). IEEE recommended practice for speech quality measurements, *IEEE Trans. Audio-Electroacoust.* **AU-17,** 226–246.

Stevens, S. S., Miller, J., and Truscott, I. (1946). The masking of speech by sine waves, square waves, and regular and modulated pulses, *J. Acoust. Soc. Am.* **18,** 418–424.

5

Noise and Efficiency

D. R. DAVIES
Applied Psychology Division
University of Aston in Birmingham
Birmingham
England

D. M. JONES
Department of Applied Psychology
University of Wales Institute of Science and Technology
Penylan, Cardiff
Wales

I. INTRODUCTION

Noise is usually defined as unwanted sound, and three main behavioural effects of exposure to it may be distinguished. First, noise can impair hearing and make spoken communication more difficult to achieve [see Davies and Jones (1982) for a review]; second, noise may produce feelings of annoyance and irritation, although there are considerable individual differences in the response to noise [see Jones and Davies (1984) for a review]; third, noise may affect the efficiency with which tasks are performed, usually for the worse but occasionally for the better. The effects of comparatively high levels of noise, presented either intermittently or continuously, upon task performance are the concern of this chapter, in which we focus upon research conducted since 1950 and, in view of the comparative dearth of industrial studies carried out during this period, primarily upon the findings of laboratory experiments.

A. Historical Background

As Hockey (1984) points out, it is now fairly simple to demonstrate that noise can affect task performance. Yet up until the early 1950s there was considerable doubt as to whether noise exerted any significant influence on performance at all, as two early reviews of the effects of noise on behaviour indicate. Berrien (1946), in a review mainly concerned with industrial research, concluded that although the results of the majority of studies were inconclusive and ambiguous, there were nevertheless some indications that noise exerted a slight effect upon output and upon speed of work. Kryter (1950) in a more extensive review, which placed much greater emphasis upon the results of laboratory studies, concluded that such studies could be divided into three categories. Kryter considered that findings from the first category, comprising studies demonstrating adverse effects of noise, could be largely discounted on methodological grounds and that findings from the remaining two categories, containing studies either demonstrating slight or inconclusive effects of noise or showing that mental and motor performance were unimpaired by noise, were rather more reliable. Thus, around 1950, the general view of the effects of noise can be broadly summarised as being that noise had not been shown to have any effect on efficiency of practical importance and that industrial noise reduction programmes could therefore be regarded as likely to produce little significant improvement in output.

The military and economic pressures of the Second World War resulted in many psychologists becoming involved with the selection and training problems of the armed forces, one of whose main priorities was to train their personnel as quickly and effectively as possible to perform novel and complex tasks often requiring high levels of skill. At the end of the war, therefore, research on human skilled performance had developed considerably and had become one of the central concerns of experimental psychology. In addition to the fundamental reappraisal of theories of skilled performance that took place during the late 1940s, due mainly to the work of Craik and Bartlett in England and of Fitts in the United States, new tasks, which psychologists had not previously utilised, became available for the investigation of human performance. Many of them simulated actual military tasks, for example, the job of the aircraft pilot or that of the airborne radar operator, and during the 1950s performance at several of these newly developed tasks was investigated under conditions of loud noise by psychologists working in both military and civilian research settings. The principal features of these tasks were that they involved sustained attention, were performed continuously without a break and lasted for at least 30 min and frequently for longer. Examples are vigilance, tracking and the five-choice serial reaction task, which are further discussed later. The model of skilled performance guiding the selection of such tasks for investigation was that of a self-regulatory system which could compensate for below average efficiency at certain times by greater than average efficiency at others, thereby maintaining a stable level of performance until the regulatory process itself became disrupted through fatigue. Noise was assumed to increase the number of periods of inefficiency but would only affect the overall level of performance in continuous tasks of long duration, due to poorer performance towards the end of the task resulting from the accumulation of fatigue.

During the 1960s a number of studies were conducted which made detailed comparisons of the effects of noise and those of other stressors, such as sleep deprivation and environmental warmth, and of motivational factors such as financial incentives and knowledge of results upon the performance of vigilance and serial reaction time tasks. In addition the changes in efficiency that occurred with the simultaneous administration of two stressors, or of a stressor and a motivational factor, were examined (Poulton, 1966; Wilkinson, 1969); some of the main findings from studies using noise as one of the stressors are described in this section. Out of this work came the 'arousal theory of stress' (Broadbent, 1963, 1971), which assumes that there is a general state of arousal or reactivity which is increased by noise or by incentives and reduced by boredom or loss of sleep. The application of arousal theory to the explanation of the effects of various stressors and motivational factors rests on two main assumptions:

(1) *The Inverted-U Hypothesis:* This hypothesis assumes that the relation between the level of arousal and the level of performance follows an inverted U, so that performance is optimal at some intermediate level of arousal and poor when arousal level is either too high or too low [see Duffy (1957, 1962) and Malmo (1959)].

(2) *The Yerkes–Dodson Law:* This law is derived from work by Yerkes and Dodson (1908) concerning the relation between the intensity

of electric shock and the difficulty of a discrimination problem upon learning in mice; the law implies that the optimal level of arousal for the
performance of a particular task is inversely related to task difficulty, so
that the performance of more difficult tasks will be impaired at lower
levels of arousal than will the performance of easier tasks [see Easterbrook (1959)].

For a variety of reasons [see Broadbent (1971), Eysenck and Folkard
(1980), Gale (1977), Hockey (1979, 1984), Hockey and Hamilton (1983) for
detailed discussion] the unidimensional arousal model has come to be
regarded as insufficiently robust to account either for the effects of all
stressors, whether administered singly or in combination, or for all aspects of the experimental results obtained with a particular stressor such
as noise. For instance, as noted earlier, noise is considered to raise the
level of arousal, a view for which there is some support from psychophysiological studies [see Davies (1968, 1976) for reviews]. However, the
effects of noise on performance in different task situations are not entirely
consistent with those of other variables which are also assumed to increase arousal level, such as stimulating drugs, incentives of various
kinds or late afternoon and evening testing [see Hockey and Hamilton
(1983)]. Increasingly, therefore, arousal is regarded as a multidimensional
rather than as a unidimensional state. For example, the 'variable state'
approach proposed by Hockey (1984) suggests that there are a number of
specific states each associated with the presence of a particular stressor.
An advantage of this approach is that it focusses upon differences *within*
the categories of 'high-arousal' and 'low-arousal' variables in their effects
on performance as well as upon differences between the two sets of
variables. Such an approach is, in principle, able to provide detailed
'maps' of the various stress states (for instance, the 'noise' state) through
an examination, in as wide a range of task situations as possible, of the
patterns of performance change typically associated with each state.

Two complementary research strategies have been employed in the
investigation of the effects of stress upon performance. The first examines
the effects of a single stressor upon the performance of a wide range of
tasks; the principal results obtained with this approach using noise as the
stressor are described in Section II. The second strategy compares the
effects of a number of different stressors upon the performance of a single
task or upon the performance of a narrow range of similar tasks; the main
findings obtained with this approach are outlined in Section III, which
focusses upon the interaction of noise with stressors such as sleep deprivation and environmental warmth and with motivational variables such as
incentives and knowledge of results, although a brief account of the simi-

larities and differences among the effects of various stressors administered singly is also provided. Data accumulated through the use of these
two research strategies has enabled a sketch-map of the 'noise state' to be
gradually built up, although, as will become apparent, a detailed picture of
the terrain remains to be drawn. In Section IV we consider the influence
exerted by situational and contextual variables upon the effects of noise,
placing particular emphasis upon the perception of control. In Section V
we examine the relatively few studies of the effects of noise upon performance in real-world situations, and in Section VI we discuss the various
explanations of the effects of noise upon performance that have been
advanced. The final section, Section VII, provides a brief summary of the
chapter.

II. EFFECTS OF NOISE ON TASK PERFORMANCE

In this section we focus upon laboratory studies of the effects of noise
upon task performance, and it is worth indicating at the outset that the
interpretation of data from such studies is frequently limited by methodological considerations. First, as reviewers of the effects of noise have
often emphasised, the noise levels selected for experimental (noise) and
control (quiet) conditions of noise experiments vary widely among studies; for instance, in some experiments the sound level of the noise condition may be as low as 50 dB, while in others that of the quiet condition
may be as high as 80 dB. Sometimes, too, the frequency composition of
the noise used is unspecified. Discrepancies among the results of different
studies, in which performance on similar tasks has been examined, may
thus be attributable to the use of different kinds or levels of noise. The
type of noise used in experiments on noise and performance can vary
from relatively infrequent bursts of noise, presented during particular
phases of task performance, through intermittent noise presented on a
regular or an irregular schedule, to noise that is presented continuously,
although the level of continuous noise may occasionally be varied upwards or downwards within the same work session. In the majority of
laboratory studies of the effects of noise, investigators have used 'white'
noise, in which the constituent frequencies, taken from a wide range of
the audio-frequency spectrum, are of equal intensity. 'Pink' or 'green'
noise, in which the constituent frequencies are predominantly high or
low, respectively, have been employed comparatively rarely. As far as
noise intensity is concerned, virtually all the studies to be discussed here
have used for the noise condition noise levels within the range 80-dB SPL
(sound-pressure level) to 110 dB-SPL, although the utilisation of sound-

level meters incorporating different weighting networks, thus yielding most commonly, intensity levels in terms of decibels(A) or decibels(C) is a complicating factor in the comparison of different studies [see Bruel (1976), Davies and Jones (1982) and Chapter 1 of this volume for elaboration].

Second, when noise seems to exert no reliable effect upon the performance of a particular task, it may be that performance is genuinely unaffected by noise or that the measure of performance selected, or the task yielding the measure, is insufficiently sensitive to the chosen level of noise [see Poulton (1965), Wilkinson (1969)]. However, the absence of a change in performance does not preclude a change in the way in which the task is carried out. Third, the choice of experimental design can influence the outcome of noise experiments and the employment of a repeated measures design, in which the same people perform the task in both noise and quiet conditions, may result in incorrect inferences being drawn with respect to the effects of noise in a particular task situation (Poulton and Edwards, 1979; Poulton and Freeman, 1966). It is customary to attempt to control for the effects of transfer, that is, the effect of a first attempt at a task upon performance at the second attempt, by balancing the order in which people experience the two conditions (say, noise and quiet). However, this control is ineffectual if transfer is asymmetrical, in that the effect of performing first in noise upon later performance in quiet is not equal to the effect of performing first in quiet upon later performance in noise. In such a case the only valid comparison that can be made between conditions A and B is to compare them when each was performed first, that is, to use in effect an independent groups design. However, the presence of asymmetrical transfer can usually be detected fairly easily, and where necessary, conclusions concerning the effect of a particular stressor can be modified. It is useful, nevertheless, to make certain that sufficient practice at the task has been given for performance levels to stabilize before a stressor is introduced. Such a procedure does not seem to reduce the chances of obtaining reliable effects of noise (Cohen *et al.*, 1973) and has the advantage of ensuring that where a treatment effect is obtained it is indicative of an effect upon *performance* rather than upon learning. In addition to the effects of previous experience with the task, those of acclimatization or adaptation to the stressor can also be mentioned since, for a number of stressors, prior adaptation has been shown to reduce the effects of stress on performance (Wilkinson, 1969).

For most areas of research on human performance, adequate task classification schemes do not exist, although important contributions to the development of such taxonomies have been made by Fleishman and his associates [for example, Fleishman (1975)] and these seem likely to prove useful in assessing the effects of noise on performance (Theologus *et al.*, 1974). One consequence of the lack of adequate task taxonomies is that in many cases it is difficult to specify with any confidence the practical implications of laboratory findings, since it is often unclear what aspect of general efficiency is being tapped by a particular task. Any task classification scheme adopted for review purposes is thus inevitably somewhat arbitrary and will be selected on grounds of convenience as much as anything else. We have classified the tasks surveyed in this section under four main headings: (1) perceptual-motor performance, (2) selective and sustained attention, (3) learning and memory and (4) mental performance.

A. Perceptual-Motor Performance

Under this heading we examine the effects of noise upon reaction time and serial responding, upon tracking and upon psychomotor performance.

1. Reaction Time and Serial Responding

Very few experiments have been conducted on the effects of noise upon discrete reaction time (RT), that is, where following the response to an RT stimulus, there is a short pause before the presentation of the next trial. In serial or continuous RT tasks, such as the five-choice serial reaction task described later, the individual's response to one stimulus triggers the presentation of another. Reaction time tasks can be further divided into simple RT tasks, in which there is only one stimulus, and disjunctive, discrimination or choice RT tasks, in which an appropriate response must be made as quickly as possible to one of a number of possible stimuli. In some such tasks the response must be withheld rather than emitted when a particular stimulus is presented; for example, individuals may be required to respond to a green light but not to a red one.

It is clear that noise presented at quite moderate levels can significantly lengthen simple RT, although it does not invariably do so. Kallman and Isaac (1977), using a repeated measures design in which the same people experienced all possible combinations of two noise conditions and two illumination conditions, found that continuous noise at a level of 69 dB(A) reliably increased simple tactile RT, although only when the ambient illumination was high. In an extremely thorough study, Theologus *et al.* (1974) tested independent groups of people for three sessions, the first of which was spent in task familiarisation and practice, on a 20-min simple visual RT task. Individuals in the noise group were exposed to 85-dB noise bursts which were either random (session 2) or patterned (session 3). In the random condition the burst duration and the inter-burst interval were varied, while in the patterned condition these parameters were fixed at 5 and 2 sec, respectively. Theologus *et al.* found that performance showed considerable improvement across sessions with the noise group showing the greater improvement, although, as they themselves note, their experimental design confounded the effects of practice across sessions with those of intermittent noise type. Compared to quiet, performance during exposure to random intermittent noise was reliably worse in the first half of the task but not in the second, suggesting that some adaptation to noise had occurred. The main effect of noise was upon performance rate, as assessed by the mean response latency (the mean of all RTs) and the mean reaction time (the mean of all RTs excluding those two or more times the mean of the 20 fastest response latencies, designated as 'response blocks'). There was no effect of noise upon the number or duration of such blocks: hence the effect of noise was to make all responses slightly slower, rather than to alter the distribution of response latencies by increasing the number of unduly long response times.

When two stimuli are presented for response in rapid succession, that is, at inter-stimulus intervals (ISIs) ranging from about 20 to 200 msec, the response to the second stimulus tends to be unduly delayed, a phenomenon known as the psychological refractory period [see Smith (1967)]. Goolkasian and Edwards (1977) found that a noise level of 95 dB(C), compared to a quiet condition of 55 dB(C), significantly increased the psychological refractory period, as indicated by the RT to the second of two successive visual stimuli, while minimally affecting the RT to the first. This result was obtained with ISIs of 20, 50, 100 and 200 msec, each being employed 60 times. The rapid presentation of two stimuli, each requiring a response, thus seems to make performance vulnerable to noise.

A similar effect can be observed in the performance of serial RT tasks, and the five-choice serial RT task (Leonard, 1959) has probably been the task most extensively studied under conditions of loud noise. The task consists of five small light bulbs and five brass discs, each bulb corresponding to a disc. The individual performing the task is required to tap with a metal stylus the disc corresponding to a lighted bulb, whereupon the bulb is extinguished and another bulb is lit. The lighting of a bulb is contingent upon a disc, not necessarily the correct one, being tapped. Three performance scores can be derived from the serial reaction task: the number of correct responses, the number of errors and the number of long response times (1500 msec or longer), which are known as 'gaps'. The task duration generally ranges from 25 to 40 min and performance scores are usually obtained for 5- or 10-min time blocks in order to assess the effects of time on task. The five-choice serial reaction task thus lasts for a relatively long time, is performed continuously without a break and is self-paced (that is, the rate of work is determined by the individual performing the task rather than by the experimenter).

We now consider the results of 10 experiments employing the five-choice serial reaction task and 1 employing a four-choice serial reaction task, which is described in more detail later. Most of these experiments (9 out of 11) have employed repeated measures designs, in which the same individuals have been exposed to more than one treatment and in which it is thus possible that asymmetrical transfer may have influenced the results obtained. While the results of studies employing repeated measures designs are fairly consistent [with one exception, see Poulton and Edwards (1979)] and similar findings have also been obtained from studies employing independent groups designs (in which different groups of individuals are exposed to different treatment conditions), little attempt has been made to test for the presence of asymmetrical transfer in those experiments in which it may be operating.

Of 10 experiments in which the five-choice serial reaction time task has been used, 6 (Broadbent, 1953, 1957a, experiment 1; Hartley, 1973, 1974, experiment 1; Hartley and Carpenter, 1974; Wilkinson, 1963, experiment 1) find that errors are reliably increased in continuous noise at levels of either 95 dB(C) or 100 dB. Only 3 (Hartley, 1973, 1974, experiment 2; Hartley and Carpenter, 1974) find that gaps are reliably increased in continuous noise, and in these experiments the light display was separated from the response panel and reclined at an angle of about 30° from the vertical. People work more slowly at this task, producing around 100 responses per minute, compared to 170 responses per minute when the bulbs and discs are set into the same board (Poulton, 1977a). It seems likely that the slower rate of responding increases the probability of emitting gaps, provided that the distribution of fast and slow responses remains much the same in the two versions of the task, since the cut-off point (1500 msec) for a gap is fixed. It would perhaps be more useful to examine instead the number of response times that are greater than two standard deviations above the mean for each individual's responses. None of the 10 experiments finds any significant effect of noise upon the number of correct responses.

Thus the main way in which continuous noise impairs the performance of the five-choice serial reaction task is by increasing errors. Loud noise in which the higher frequencies predominate is more likely to impair performance than is noise containing predominantly lower frequencies. Broadbent (1957a) compared the effects of low-frequency noise containing equal energy in the 100–2000-Hz range with those of high-frequency noise containing equal energy in the 2000–5000-Hz range at levels of 80,

90 and 100 dB. Errors were significantly increased only in the 100-dB high-frequency noise. Hartley (1974) found that intermittent noise with an on-time of roughly 60%, and alternating between 95 and 70 dB(C), increased the number of errors by about the same amount as did continuous 95-dB(C) noise, relative to the 70-dB(C) quiet condition. Hartley also found that ear defenders did not reliably influence the effects of noise upon errors. Hartley and Carpenter (1974) found that more errors were made in 95-dB(C) free-field noise than in 95-dB(C) headphone noise, compared to a 70-dB(C) quiet condition, but this difference became significant only in the second half of the task. Fisher (1972, 1973) examined the effects of 80-dB noise bursts, each lasting for 2 sec, presented randomly during performance of the five-choice task. She found that performance was only impaired when the onset of a noise burst occurred during the production of a response, although the impairment was shortlived.

It seems, therefore, that when a sequence of rapid actions has to be carried out over a relatively long period of time in response to an unpredictable sequence of inputs, the accuracy of performance is reduced in noise (in that errors increase), while the speed of work is unchanged (in that the number of correct responses remains about the same). Noise may thus be described as affecting the balance, or trade-off, between the speed and accuracy of performance. People attempt to maintain their normal rate of working in noise, but this can only be achieved at the cost of making mistakes. Three main explanations have been proposed as to why this shift occurs. First, and most obviously, noise may act as a distractor, interfering either with the intake of task information or with the selection of an appropriate response, or both. Broadbent (1957b, 1958a) suggested that loud noise produced involuntary interruptions in the intake of task information, referred to as 'internal blinks' resulting in brief periods of perceptual inefficiency in which attention is diverted away from the task. In repeated measures designs the same individuals perform the same task several times and become highly practised. Highly practised individuals will continue to respond even when the intake of task information is interrupted since they can anticipate what is likely to happen. People working at the five-choice task in noise will occasionally fail to pay attention to the information provided by the task and their responses will instead be based upon their estimate of which light will come on next. These guesses will sometimes be wrong and will result in errors. The longer people work in noise, the more easily distracted they become, since the task has lost its novelty, and hence errors tend to increase in noise as a function of time at work.

The 'distraction' hypothesis just briefly outlined assumed that the occurrence of both errors and gaps in serial responding could be explained in terms of failures of attention. The hypothesis was abandoned when it was discovered that the application of other stressors to the five-choice task, in particular heat and loss of sleep, suggested that gaps and errors were not caused in the same way [see Broadbent (1971, p. 405)]. The arousal hypothesis replaced the distraction hypothesis as the principal explanation of the effects of noise and other stressors on performance but encountered difficulties in explaining why, if noise increases arousal level while prolonged work decreases it, the adverse effects of noise should be greatest towards the end of the task. More recently, an early explanation of the adverse effects of noise [see Kryter (1950, 1970) and Stevens (1972)] has been revived by Poulton (1976, 1977b, 1978); this explanation suggests that noise impairs perceptual-motor performance by masking task-relevant cues which provide helpful feedback to the individual performing the task. For example, in the five-choice serial reaction task, when the metal disc is tapped by the metal stylus 'clicks' are inevitably produced as a side effect of responding. These clicks act as cues that indicate whether or not the metal disc has been tapped correctly. Such cues are masked by loud noise and hence the feedback received is degraded. Since both visual and auditory feedback can be considered to play an important part in the maintenance of the speed and accuracy of responding in this task, if one source of feedback is degraded, performance is likely to be impaired. Detailed arguments for and against the masking hypothesis can be found in Poulton (1977b, 1978, 1981), Broadbent (1978) and Hartley (1981a,b).

The most obvious and direct way of clarifying the role of masking in the impairment of trial reaction performance by noise is to eliminate acoustic cues in both noisy and quiet conditions and then to determine whether the typical impairment is obtained. Jones (1983) used a virtually silent, four-choice, 'piano-key' type of response panel with the individual's fingers resting on the keys, so that the large ballistic movements required by the usual versions of the task were no longer necessary. At the distal end of each key was a stimulus light, to which the appropriate response was made. A further way in which acoustic cues were reduced was through the use of ear-defender headphones, which permitted white noise to be presented at the ear while providing attenuation in excess of 40 dB to ambient sounds. Jones compared performance under noise [90 dB(C)] and quiet [60 dB(C)] conditions. The task lasted for 40 min, although for the final 10 min of the noise condition the sound level was reduced to 60 dB(C) in order to examine 'aftereffects' of noise (see Section IV). As Table I indicates, Jones obtained the typical finding with this task. Noise significantly increased errors, and errors rose significantly from the first period of the task to the third in noise but not in quiet. Noise did not

TABLE I. The Mean Number of Errors in Serial Reaction Performance When Performed in Soft [60-dB(C)] and Loud [90-dB(C)] Noise in Four Quarters of a 40-Min Period of Work[a,b]

	Quarters			
Noise	1	2	3	4
Loud	14.9	20.5	27.2	30.3
Soft	13.8	14.5	15.8	18.5

[a] Adapted from Jones (1983).
[b] Note that in the last quarter the sound was switched off for both groups.

reliably affect either correct responses or gaps, however. It does not seem, therefore, that masking can be responsible for the impairment of serial responding by noise.

Jones found that in his task noise increased the likelihood that a fast response would be an error, even though noise exerted only a modest effect overall upon the speed of responding, as inferred from the distribution of response latencies in noise and quiet (see Fig. 1). This suggests that noise influences the way in which the speed of responding is regulated and controlled by the system responsible for selecting appropriate responses to environmental stimuli. While in simple RT tasks noise slows responding early in the session, and the amount of impairment decreases with time at work, presumably as a result of adaptation, in serial responding accuracy is impaired by noise towards the end of the task. But for serial responding to be impaired the task must be fairly prolonged, and probably at least 20 min of continuous work is required; if the task is not particularly sensitive, the noise must be intense if performance is to be affected, probably around 95 dB, although more sensitive tasks, such as that used by Jones, will show effects at lower noise levels. Thus Conrad (1973) did not obtain any reliable effect of either 93 dB(A) continuous or intermittent noise on the performance of a rapid serial decoding task lasting for only 5 min, and similarly Viteles and Smith (1946) found that 90-dB noise did not significantly affect performance at a 'discrimeter' task involving four-choice serial responding, which lasted for 30 min.

2. Tracking

Many perceptual-motor skills involve a smooth and integrated response to a continuously or periodically varying input, rather than discrete or

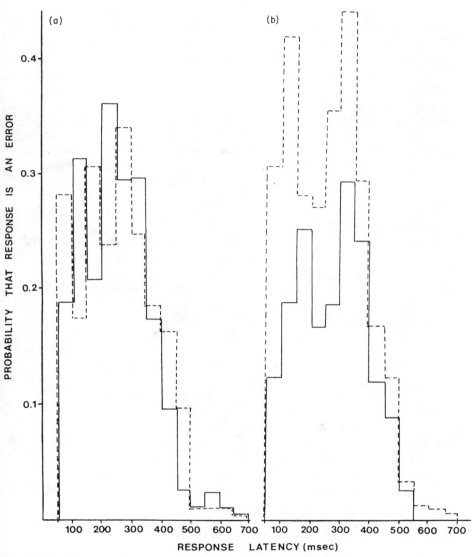

Fig. 1. The speed of errors expressed as a proportion of the speed of all responses in a serial reaction task: results for (a) the first and (b) third quarters of a 40-min period of work in soft [60 dB(C)] (———) and loud [90 dB(C)] (------) noise. [After Jones (1983).]

serial responding. One of the best-known examples of such continuous tasks is the tracking task, which began to be studied in the 1940s in the context of gunlaying. The two main components of a tracking task are a target, which usually moves both vertically and horizontally, and a cursor, which is used to track the target. People performing tracking tasks have to keep the cursor on the target as much as possible, any discrepancy between the two being recorded as an error. The error may be directly observable, as in pursuit tracking, or detectable only indirectly, as in compensatory tracking. The most frequently employed measure of tracking performance is the 'time on target' measure, that is, the length of time for which the cursor is not on the target or some specified distance away from it. This measure provides no information about the extent of the discrepancy between target and cursor and is thus somewhat unsatisfactory. A somewhat more sensitive index of tracking performance is the root-mean-square (rms) error.

A number of different tasks have been used in the investigation of the effects of noise upon tracking. Although it is clear that noise can sometimes impair tracking performance, in the majority of studies no reliable effect of noise has been obtained. For example, Plutchik (1961) showed that high-intensity intermittent noise ranging between 105 and 122 dB had no effect either upon compensatory tracking performance or upon performance at a mirror-tracing task, although in the later case, the variability of performance increased in noise. Hack et al. (1965) found that intermittent loud noise produced an initial decrement in tracking performance but that adaptation soon occurred. Stevens (1972) found no effect of 115-dB continuous noise, compared to 90-dB quiet, on the performance of either a 10-min coordinated serial pursuit task or a fast-speed pursuit motor task, each of which was performed several times over a 4-week period. Theologus et al. (1974) showed that 85-dB intermittent noise did not impair performance at a 20-min 'rate control' task (a compensatory tracking task), and no significant effects of 105-dB noise were observed with the tracking task used by Grether et al. (1971). Finally, Jones et al. (1982), using a pursuit tracking task, sampled the rms error 25 times per second over the 30-min period of the task. Although the increase in tracking error from the first 5 min of the task to the last was higher in continuous noise, relayed at a level of 90 dB(C) than in quiet [55 dB(C)], this difference was not reliable and no overall effect of noise intensity on tracking performance was observed.

In some studies, however, noise has been found to exert a significant effect on tracking performance, occasionally producing an improvement (Gawron, 1982) but more frequently an impairment (Eschenbrenner, 1971; Grimaldi, 1958; Simpson et al., 1974; Stevens, 1972; Viteles and

Smith, 1946). Although as noted previously Stevens obtained no reliable effect of noise on the performance of two tracking tasks, he did find that continuous loud noise at a level of 115 dB significantly impaired performance at a coordinated serial reaction time task requiring a rectangular beam of light to be directed along a series of narrow, curved pathways from one target to another, using a joystick and rudder bar. Since this task appeared to be virtually indistinguishable from several other tasks which did not show impairments of performance in noise, Stevens suggested that the result obtained with the coordinated serial pursuit task may have been artefactual, the impairment being attributable to apparatus factors. The tasks used by Grimaldi and Viteles and Smith were similar, simulating the operation of a lathe, and were self-paced, the individual performing the task determining the rate of work; this makes the nature of the impairment due to noise difficult to specify because of the trade-off between speed and accuracy. However, it is difficult to account for the results of Eschenbrenner (1971), who employed a complex manual image motion compensation task, or those of Simpson et al. (1974), who obtained a reliable decrement in pursuit rotor performance with a noise level of only 80 dB(A). One of the few clues as to how noise may impair tracking performance comes from an unpublished study by Broadbent and Gregory [cited by Broadbent (1971, p. 404)]. In this study noise was found to produce an increase in the number of 'reversals', that is, the number of corrective movements during the period of tracking, suggesting that errors are being overcorrected in noise. It is possible that noise makes people over-reactive or more attentive to errors in tracking situations. Further studies of this kind involving an analysis of tracking performance in control theory terms would seem to be required before any advances can be made in the understanding of the way in which noise impairs tracking, and why performance is sometimes enhanced, sometimes impaired and more often unaffected by noise.

Perhaps the most common everyday example of a tracking task is driving a car. Finkelman et al. (1977) investigated the effects of 93-dB(A) intermittent white noise on driver performance over a prescribed route marked out by rubber pylons. While noise alone did not reliably affect either the time taken to complete the circuit or the number of driver errors (as indicated by the number of times a pylon was hit or knocked down), when both noise and a subsidiary task (delayed digit recall) were present, driving time and errors were significantly increased.

3. Psychomotor Tasks

Investigations of the effects of noise on a wide variety of psychomotor tasks indicate that noise seldom impairs performance in such tasks, and indeed occasionally improves it, if the task is a simple one. Weinstein and MacKenzie (1966), for example, found that performance at the Minnesota Rate of Manipulation test was significantly better in 100-dB white noise than in a control condition. The task required subjects to turn over as many blocks as possible using both hands in a series of 5-sec trials. Tasks in which performance is unaffected by noise include a rowing task producing considerable physical fatigue (Loeb and Richmond, 1956), hand steadiness and marksmanship (Stevens, 1972) and a complex coordination task (Key and Payne, 1981). Bachman (1977) found that a 1000-Hz tone presented through headph nes at a level of 105 dB did not reliably impair the rate at which two psychomotor tasks were learned, compared to a no-noise condition. The first task, the stabilometer task, involved learning to balance on a horizontally pivoted platform, while the second, the ring–peg task, involved learning to transfer washers from one set of pegs to another by using mirror vision only.

At extremely high levels of noise, however, performance can be impaired. For example, Harris in a number of studies [see Harris (1968)] has examined the effects of very high-intensity noise (120–140 dB) on the performance of discrimination and psychomotor tasks. At these intensities impairments were usually obtained, particularly when the noise exposure was asymmetrical, one ear being protected by an ear protector. Harris discusses his results as a possible effect of high intensity on the vestibular system. Such an effect is rather different from the other effects discussed here. Similarly, bursts of noise presented at high intensities can produce adverse effects upon performance. Occasional bursts of noise produce an involuntary 'startle' response, which diminishes with repeated presentation although it is fairly resistant to complete extinction. For example, the eye-blink and head-jerk components of the startle response tend to recur, and to the extent that they are involved in task performance, efficiency is impaired. May and Rice (1971), for instance, found that motor performance involving a high degree of precision was impaired in the period immediately following a pistol shot (with a peak level of 124 dB). While this effect diminished over the 100-min task, during which a total of 16 shots were presented at random intervals, it did not disappear entirely. Generally the effects of noise bursts on performance tend to be localized to at most the 30 sec following onset, although periods of disruption as short as 2 or 3 sec are more commonly reported.

B. Selective and Sustained Attention Tasks

Selective attention tasks can be broadly classified as involving either focussed or divided attention [see Treisman (1969) and Kahneman (1973) for classification schemes]; the former require attention to be focussed on one source or kind of information to the exclusion of others, while the latter require attention to be divided or shared between two or more sources or kinds of information or two or more mental operations. The effects of noise upon the performance of several focussed attention tasks have been examined, among them the Stroop test, visual search and speeded classification and cancellation and proofreading. Divided attention, or time sharing, has been investigated in dual-task situations, in which people carry out two tasks concurrently; in noise studies the two tasks have usually been tracking and monitoring. The vigilance task has been described as providing "the fundamental paradigm for defining sustained attention as a behavioural category" (Jerison, 1959, p. 29) and numerous studies have assessed the effects of noise upon vigilance performance. Vigilance tasks require attention to be directed to one or more sources of information over long, unbroken periods of time for the purpose of detecting infrequently occurring small changes in the information being presented [see Davies and Parasuraman (1982)]. Such tasks, which generally last for at least 30 min, and often for much longer, are also known as "monitoring" or "watchkeeping" tasks and began to be studied during the Second World War in the context of radar and sonar monitoring.

1. The Stroop Test

There is no standard version of the colour–word interference test generally named after J. R. Stroop, who introduced the technique to the American experimental literature in 1935 (Stroop, 1935). Although there are many variations of the Stroop test, notably those involving tachistoscopic presentation, key-pressing and card-sorting responses, the 'traditional' Stroop procedure utilises three charts: a word chart (W) bearing colour words printed in black, a colour chart (C) bearing colour patches and a colour–word chart (CW) on which colour words are printed in incongruent ink colours, for example, the word RED printed in green. The task is to read the colour words on W, to name the colours on C and to name the ink colours, ignoring the words, on CW. It has consistently been found that CW completion time is longer, frequently by as much as 50 to 100%, than the completion time for C. This 'interference' phenomenon, usually measured by CW − C or CW/C [see Jensen (1965)] has been taken to indicate an individual's general 'interference proneness' or, conversely, the ability to focus attention on a relevant stimulus dimension (ink colour) and to ignore an irrelevant one (word meaning). Thus if noise impairs the ability to focus attention on a relevant stimulus dimension, interference scores should be greater in noise than in quiet.

Moderate levels of noise, that is, up to about 85 dB, have been shown to bring about a significant *reduction* in interference (Houston, 1969; Houston and Jones, 1967; O'Malley and Poplawski, 1971), although in one experiment Hartley and Adams (1974) observed a tendency in the direction of reduction with a noise level of 95 dB(C). However, in another experiment Hartley and Adams found that 100-dB(C) noise reliably *increased* interference, indicating an impairment of focussed attention. A longer exposure to noise (30 min) was also found to produce greater interference than a shorter one (10 min). Work with the Stroop test seems to suggest therefore that while moderate noise levels may improve focussed attention, higher levels are likely to impair it, and the impairment is likely to increase with the duration of exposure.

2. Speeded Classification and Visual Search

Speeded classification tasks also require relevant to be selected from irrelevant information; an example of such a task would be sorting cards bearing symbols varying in colour, size and shape into different piles on the basis of colour alone while ignoring size and shape. Thus performance, which is usually assessed in terms of sorting time, can be compared when only the relevant dimension (colour) is present and when one or two irrelevant dimensions (size and shape) are added. Generally sorting times for the relevant dimension increase when irrelevant dimensions are present. In a study of the effects of 95-dB(A) noise on the speeded classification performance of adults and children, von Wright and Nurmi (1979) found that noise affected their performance differentially. While noise increased the sorting times of adults significantly, it exerted no effect upon errors. But noise slightly improved the performance of 9-year-old children and significantly reduced sorting times for 6-year-old children when irrelevant dimensions were present, although error rates were reliably increased. This result suggests that noise affects the speed–accuracy trade-off in children and adults in different ways; young children sacrifice accuracy for speed while adults maintain accuracy at the expense of a reduction in the rate of working.

The effects of noise on the search for, and identification of, targets in visual arrays have also been investigated. In a series of experiments with intermittent and continuous noise, Warner and Heimstra (1971, 1972, 1973) [see also Warner (1969)] employed a task in which people searched rows of identical letters for the occurrence of a different letter, which appeared on average in every fifth display searched. The task could be made more difficult by increasing the number of letters to be searched, from 7 to 15 to 31. In different experiments Warner and Heimstra used noise levels of 90 and 100 dB and the on–off ratios for the intermittent noise were also varied. Although significant effects of noise were obtained in most of their experiments, noise sometimes reduced response times to targets and sometimes increased them, and there was no consistent relationship between the effects of noise and task difficulty. Since repeated measures designs were used in this series of experiments, the somewhat confusing results obtained are probably attributable to asymmetrical transfer between conditions [see Poulton (1977a, p. 447)].

Teichner (1963) used a task in which 10 letters of the alphabet were displayed following the release of a shutter; the individual performing the task controlled the length of time for which each display was presented while searching for particular letter combinations memorised beforehand. The task comprised 200 displays in all and lasted for approximately 15 min. Teicher *et al.* found that performance tended to improve with time at work but that five different levels of white noise (57, 69, 81, 93 and 105 dB) exerted no differential effect on performance, the performance measure being a transformation into bits per second of the time taken to search the displays. However, when other groups of people worked through the first 50 displays in 81-dB white noise and through the last 50 displays either at the same noise level or at one changed without warning to 57, 69, 93 or 105 dB, the best performance on the last 50 displays was found to occur in the group for whom the noise level remained the same and who were therefore exposed to a noise level of 81 dB throughout the task. But in this study the overall task duration was relatively short and performance improved rather than deteriorated with time at work; it would be interesting to see whether, in a longer task in which performance was deteriorating, an increase in noise level towards the end of the task would produce a beneficial effect on performance. In this connection, it is worth noting that Teichner *et al.*, using 100-dB noise bursts with varying on-times within a 5-sec cycle, found that in the early stages of their search task intermittent noise appeared to act as a distractor, with shorter bursts being more distracting, while later in the task, the same noise appeared to serve an alerting or arousing function.

3. Cancellation and Proofreading Tasks

In cancellation and proofreading tasks, people search through rows of random letters or digits, or passages of prose, in order to detect specified target items or to detect and correct textual errors. Such tasks thus have strong affinities with visual search tasks. Usually people carry out cancellation and proofreading tasks at their own pace, so that measures of both the speed and accuracy of working are available; assessments can therefore be made of the extent to which the speed–accuracy trade-off is affected by noise. Davies and Davies (1975), for example, found that in 95-dB(A) broad-band noise compared to a quiet condition of 70 dB(A),

TABLE II. Speed of Performance: The Mean
Number of Lines Checked during the
15-Min Cancellation Task by Old and
Young Subjects Working in Four
Conditions[a]

Condition	Old		Young	
	A.M.	P.M.	A.M.	P.M.
Noise	84.8	78.9	91.6	91.4
Quiet	73.0	72.4	84.7	89.5

[a] From Davies and Davies (1975).

men aged 65 years and over significantly increased their rate of working at
a cancellation task requiring the detection of the letter 'e' in passages of
English prose, although their accuracy was unaffected. For men aged 30
years and under, however, noise exerted no reliable effect on either speed
or accuracy (see Tables II and III). Overall, noise did not significantly
affect either the speed or the accuracy of performance. Bailey *et al.* (1978)
also obtained no effect of either 95-dB intermittent periodic or 95-dB
intermittent aperiodic white noise on the performance of a similar task.
Jones and Broadbent (1979), however, found that 80-dB(C) low-frequency

TABLE III. Accuracy of Performance: The Mean Number of Omission
and Commission Errors and the Mean Percentage of *e*s
Correctly Cancelled Overall and for 'Easy' and 'Difficult'
Signals Separately, by Old and Young Subjects in Four
Conditions[a]

	Old/Noise		Young/Noise	
	A.M.	P.M.	A.M.	P.M.
Percentage total *e*s	79.6	90.5	95.2	91.5
Percentage *e*s (easy signals)	83.8	91.5	96.1	94.0
Percentage *e*s (difficult signals)	55.2	85.1	90.3	80.0
Commission errors	1.2	0.6	0.3	1.4
	Old/Quiet		Young/Quiet	
	A.M.	P.M.	A.M.	P.M.
Percentage total *e*s	88.5	88.9	93.8	95.8
Percentage *e*s (easy signals)	92.8	91.5	95.2	96.6
Percentage *e*s (difficult signals)	72.2	80.9	85.5	91.0
Commission errors	0.6	0.6	0.4	0.2

[a] From Davies and Davies (1975).

office noise significantly reduced the speed of working of middle-aged housewives at a 30-min proofreading task, compared to a quiet condition of 55 dB(C). Noise also significantly increased the overall error rate, that is, false detections of errors and failures to notice errors. Jones and Broadbent used a somewhat more complex situation than the usual proofreading task; the housewives worked in pairs, one acting as a 'sender' and the other as a 'receiver', the former reading aloud the correct version of the text and the latter listening and checking the incorrect version. The speed and accuracy scores referred to above thus represent the joint performance of senders and receivers rather than the performance of individual proofreaders. Senders apparently found reading to be no more difficult in noise than in quiet, but found it more difficult to remember or to understand what they had read. Jones and Broadbent therefore suggested that processing of the surface structure of the material to be checked is unaffected by noise but that comprehension of the material is likely to be impaired. Support for this view comes from two experiments by Weinstein (1974, 1977). Weinstein used a proofreading task in which both contextual errors (grammatical mistakes, omissions and the presence of incorrect or inappropriate words) and non-contextual errors (misspellings and typographical errors) had to be detected. In his first experiment Weinstein employed 70-dB(A) intermittent noise consisting of a recording of the operation of a teletype machine, and in his second 68-dB(A) intermittent noise consisting of a recording of radio news items. These are, of course, comparatively low levels of noise, frequently used as quiet or control conditions. In both experiments Weinstein found that noise significantly impaired the detection of contextual errors but not that of non-contextual errors. In neither experiment was the speed of working affected by noise. Table IV shows the results from Weinstein's (1977)

TABLE IV. Mean Performance on Proofreading Passages[a]

Passage and condition	Contextual errors[b]	Noncontextual errors[b]	Speed (line/s)
Passage 2			
Quiet	56.1	45.6	0.273
Noise	60.1	45.9	0.274
Passage 3			
Quiet	47.2	45.1	0.282
Noise	58.1	46.3	0.281

[a] From Weinstein (1977). Copyright 1977 by the American Psychological Association. Reprinted by permission of the author.

[b] Percentage undetected.

experiment. It thus appears that even comparatively low levels of familiar kinds of noise are likely to affect the detection of textual errors requiring comprehension of textual units larger than the individual word, and hence. involving semantic processing, whereas the detection of errors in individual words, for which semantic processing is less necessary, remains unimpaired. The overall rate of working is likewise unaffected. On the basis of responses to a post-task questionnaire, Weinstein (1977) suggested that people expect noise to impair their performance and that they are aware of changes in their rate of working, and hence are able to adjust it, but are unaware of changes in accuracy. Consequently, lacking the feedback which would inform them that their performance was deteriorating, individuals working in noise fail to make the necessary compensatory adjustments.

4. Time Sharing

Time sharing must occur, to a greater or lesser extent, in situations in which two tasks are performed concurrently or in which two or more displays are to be monitored at the same time. In dual-task paradigms, one task is frequently designated as the 'primary' task and the other as the 'secondary' task. The tasks involved may also possess either many or few elements in common.

Working in noise invariably influences dual-task performance, the usual finding being that primary task performance is relatively unaffected while secondary task performance is impaired. Boggs and Simon (1968), for instance, compared the effects of 92-dB(A) intermittent noise on the concurrent performance of a primary task (four-choice serial responding) and a secondary task (digit sequence monitoring) lasting for 10 min. Simple and complex versions of the primary task were used, complexity being varied by manipulating stimulus–response compatibility. Boggs and Simon found that noise exerted no effect on primary task performance but that it reliably impaired secondary task performance; moreover the impairment of secondary task performance was significantly greater in noise when the complex version of the primary task was used (see Table V). This result can be explained by assuming that task performance absorbs mental effort or capacity and that complex tasks absorb more capacity than do simple ones. Since the supply of capacity is limited, when a complex task is being performed, there is less residual capacity available for the performance of the secondary task, so that secondary task performance is further impaired. Resisting the distracting effects of noise also absorbs capacity, so that secondary task impairment in noise is greater than in quiet.

TABLE V. Effect of Noise on Secondary Task Performance under Simple and Complex Task Conditions[a]

	Mean errors			
Task	Quiet	Noise	Overall M	Difference (noise minus quiet)
Simple	8.50	11.75	10.13	3.25
Complex	11.25	18.54	14.90	7.29

[a] From Boggs and Simon (1968). Copyright 1968 by the American Psychological Association. Reprinted by permission of the author.

The allocation of capacity to high- and low-priority components of the dual-task situation also seems to be affected by noise. When working in loud noise people appear to allocate relatively more attention to high-priority task components and relatively less to low-priority ones. In other words, the allocation of attention, or capacity, becomes more selective in noise. Support of the selectivity hypothesis comes both from dual-task situations and multi-source monitoring tasks.

Hockey (1970a) required subjects to perform a combined tracking and multi-source monitoring task for 40 min. In the instructions given to subjects, the tracking task was designated as the 'high-priority' task and the monitoring task, which required the detection of the onset of lights at different spatial locations, as the 'low-priority task'. Tracking performance was unaffected by 100-dB broad-band noise, while signals appearing at central sources were detected more frequently. However, signals appearing in peripheral locations were detected less frequently in noise. Hockey (1970b) showed that this differential detection of signals at different spatial locations resulted from the expectation that signals were more likely to appear at central locations than at peripheral ones. In a third experiment, Hockey (1970c) found that sleep deprivation produced changes which could be interpreted as being the opposite of those obtained with noise, impairment of performance being greater on the high-priority task (tracking). Loud noise thus appears to produce a reallocation of attentional resources to task components that the individual regards as important, either because the instructions provided by the experimenter make the task priorities explicit or because on the basis of experience with the task the individual's own set of task priorities is determined. Various attempts have been made to replicate Hockey's original results, although none is an exact replication, and other studies of a broadly similar kind have been conducted. The results of these studies furnish mixed support for the attentional selectivity hypothesis. Hamilton and Copeman (1970) obtained findings broadly similar to those of Hockey (1970a) in that tracking performance improved in 100-dB white noise while the monitoring of peripheral lights was reliably impaired. Bell (1978) found that 95-dB white noise exerted no effect on tracking performance but significantly impaired the performance of a secondary number-checking task. Finkelman and Glass (1970) observed that intermittent 80-dB noise did not affect tracking but that performance on a subsidiary short-term memory task deteriorated significantly. In contrast, Loeb and Jones (1978) found that tracking was impaired but a secondary monitoring task unaffected, Finkelman et al. (1979) reported that primary and secondary task performances were both reduced in noise, and Forster and Grierson (1978) observed no effect of 92-dB(A) noise on either tracking or monitoring. Hartley (1981a) found that 95-dB(A) noise reliably impaired tracking performance but, as in Hockey's (1970a) experiment, increased response latencies to peripheral lights in a monitoring task while reducing latencies to central lights. Such a diversity of findings is, perhaps, not entirely unexpected, if it is assumed that the phenomenon first described by Hockey is not principally due to the functional demands made by the respective task components but to the individual's perception of the priorities to be followed when carrying out the task [see Forster (1978), Hartley (1981a,b), Hockey (1978) and Poulton (1981) for detailed discussion]. Even so, the generalisability of the attentional selectivity phenomenon should not be taken for granted (Smith and Broadbent, 1980).

A clearer picture of the way in which noise biases the intake of information comes from studies of multi-source monitoring tasks. Broadbent (1950, 1951, 1954) used two multi-source monitoring tasks, the 20 dials task and the 20 lights task, and compared performance in noise (100 dB) and quiet (70 dB). In the 20 dials task he found that noise had no effect on the total number of signals detected but did reduce the number of 'quick founds', that is, signals detected within 9 sec of onset. Some impairment due to noise was also found with the 20 lights task, but in both cases this impairment tended to be restricted to those areas of the display which were not in the direct line of vision, the displays being arranged to form three sides of a square, with the subject facing the central position. Hockey (1973), using a three-source monitoring task developed by Hamilton (1969) which required individuals to make a sampling response in order to obtain a brief glimpse of the present state of one of the three sources, which might or might not contain a signal, found that 100-dB broad-band noise produced increased sampling of the source on which signals had a high probability of appearing. In these experiments it seems that attention is unequally distributed to the displays being monitored;

some displays receive more attention while others receive less. Noise accentuates this unequal distribution of attention and Hockey's (1973) result indicates that this accentuation results from people's expectations about where significant events requiring action are most likely to occur. In noise there is a greater shift in the pattern of information intake in the direction of perceived task priorities. Examples of the effects of noise on attentional selectivity in memory tasks are discussed later in Section C.

Two experiments have examined the effects of noise on time estimation while another task was being performed. Loeb and Richmond (1956) asked people to estimate 10-min intervals while working at a jigsaw puzzle in noise (110 dB) and quiet (80 dB). The estimated time was longer in noise than in quiet, although the difference just failed to be reliable. Using a longer task, which was extremely attention demanding, Jerison and Smith (1955) found that the average estimate of a 10-min interval was 7.16 min in noise (111.5 dB) and 8.79 min in quiet (77.5 dB), a highly significant difference. Noise thus appears to affect the perception of elapsed time during task performance, but in what way is uncertain, and little work relevant to this topic has been considered in recent years.

5. Vigilance

Efficiency in vigilance situations is most commonly assessed by recording the number of occasions on which a change in the state of the display being monitored, known as a 'signal', is correctly reported, and this measure is called the 'detection rate' or sometimes the 'hit rate'. A second measure is the time taken to detect a signal, the detection latency. A third measure is the number of occasions on which a signal is reported when none, in fact, has been presented. Such errors are called 'commission errors' or false alarms. The application of decision theory analyses to vigilance situations [see Davies and Parasuraman (1982) and Swets (1977)] yields two further ways of assessing performance: in terms of response bias (broadly, the observer's readiness to make a positive response) and perceptual sensitivity (broadly, the efficiency with which the observer is able to discriminate between signal and non-signal events). Typically, performance in vigilance situations deteriorates with time at work, sometimes as a result of a decline in perceptual sensitivity and sometimes as a result of a change in response bias. Which type of change occurs depends on the nature of the demands made by the task on time pressure and working memory (Davies and Parasuraman, 1982).

Since vigilance tasks generally require subjects to make rapid observations of a constantly changing display in order to detect the presence of faint signals occurring at unpredictable times, it might be expected that detection efficiency would also be impaired by loud noise. But in general noise exerts a negligible effect on performance at single-source monitoring tasks, as several studies attest (Blackwell and Belt, 1971; Davies and Davies, 1975; Davies and Hockey, 1966; Jerison, 1959; Jerison and Wallis, 1957; Poulton and Edwards, 1974; Tolin and Fisher, 1974).

There are, however, two exceptions to the general finding that noise exerts little or no effect on single-source vigilance performance. First, noise does appear to affect the confidence with which detection responses are made, there being a fall in the proportion of doubtful responses while the proportion of confident responses rises (Broadbent and Gregory, 1963, 1965; Poulton and Edwards, 1974). Subjects are thus much more likely in noise to report that a signal definitely was or was not presented. Since very frequent and easily discriminable signals encourage confident responses, the effect of noise on tasks containing such signals will be to increase the number of reports of signals, some of which may well be false alarms. In tasks containing infrequent signals, which are difficult to discriminate from background events, the reluctance to make uncertain reports in noise will tend to decrease the number of reports of signals, leading to a possible increase in omission errors. Noise thus appears to bias the weighing of evidence in some way, possibly by affecting the costs and payoffs associated with hits and false alarms.

Adverse effects of noise on performance at single-source vigilance tasks can also be obtained with comparatively low noise levels when 'cognitive' or 'verbal' vigilance tasks are employed (Benignus et al., 1975; Jones et al., 1979). Instead of requiring subjects to detect changes in the brightness of a light or the loudness or duration of a tone, cognitive vigilance tasks require them to detect specified sequences of digits or letters. Benignus et al. found that 80-dB low-frequency noise reliably increased omission errors in a visual task requiring the detection of digit sequences, and Jones et al. reported that broad-band noise, relayed at levels of 80 or 85 dB(C) increased omission errors in a similar task. Why noise presented at relatively low intensities should exert a greater effect on the performance of cognitive vigilance tasks is uncertain. Finally, while continuous high-intensity broad-band noise does not appear to affect the extent of the vigilance decrement (Jerison and Wallis, 1957), varied noise at moderate intensity levels does seem to attenuate the deterioration in performance with time at work (Davies et al., 1973; McGrath, 1963), an effect probably attributable to the maintenance of arousal. An example of the beneficial effects of music, compared to white noise of the same intensity (75 dB), on the deterioration with time in the speed of responding to signals in a vigilance task is shown in Fig. 2.

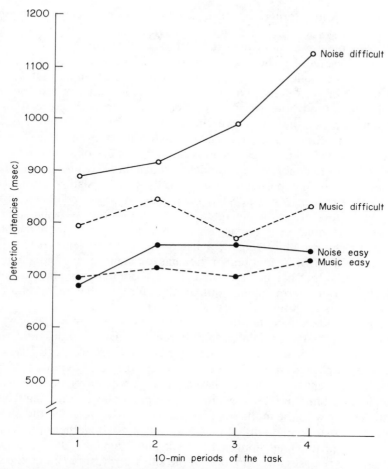

Fig. 2. Detection latencies for each 10-min period in a vigilance task. [From Davies *et al*. (1973).]

C. Verbal Learning and Memory

There are three components of memory which need to be distinguished in principle, although they are not always easily separable in practice. First, if incoming information is to be remembered, it must be registered, that is, learned, or in some sense 'put into storage'. Second, information that has been registered must be retained in storage rather than lost from it, or forgotten. Third, when the information is needed, it must be able to be retrieved from storage; that is, it must not only be available in that it has been put into storage, but also accessible, in that it can be readily located.

Experiments on memory tend to be of at least two different kinds: the first category involves the short-term recall of a series of simple items such as letters, digits or words, items which are so simple that a brief presentation of each item is considered to be sufficient to ensure registration. Recall of the whole list of items either in any order (free recall) or in the order in which they were presented (ordered recall) is usually asked for as soon as the presentation of the list has been completed. The rehearsal of individual items can be prevented by the use of a 'distractor' or 'filler' task, for instance, counting backwards in threes from a given three-digit number. Such experiments are generally known as short-term memory (STM) experiments, and STM studies are thought to be concerned with the factors affecting the retention of under-processed and under-rehearsed information (Cermak 1972). In contrast, long-term memory (LTM) studies normally require a list of items to be learned to some criterion, thus ensuring that registration has taken place, and the retention of the material that has been learned is tested some hours, days or even weeks later. Long-term memory studies therefore involve the retention of material that is both fully processed and well rehearsed.

The most widely accepted theoretical approach to memory, at least up until the early 1970s, postulated that there are at least two different storage systems, a short-term store and a long-term store, respectively, and sought to emphasise the differences between them. The apogee of this approach is, perhaps, the multi-store model of memory developed by Atkinson and Shiffrin (1971). More recent approaches, stemming from a dissatisfaction with the multi-store model, have either rejected the idea of separate short-term and long-term storage systems altogether in favour of a 'levels of processing' approach (Craik and Lockhart, 1972) or have attempted to re-assess the function of the short-term store (Baddeley and Hitch, 1974). The levels of processing approach suggests that there is a continuum of processing levels ranging from shallow (the processing of physical attributes) to deep (the processing of semantic attributes) and that the more deeply the information is processed, the more likely it is to be retained. Baddeley and Hitch (1974) expressed the view that the concept of the short-term store should be replaced by the concept of 'working memory', consisting of a limited capacity, modality-free central processor (essentially an attentional system) and an articulatory loop, which enables information to be briefly stored in a phonemic code without demands being made on the central processor.

Both loud and moderate levels of noise seem to exert various effects on performance in memory tasks [see Jones (1979)], although such effects have not been obtained in all the relevant studies. Short-term memory studies are considered first, followed by experiments that have examined the effects of noise on long-term retention.

1. Noise and Short-Term Retention

At first sight the effects of noise on short-term retention are somewhat inconsistent; in some studies no effect of noise has been obtained (Berlyne *et al.,* 1966; Davies and Jones, 1975; Sloboda, 1969; Sloboda and Smith, 1968); in others an impairment has been reported (Berlyne *et al.,* 1965; Hamilton *et al.,* 1977; McLearn, 1969; Salamé and Wittersheim, 1978); and in others still an improvement in noise has been observed (Archer and Margolin, 1970; Hockey and Hamilton, 1970). Some of these discrepancies are probably attributable to differences in the noise conditions and task procedures employed in different studies, although any difference, or the lack of one, between performance scores in noise and quiet may also conceal changes in the structure of performance. Both at loud and moderate noise levels the performance of tasks requiring short-term retention can alter in a variety of ways, whether or not any difference is observed in the overall level of efficiency.

Noise seems to produce an improvement in the retention of order information (Daee and Wilding, 1977; Hamilton *et al.,* 1972; Hamilton *et al.,* 1977; Hockey and Hamilton, 1970). In noise the tendency for people to 'parrot back' lists of items in the order in which they were presented is enhanced (Dornic, 1975). This effect persists when the requirement for item retention is either reduced to a minimum by using already familiar items (Wilding and Mohindra, 1980) or dispensed with altogether by supplying the items contained in the list with the instruction to place them in their original order (Daee and Wilding, 1977). It also appears when rehearsal is prevented by articulatory suppression (Millar, 1979).

An improvement (Hockey and Hamilton, 1970) or at least no decrement in ordered recall (Davies and Jones, 1975) has also been observed in multi-component memory tasks. The procedure employed in both these studies involved the serial presentation of a list of eight words on a projection screen, with each word being presented at one of the four corners of the screen. Instructions emphasised ordered recall and once this had been completed the recall of presentation location was asked for. The results from the two experiments are shown in Table VI, from which it is apparent that noise either improves or leaves unaffected ordered recall scores while reducing location scores. Such results have been interpreted as providing further evidence for increased selectivity in noise. This conclusion is supported by a result obtained by Smith (1982), who found that noise benefitted whichever of the two tasks, the recall of words or of

TABLE VI. Mean Percentage Recall of 'Relevant' and 'Irrelevant' Items in Noise and/or Incentive[a,b]

	Relevant task		
Condition	Percentage of words recalled in correct order	Percentage of words recalled irrespective of order	Irrelevant task— percentage of locations recalled
Control	42.50 [43.75]	73.75 [73.12]	60.12 [48.50]
Noise	46.25 [54.12]	80.00 [69.00]	33.33 [32.00]
Incentive	57.50	79.37	47.04
Noise and incentive	50.00	80.00	44.94

[a] From Davies and Jones (1975).
[b] Where appropriate, Hockey and Hamilton's (1970) results are shown in brackets for comparison.

locations, was given priority in the instructions. However, Smith argued that assigned task priority and the order of recall of the two types of item (words and locations) were confounded. It is possible that noise merely improves recall for items recalled first, whether they are words or locations, priority being irrelevant. In two further experiments, therefore, Smith independently varied task priority and order of recall and found that priority instructions were necessary for an effect of noise to appear. Increased selectivity in noise in multi-component memory tasks thus depends upon one component of the task being emphasised by the instructions.

Noise also appears to reduce category clustering in free recall and hence to affect the organisation of memory. When the material to be remembered consists of words drawn from a small number of semantic categories (for example, animal names, the names of trees, the names of countries and so on) scattered through a long list, the usual tendency is for words from the same category to be recalled together, a phenomenon known as 'category clustering'. Clustering tends to be more fragmented in noise (Daee and Wilding, 1977), and detailed analyses have revealed that even at quite moderate levels (85 dB and below) noise increases the tendency to make an initial perfunctory report of a few words from each category followed by individual words that are not clustered (Smith *et al.,* 1981). Although the number of words recalled is the same in noise and quiet, it is their organisation which is disrupted. The effect of noise on clustering can be attenuated by changing the nature of the list, so that when the list contains all the members of a particular category or when the words employed are poor exemplars of a category, the effect of noise becomes rather small.

Reports of decreased category clustering in noise suggest the possibility that noise discourages 'deeper', more elaborate levels of processing verbal information so that less attention is devoted to semantic properties of the material to be remembered. If less attention is paid to semantic cues in noise, it is also possible that more attention is paid to physical properties of the items to be retained, including perhaps the order of their presentation. Evidence indicating that noise improves the retention of phonemically related, as opposed to semantically related, material (Baddeley, 1968; Schwartz, 1975; Wilding and Mohindra, 1980) provides some support for this possibility. However, an attempt to manipulate the type of encoding (physical versus semantic) employed by individuals working in loud or soft noise did not produce any shift in the level of processing verbal material in loud noise (Smith and Broadbent, 1981); moreover the improved retention of phonemically related items in noise appears to be the result of an enhancement of the quality of information held in the articulatory loop, attributable to the slowing of rehearsal rate in noise (Mohindra and Wilding, 1982; Wilding and Mohindra, 1980). The reduction in the rate at which individual items are rehearsed would result in fewer items being held in the articulatory loop at any one time, thus decreasing the possibility of confusion among them. Noise also seems to reduce the 'holding capacity' of working memory, improving the retention of very recently presented items at the expense of earlier ones (Hamilton *et al.*, 1977; Hockey *et al.*, 1981) and, in the absence of instructional bias towards alternative strategies, promotes a maintenance rehearsal strategy involving increased reliance on the articulatory loop (Wilding and Mohindra, 1980; Wilding *et al.*, 1982).

The principal effects of noise in STM situations calling for the retention of verbal materials, therefore, are on the encoding of information: order information is better preserved, perhaps because of the shift towards a maintenance rehearsal strategy emphasising the articulatory loop, but semantic information is less well utilised since elaborative rehearsal is less likely to be employed, perhaps because the storage space available to working memory as a whole is reduced, thus decreasing the efficiency of the executive component. In multi-component memory situations, increased selectivity is apparent, resulting from a re-allocation of attentional resources.

2. Noise and Long-Term Retention

Several studies have indicated that the long-term retention of verbal materials originally presented in noise is enhanced (Berlyne and Carey, 1968; Berlyne *et al.*, 1966; Haveman and Farley, 1969; McLean, 1969) and the interaction between noise and retention interval has been explained in terms of attentional processes at input, with Tulving's (1979) encoding specificity principle receiving particular emphasis [see Eysenck (1982, p. 160)]. But perhaps the major focus of studies concerned with noise and long-term retention has been upon the retrieval of information from semantic memory. Semantic memory has been defined by Tulving (1972, p. 386) as "a mental thesaurus, organized knowledge a person possesses about words and verbal symbols, their meanings and referents, about relations among them, and about rules, formulas and algorithms for the manipulation of these symbols, concepts and relations".

A typical semantic memory task resembles a test of 'general knowledge' and may require people to produce the names of as many European countries as they can in a given time period or to produce the names of as many fruits beginning with the letter A, or synonyms for a given word and so on. von Wright and Vauras (1980) found that 95-dB(A) intermittent white noise significantly reduced the number of European, American and Asian countries that could be recalled in an 8-min period, the difference being almost entirely attributable to the reduction in the number of non-European countries named. Two further experiments, in which shorter recall periods were used, and in one of which some different semantic memory tasks were employed, yielded similar findings. It appears that the performance of individuals scoring high on 'neuroticism' was particularly adversely affected by noise. Semantic memory is sometimes considered to be organised hierarchically, with some words or names being particularly dominant or accessible, and others being relatively inaccessible or non-dominant, and it has been hypothesized that noise enhances the retrieval of dominant items from semantic memory.

Eysenck (1975) found that continuous 80-dB white noise facilitated the speed of recall from semantic memory of dominant items but exerted a detrimental effect on recall speed for non-dominant items, particularly in individuals who reported themselves on a checklist as being relatively highly activated at the beginning of the experiment (see Fig. 3). In a similar experiment, Millar (1980) found that continuous broad-band noise relayed through headphones at a level of 95 dB(A) produced faster recognition times for semantically 'high-dominant' words than did the same noise at a level of 70 dB(A), but had no effect on recognition times for 'low-dominant' words. It thus appears that noise may differentially affect the retrieval of dominant and non-dominant items from semantic memory.

It is possible that this effect may be accounted for in terms of a shift in criterion placement. In partial replication of a study by Schwartz (1974), Jones *et al.* (1982) visually presented short stories containing characters possessing rare or common surnames obtained from the local telephone directory. Subsequently they administered a recognition test for both

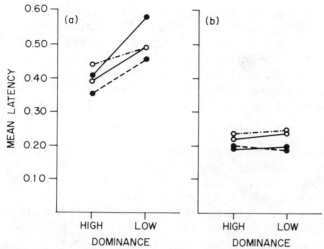

Fig. 3. Effects of mood on retrieval speed (log seconds): (a) recall and (b) recognition (●——●, high activation, white noise; ●------●, high activation, no white noise; ○——○, low activation, white noise; ○—.—○, low activation, no white noise.) [From Eysenck (1975). Copyright 1975 by the American Psychological Association. Reprinted by permission of the author.]

types of surname. When the stories were presented in 85-dB(A) white noise, people became more cautious about reporting rare surnames as having occurred in the stories. Since such surnames were, presumably, 'non-dominant', this result supports Eysenck's (1975) findings and suggests that, as in other task settings such as vigilance, noise affects pigeon-holing rather than filtering strategies (Broadbent, 1971); that is, it influences response rather than stimulus selection.

D. Intellectual Tasks

The tasks to be considered under this classification fall into four main categories: (1) tasks drawn from tests of clerical ability, (2) coding and sorting tasks, (3) mental arithmetic tasks and (4) standard intelligence tests. As Grether (1971) has pointed out, the first category contains many tasks whose demands appear to be primarily perceptual rather than intellectual, although since investigators using such tasks have usually labelled them as intellectual tasks, this classification has been retained here. But in many respects such tasks as letter, name or number checking or pattern matching can be considered as more closely resembling the cancellation tasks discussed in Sub-section B earlier. The second category includes a number of tasks, such as card sorting, which have been shown to be sensitive to the effects of such stressors as hypoxia (Poulton, 1970), while tasks in the third category, mental arithmetic, probably make greater demands on information processing than do tasks in the first two categories. Mental arithmetic tasks are also fairly sensitive to the effects of hypoxia (Grether, 1971). The fourth category contains standard tests of intelligence or general mental ability, such as the California Capacity Questionnaire and the Otis Self-Administering Tests of Mental Ability.

1. Tests of Clerical Ability

The effects of noise on the performance of tasks involving clerical ability appear to be slight, although studies employing continuous and intermittent noise have produced somewhat conflicting results. Wilbanks *et al.* (1956) examined the effects of recorded aircraft noise presented continuously at a level of 110 to 114 dB upon the performance of four tests drawn from the Differential Aptitudes Test. The noise was relayed through headphones at a level of 106 dB. The tests were mechanical reasoning, abstract reasoning, clerical speed and accuracy and numerical ability. Two groups of 46 naval cadets performed these tests in both noise and quiet (consisting of "reasonably normal sound conditions"), one group receiving the noise condition first and the other quiet. Testing took about three hours to complete. Only on the clerical speed and accuracy test was any effect attributable to noise obtained, performance being significantly *enhanced* under noise conditions.

Smith (1951), on the other hand, investigating the effects of intermittent noise at a level of 100 dB with an on–off ratio of 50% on performance at the name and number checking subtests of the Minnesota Clerical Test and at the Minnesota Paper Form Board Test, found that in noise subjects consistently worked more quickly than in quiet but made more errors. However, these differences were only occasionally significant. Using continuous noise at a lower intensity (80-dB speech), but essentially the same kind of task as Smith, although of longer duration, Auble and Britton (1958) found no effect of noise on clerical performance. Tests of clerical ability, therefore, appear to be little affected by noise, although both intermittent and loud continuous noise may produce some increase in speed of work.

2. Coding and Sorting Tasks

The Digit Symbol subtest of the Wechsler Adult Intelligence Scale (Wechsler, 1958) is a coding task which requires the substitution of symbols for digits in accordance with a code. Houston (1968) found that performance of this 1.5-min task (assessed in terms of the number of correct substitutions) was slightly but significantly improved by a low level of varied noise (78 dB), which subjects were instructed to ignore, compared to a quiet condition consisting of normal room noise. Stevens (1972), however, found no effect of 115-dB aircraft noise, compared to a level of 90 dB, upon either card sorting or coding in highly practised individuals.

3. Mental Arithmetic

A number of investigators have examined the effects of noise upon the performance of tasks involving mental arithmetic. Low-intensity noise has consistently been found to have no effect on performance at such tasks (Angelino and Mech, 1955; Kurz, 1964; Mech, 1953), and at higher noise levels little effect has been observed in the majority of studies. In a carefully designed study, Park and Payne (1964), for example, investigated the effects of low-frequency noise upon the performance of an easy and a difficult mental division task. Four independent groups of subjects, matched for ability, were tested, two working in noise at a level of 50 to 70 dB. No effect of noise upon performance was observed, and there was no interaction between noise and task difficulty. However, noise was found to increase the variability of performance for subjects working with easy problems, although no such effect was obtained with subjects working with difficult problems.

It seems, therefore, that noise also exerts little influence on the performance of conventional mental arithmetic tasks. However, when such tasks are made more difficult by the inclusion of a short-term memory component, adverse effects of noise have been consistently obtained (Broadbent, 1958b; Frankenhaeuser and Lundberg, 1976; Woodhead, 1964). Harris and Sommer (1971), who used a shortened version of the task employed by Broadbent and Woodhead, found that a combination of loud noise (110 dB) and vertical vibration produced a significant reduction in the number of correct responses, although lower levels of noise (80 and 90 dB) in combinations with vibration did not. In a second experiment it was found that high-intensity noise (107 dB) and vibration combined to produce a greater decrement in performance than either condition alone.

Thus when mental arithmetic tasks involve a short-term memory load, they appear to become susceptible to the effects of noise. It might also be expected that performance at multi-source tasks would be more adversely affected by noise than performance at single-source tasks. In a test of this hypothesis, Samuel (1964) examined the effects of 110-dB noise on performance at a dual-source mental addition task. Four independent groups of five men and five women worked in noise or quiet (80 dB) and performed a 21-min mental addition task in which the two digits to be added were either presented at the same source or on two spatially separated sources.

Contrary to expectation, the percentage of errors was significantly lower in noise in the spatially separated condition and fewer omissions were made. No difference was found between noise and quiet in the same source condition. Noise did not appear to impair the process of calculation itself, and Samuel concluded that "either the perceptual or the effector mechanism . . . is affected by noise" (p. 266). An incidental finding was that in noise women completed significantly more additions than did men without making proportionately more errors.

Samuel pointed out certain differences between his task and multi-source tasks in which performance is impaired by noise. He suggested that his task was simpler, less highly paced and of briefer duration and that it placed a smaller load on short-term memory. Nevertheless, it is not entirely clear why Samuel's results are so different from those normally obtained with multi-source tasks.

4. Standard Intelligence Tests

Stambaugh [as reported by Corso (1952)] examined the effects of noise upon performance on the California Capacity Questionnaire (CCQ). The same subjects completed two forms of the CCQ, one in noise (105 dB) and the other in quiet (normal room noise). Each testing session lasted approximately 50 min, and a 24-hr period separated them. Noise was found to exert no significant effect upon performance, although some subjects improved in noise while others deteriorated. Subjects who showed most deterioration in noise appeared to be more highly aroused, as inferred from physiological measures recorded during performance. Stambaugh's failure to find any effect of noise on intelligence test performance has been confirmed by other investigators [see Grether 1971)].

In conclusion, therefore, noise seems to have little or no adverse effect on the performance of intellectual tasks, except where a load on short-term memory is introduced. Performance appears to improve significantly in noise when multi-source tasks are used, although this conclusion is based on the results of one experiment only.

III. NOISE AND OTHER STRESSORS

Outside the laboratory environmental stressors rarely occur in isolation. Generally, several stressors are simultaneously encountered, the most frequently occurring combination probably being noise and heat. In addition, 'host' factors such as sleep deprivation, fatigue or drugs may be present and may combine with externally imposed stressors to produce effects on performance. Moreover, jobs in which loud noise is encountered are also likely to be dangerous, and this may increase the level of anxiety. As several authors have emphasised (Grether, 1971; Poulton, 1966, 1970; Wilkinson, 1969), different stressors probably exert their effects through quite different mechanisms and the effects on performance of two or more stressors acting in combination cannot be predicted from a knowledge of the effects of each stressor administered alone. Two stressors may combine to produce effects on performance which are additive (where the effect on performance of the two stressors in combination is the sum of their independent effects), greater than additive (synergistic) or less than additive (antagonistic or interactive). Most combinations of environmental stressors appear to be additive in their effects on performance, a few are antagonistic and scarcely any are synergistic (Grether, 1971). As noted in Section I, the effects of stressors administered in isolation can also be compared with respect to performance on a particular task, for example, serial responding.

The stressors most often compared with noise and most frequently studied in combination with noise are environmental warmth, sleep deprivation, drugs such as alcohol and chlorpromazine and vibration. In addition, the effects of knowledge of results (KR) and monetary incentives have been examined in conjunction with those of noise. The effect of environmental warmth on serial responding is quite different from that of noise (Pepler, 1959), while on a multi-component task the effects of warmth and noise are similar (Bursill, 1958). Poulton and Edwards (1974) found that warmth on its own reliably reduced perceptual sensitivity in a vigilance task, while low-frequency noise also reduced perceptual sensitivity, but not significantly. When warmth and noise were combined, noise reduced the adverse effects of warmth, since the reduction in perceptual sensitivity was no longer significant. But, in general, the effects of warmth and noise do not interact (Bell, 1978; Viteles and Smith, 1946). Hartley et al. (1977) found that 95-dB(A) white noise and chlorpromazine, when applied separately, both affected vigilance performance in similar ways, and when applied together cancelled each other out. For sleep loss and noise the picture is far more consistent. Each produces its effect at the end of the period of work on a serial reaction task. Together their effects are antagonistic. The poor performance of sleep-deprived individuals on serial reaction improves when loud noise is also present (Wilkinson, 1963). Similarly, on a vigilance task, the reduction in intermediate categories of confidence normally found in loud noise is counteracted by the joint action of sleep loss.

By one account these findings can be understood in terms of the arousing qualities of each stressor. When presented alone, noise increases and sleep loss diminishes arousal. In combination their effects are antagonistic. Some difficulties for this view are posed by the fact that in physiological terms the sleep-deprived individual may sometimes appear to be highly aroused (Malmo and Surwillo, 1960). Another way of treating the interaction of noise and sleep loss is to regard each as influencing different parts of the same system. Suppose that sleep loss increases the cost associated with the performance of a task. If the gain associated with responding is increased, then the generally soporific effects of sleep loss may be overcome. One view of noise is that it focusses attention on the adequacy of response, in that the presence of noise serves as a challenge to the individual's competence (Jones, 1983). The effects of sleep loss can be counteracted by giving the subjects an interesting task (Wilkinson, 1962, 1965) or by providing people with KR (Wilkinson, 1961, 1963) or by making the penalty for incorrect responses sufficiently high (Malmo and Surwillo, 1960). In each of these cases he gain associated with responding is increased, and noise may well act in the same way to oppose the effects of sleep loss.

The idea that noise makes us focus on the adequacy of response is further strengthened if we consider the way in which KR interacts with noise. The way in which the KR is presented is important to the outcome of serial reaction studies. If it takes the form of concurrent KR, with each error being signalled soon after its commission, then there is no interaction of KR with loud noise (Hartley, 1974). However, if this type of KR is augmented by the announcement of the individual's progress at periodic intervals during the task—of the form 'worse', 'better' or 'no difference'—then noise and KR do interact (Wilkinson, 1963). Noise does not interfere with the gathering of evidence about events which occur during the performance of the task, as is shown by the absence of an effect with concurrent KR. Noise appears to influence the weighing of evidence over the long term. Noise and KR interact because they both focus people's attention on the adequacy of the responses they produce.

Although the effect of incentive is sometimes treated in the same way as that of KR, there is no evidence that the effects of noise and monetary incentive are related. For example, their individual effects on a task requiring the recall of words and locations (described in Section II) are quite different (Fowler and Wilding, 1979), and when present together their effects do not interact (Davies and Jones, 1975). The time of day at which the test takes place also appears to be important; noise sometimes appears to improve performance if the test takes place early in the day but shows no effect later in the day (Blake, 1971; Mullin and Corcoran, 1977). Another variable which appears to counteract the adverse effects of loud noise is glucose. Simpson et al. (1974) found that pursuit motor performance reliably deteriorated at a sound level of 80 dB(A) but that a preloading of 18 mg of glucose in 100 ml of water prevented this deterioration from occurring. However, the same preloading of glucose *impaired* performance in a quiet condition of 50 dB(A). Davies and Gill (1982) found that aperiodic intermittent broad-band noise at a level of 92 dB(A) reliably increased solution times in a self-paced problem-solving task compared with a control condition but that a preloading of glucose at the same dosage as in the study of Simpson *et al.* abolished this effect.

IV. SITUATIONAL AND CONTEXTUAL EFFECTS

In this section we draw upon a growing body of evidence which suggests that the effects of noise on performance depend at least in part on the context in which it is presented. One strand of evidence comes from the work of Glass and Singer, whose influential monograph "Urban Stress" (1972) was based on the premise that "noise stressors occur in the context of cognition and social circumstances" (p. 157). In a series of laboratory studies, they investigated the effect of giving subjects a greater or lesser degree of mastery or 'control' over the noise that they heard. This was done in several ways with much the same outcome: increasing an individual's feelings of control over the noise markedly attenuated the effect of noise on performance.

Glass and Singer's experiments typically consisted of two stages. In the first, lasting for about 25 min, people were exposed to bursts of noise presented at levels of either 108 dB(A) (loud) or 56 dB(A) (soft) on periodic or aperiodic schedules. The type of noise used by Glass and Singer can best be described as conglomerate, consisting of the superimposed sounds of two people speaking Spanish, a desk calculator, a typewriter and a mimeograph machine. In the second stage of their experiments, which immediately followed the first, the noise was switched off and the same people worked at a different set of tasks, generally concerned with problem solving in some form. Early in their series of studies, Glass and Singer showed that the intensity at which the noise was presented did not affect performance during the first stage of the experiment. However, if loud noise had been presented during the first stage, then performance during the second stage tended to be impaired, persistence at problem solving tasks, for example, being markedly reduced. Perhaps the most interesting feature of Glass and Singer's research is that these 'behavioural after-effects' associated with exposure to loud noise could be attenuated by the suggestion that people could turn the noise off at any time they wished during the first stage. Only those people who did not exercise this option were included in the final data analysis, this group being deemed to have had 'perceived control' over the noise stimulus. The performance of the 'perceived control' group during the second stage closely resembled that of the group exposed to low levels of the same noise during the first stage of the experiment. In other words, perceived control effectively abolished the usual after-effect associated with exposure to high-intensity noise. The abolition or attenuation of the after-effect could also be obtained in other ways, for instance, by making the noise more predictable (either by signalling the onset of noise bursts visually or by increasing the regularity of their occurrence). It thus appears that the presence in the experimental situation of any factor which helps people to feel that they have more control over what happens reduces the likelihood that substantial after-effects of noise will be observed.

A related series of studies {using the 'learned helplessness' paradigm [see, for example, Hiroto (1974)]} has produced similar results, indicating that the degree to which people feel in control of a stimulus directly influences the extent to which they are likely to initiate various types of behaviour designed to avoid failure. There is some evidence that these phenomena are not confined to the laboratory. For example, Graeven (1975) argued, on the basis of the results of a study of community noise, that noise annoyance was mediated by the degree to which individuals thought the noise to be outside their control. Several other workers have used similar constructs in an attempt to understand individual differences in noise annoyance; for example, 'preventability' (McKennell, 1963), 'considerateness' (Borsky, 1961) and 'misfeasance' (TRACOR, 1971).

Several attempts have been made to extend Glass and Singer's (1972) findings both to other types of noise (Jones et al., 1982; Moran and Loeb, 1977; Percival and Loeb, 1980; Rotton et al., 1978) and to other kinds of behaviour (Donnerstein and Wilson, 1976; Sherrod and Downs, 1974). It is clear from such studies that certain kinds of noise—particularly conglomerate noise, combinations of aircraft noise peaks and loud speech—are more likely to produce after-effects than are other kinds of noise and that such after-effects are not confined to the degree of persistence shown in tackling insoluble puzzles, but also include changes in aggression and helping behaviour.

Cohen (1980) in an extensive review of research and theory concerned with the after-effects of various stressors, including noise, considered a number of theoretical explanations of such effects. These included the adaptive-cost hypothesis, first put forward by Glass and Singer (1972), Cohen's own information overload hypothesis [see Cohen (1978)], learned helplessness theory (Seligman, 1975), arousal theory and explanations couched in terms of coping strategies, frustration, cognitive dissonance and self-perception. Cohen concluded that most explanations of the after-effects observed following exposure to stressors such as noise are essentially variants of the original adaptive-cost hypothesis, which emphasises the psychological, and to a lesser extent the physiological, costs involved in the adaptation to a stressor. However, different explanations describe these costs in somewhat different terms.

In a less clear-cut way, a number of early experimental studies showed that the attitude of the individual towards the likely effects of noise on performance also influenced the level of efficiency. For example, Woodhead (1958) found that instructions "not to be put off" by bursts of noise reduced the effects of the bursts on performance. Broadbent (1953) found that the suggestion that noise improved performance increased the length of time before performance began to deteriorate.

Mech (1953) compared the effect of giving subjects various types of information about the likely effect of noise. One group was told that noise tended to facilitate performance, another group was told that noise usually interferes with performance and a third group was told that the effect of noise was variable, sometimes improving and at other times interfering with performance; a control group was given no information. There was indeed an effect of the type of information given in the direction suggested, but the effect was short-lived. By the fifth day of testing, the effects of suggestion had disappeared and performance was worse in the first two groups relative to the control group.

This study suggests that for the short-term at least the attitude of the individual is important in determining whether deterioration will occur. It may well be that individuals presume, despite exhortations to the contrary, that noise is damaging to performance simply because the notion that noise impairs performance is so pervasive; in time therefore they may tend to attribute any failure to perform the task adequately to the presence of noise. Gawron (1982) achieved only qualified success in an attempt to replicate Mech's results. The influence of various suggestions about the effect of noise was confined only to certain tasks and was only found for the suggestion that noise facilitated performance and not with the group that were told that noise hindered performance.

Taking these lines of evidence together, it is clear that the effect of noise on performance is not solely dependent on the acoustic properties of the noise. In previous sections it was noted that the effect of noise depended on the type of task which was undertaken and to this generalisation must be added the further complication that the way that people perceive general features of the setting in which the noise is presented also influences the likelihood that noise will affect performance. This means that the effects produced in the laboratory may not easily be generalised to the field. However, as the next section shows, some features of the laboratory response to noise have been observed in settings outside the laboratory, despite the considerable methodological difficulties which usually attend field studies.

V. FIELD STUDIES

Investigations of the effects of noise on efficiency which are conducted outside the laboratory have usually been of either productivity at work or efficiency in the classroom.

A. Effects in the Classroom

At least two types of effect may be distinguished under this heading. The first is an indirect effect on performance arising out of the power of noise to mask speech. It comes as little surprise to discover that the effects of masking include less participation in class and a shift from the use of discussion to the use of lectures (Ward and Suedfeld, 1973). It seems clear that noise disrupts teaching and listening activities and that the effects on academic achievement [see Bronzaft and McCarthy (1975) and Cohen *et al.* (1973); but *pace* Weinstein and Weinstein (1979)] are largely due to the masking effect rather than to some enduring effect on student motivation or competence (Crook and Langdon, 1974).

A second type of effect appears to be independent of the effect of masking and appears to echo the type of change just noted in our discussion of the effect of perceived control on persistence at problem solving. Cohen *et al.* (1980) examined the effect of aircraft noise on children attending school near a busy airport. This study extended the compass of measures beyond the usual ones of tests of academic ability to include tasks assessing attentional strategies, persistence at problem solving and cognitive performance which were largely free of the effects of educational attainment. It is worth noting that Cohen *et al.* took the precaution of observing performance on these tasks in a soundproofed laboratory rather than in the noise-exposed classrooms. Children were tested on two occasions a year apart, which allowed the investigators to establish whether adaptation to noise had occurred in this period.

As compared to students from schools in a less noisy area, the noise-exposed children were less able to resist auditory distraction and were less persistent in solving problems. These effects cannot be ascribed to some effect of the educational attainment in the two types of school but rather seem to be attentional and motivational deficits. Although the effects of noise were rather small, these effects showed no evidence of adaptation over the one-year interval.

B. Effects in the Workplace

Noise is only one of a multitude of factors which may influence performance in the workplace. This means that the effect of noise may be difficult to isolate. Moreover, there is the risk that loud noise will accompany other features of the work environment which also influence performance. Thus heat, vibration and danger, in greater or lesser measure, may either mask or exaggerate the action of loud noise. It is also important to recognize that a host of organizational and mechanical factors also influence output at work and that these factors are often difficult to control in field studies.

One way of understanding how noise might affect efficiency is to make a retrospective examination of accident statistics. In a study of this sort Kerr (1950) showed that mean noise level at the place of work was among the most potent of the 40 or so factors under scrutiny in predicting accident frequency (but less powerful in predicting accident severity). A similar picture emerges from an analysis undertaken by Cohen (1974, 1976). Here again the incidence of accidents was higher in noisy areas, with the incidence of accidents being especially high among inexperienced workers who were exposed to noise. These accidents might reflect some degree of momentary inefficiency rather than a slowing of the overall rate of work, although the studies just discussed, since they do not contain indices of productivity, can only be suggestive of such an effect.

A number of early studies in the weaving industry showed quite marked effects of noise by examining the effects of ear defenders (Weston and Adams, 1932, 1935). Production improved by some 12% when workers wore ear defenders (providing some 15-dB attenuation to ambient sounds). Although in some parts of the study there is the possibility that the wearing of ear defenders coincided with other conditions favourable to production, the results are unlikely to be due to bias of other kinds. Specifically, the improvement in productivity was not due to the workers having a favourable opinion of ear defenders since output improved even in those workers who did not favour the wearing of ear defenders.

The tactic of examining the effect of reducing noise at the workplace and observing the effect on production was used in a later, more methodologically robust study by Broadbent and Little (1960). They observed the rate of work and the incidence of errors in a job involving the perforation of cine film. Performance was compared at work stations with and without acoustic treatment. The acoustic treatment reduced the level of the sound to the operator by some 10 dB and care was taken that operators were moved in a systematic way from station to station. There were no effects of sound attenuation on the average rate of work, but the number of errors was consistently greater at the untreated work stations. These results are in line with those found in the laboratory with the serial reaction task (see Sub-section II.A), with a tendency for loud noise to produce occasional periods of inefficiency rather than a decline in the overall rate of work.

Despite the paucity of data, some interesting trends have emerged in results of field studies. The form of these results has tended to echo those first discovered in the laboratory, with a reduction in persistence at problem solving in the school setting and an increase in the incidence of momentary inefficiency in the case of industrial tasks. However, many more field studies are required before we may regard these findings as general and pervasive.

VI. EXPLANATIONS OF THE EFFECTS OF NOISE ON PERFORMANCE

Most accounts of the effects of noise on performance embody the notion that noise brings about some change in the 'state' of the organism. This state is usually referred to as lying on the continuum of arousal and is held to obey the so-called 'law' attributed to Yerkes and Dodson (see Subsection I.A).

Unfortunately the law is not very good at predicing outcomes. Any number of effects may be explained (including the facilitation of performance, the deterioration of performance and null effects) by imaginative assignment of the contribution of task difficulty and the effect of noise to values on the arousal continuum. Not only does this form of arousal theory fall short of predicting the value or direction of outcome, it fails completely to specify the form which strategic effects might take.

More clearly enunciated forms of theory fare little better. The best example of a theory which makes relatively clear predictions is due to Poulton (1978, 1979). However, this theory (perhaps because of its clear articulation) has encountered a substantial number of empirical stumbling blocks. In brief the theory supposes that noise has a short arousing (beneficial) effect, followed by a period when arousal subsides to the norm; finally, when the noise is switched off, the level of arousal falls below the norm (thereby accounting for after-effects). A crucial component of this theory is the masking of acoustic cues. These cues are helpful in the successful conduct of the task and the effect of noise will largely depend on the resources available to combat its effects. In addition to the masking of acoustic cues in the operation of equipment, the theory encompasses masking effects of noise on 'inner' or covert speech. Thus the effects of masking are modulated by the effects of arousal (on available processing resources). Among the several lines of evidence against such a formulation is the finding that noise has a deleterious effect on performance in tasks where no acoustic cues are present (see, for example, Subsection II.A).

A final and more tentative series of accounts have begun to emerge in the light of recent findings on the effects of noise on information processing [see Jones (1984) for an overview]. These accounts rely on the idea that the effect of noise is a response to the demands of working in more difficult environments. Among the findings that can be understood in this framework are that memory for order may be improved in loud noise and that individuals tend to become more cautious in their judgments.

The response to noise is in this case one of preserving information under difficult circumstances. Many features of this approach need to be worked out, but it does at least mark a departure from the conceptual cul-de-sac offered by the Yerkes–Dodson law.

VII. SUMMARY

The last three decades of research have brought us no closer to a simple definitive statement about the way in which noise influences performance. Research conducted in this period has suggested that the effects are much more complex than originally envisaged and that they usually involve several mechanisms that operate in a complex fashion. We now know much more about the extent of such effects: recent work has suggested that a much wider range of mental functions than hitherto supposed is influenced by noise. Moreover these effects appear at relatively low levels of noise and they sometimes extend beyond the period of exposure.

Certain features of tasks make them susceptible to the effect of noise. Among these are (1) the difficulty of the task (more difficult tasks are more susceptible to disruption by noise), (2) the demands on long-continued attention (especially when the measures of performance are chosen to be sensitive to periods of momentary inefficiency) and (3) the opportunity to do the task in several different ways (recent studies have shown that the effect of noise is the deployment of various strategies and the range of such strategies will in part dictate the range of operations open to the individual).

A number of situational variables also play a role in determining the degree of disruption of efficiency. Factors such as the degree of perceived control of the noise and temporary or enduring beliefs about the effects of noise on efficiency can serve to modulate noise effects.

Among the several pressing issues for noise research are the following topics: (1) further analysis of strategic changes in performance employing fine-grained analyses of performance rather than gross measures of efficiency such as a numerical index of errors; (2) greater understanding of the effects of the individual's attitudes towards noise and task performance on their relation to performance; (3) further work on the association between noise annoyance and task performance, especially work focussing on the effects of spectral parameters of the noise rather than on its intensity; (4) analysis of the extent to which after-effects of performance persist; at present there is very little indication of how long such after-effects last; (5) further work on the effects of noise on industrial efficiency. This last ranks as perhaps the most important of all these issues, for it is here that we must look for the application of experimental findings and theoretical formulations.

REFERENCES

Angelino, H., and Mech, E. V. (1955). Factors influencing routine performance under noise: An exploratory analysis of the influence of 'adjustment', *J. Psychol.* **40**, 397–401.

Archer, B. U., and Margolin, R. R. (1970). Arousal effects in intentional recall and forgetting, *J. Exper. Psychol.* **86**, 8–12.

Atkinson, R. C., and Shiffrin, R. M. (1971). The control of short-term memory, *Scientific Am.* **225**, 82–90.

Auble, D., and Britton, N. (1958). Anxiety as a factor influencing routine performance under auditory stimuli, *J. Gen. Psychol.* **58**, 111–114.

Bachman, J. C. (1977). The effect of non-traumatic audible stress on learning of two motor skills, *J. Motor Behav.* **9**, 243–245.

Baddeley, A. D. (1968). A 3-minute reasoning test based on grammatical transformation, *Psychonomic Sci.* **10**, 341–342.

Baddeley, A. D., and Hitch, G. (1974). Working memory, *in* "The Psychology of Learning and Motivation: Advances in Research and Theory," Vol. 8 (G. H. Bower, ed.). Academic Press, New York.

Bailey, G., Patchett, R. G., and Whissell, C. M. (1978). Effects of noise on accuracy, visible volume, and general response production for human subjects, *Perceptual Motor Skills* **46**, 76–78.

Bell, P. A. (1978). Effects of noise and heat stress on primary and subsidiary task performance, *Human Factors* **20**, 749–752.

Benignus, V. A., Otto, D. A., and Knelson, J. H. (1975). Effect of low frequency random noises on performance of a numeric monitoring task, *Perceptual Motor Skills* **40**, 231–239.

Berlyne, D. E., and Carey, S. T. (1968). Incidental learning and the timing of arousal, *Psychonomic Sci.* **13**, 103–104.

Berlyne, D. E., Borsa, D. M., Craw, M. A., Gelman, R. S., and Mandell, E. E. (1965). Effects of stimulus complexity and induced arousal on paired-associate learning, *J. Verb. Learn. Verb. Behavior* **4**, 291–299.

Berlyne, D. E., Borsa, D. M., Hammacher, J. H., and Koenig, I. D. V. (1966). Paired-associate learning and the timing of arousal, *J. Experiment. Psychol.* **72**, 1–6.

Berrien, F. K. (1946). The effects of noise, *Psychological Bull.* **43**, 141–161.

Blackwell, P. J., and Belt, J. A. (1971). Effect of differential levels of ambient noise on vigilance performance, *Perceptual Motor Skills* **32**, 734.

Blake, M. J. F. (1971). Temperament and time of day, *in* "Biological Rhythms and Human Performance," (W. P. Colquhoun, ed.). Academic Press, London.

Boggs, D. H., and Simon, J. R. (1968). Differential effect of noise on tasks of varying complexity, *J. Appl. Psychol.* **52**, 148–153.

Borsky, P. N. (1961). Community reactions to air force noise, *WADD Tech. Rep.* **6-689**.

Broadbent, D. E. (1950). The twenty dials test under quiet conditions, *Med. Res. Council Appl. Psychol. Unit Rep.*, No. 130/50.

Broadbent, D. E. (1951). The twenty dials and twenty lights test under noise conditions, *Med. Res. Council Appl. Psychol. Unit Rep.*, No. 160/51.

Broadbent, D. E. (1953). Noise, paced performance and vigilance tasks, *Brit. J. Psychol.* **41**, 295–303.

Broadbent, D. E. (1954). Some effects of noise on visual performance, *Quart. J. Experiment. Psychol.* **6**, 1–5.

Broadbent, D. E. (1957a). Effects of noises of high and low frequency on behaviour, *Ergonomics* **1**, 21–29.

Broadbent, D. E. (1957b). Effects of noise on behavior, *in* "Handbook of Noise Control," (C. M. Harris, ed.). McGraw-Hill, New York.

Broadbent, D. E. (1958a). "Perception and Communication." Pergamon, London.

Broadbent, D. E. (1958b). Effect of noise on an 'intellectual' task, *J. Acoust. Soc. Am.* **30**, 824–827.

Broadbent, D. E. (1963). Differences and interactions between stresses, *Quart. J. Experiment. Psychol.* **15**, 205–211.

Broadbent, D. E. (1971). "Decision and Stress." Academic Press, London.

Broadbent, D. E. (1978). The current state of noise research: Reply to Poulton, *Psychologic. Bull.* **85**, 1052–1067.

Broadbent, D. E., and Gregory, M. (1963). Vigilance considered as a statistical decision, *Brit. J. Psychol.* **54**, 309–323.

Broadbent, D. E., and Gregory, M. (1965). Effects of noise and of signal rate upon vigilance analysed by means of decision theory, *Human Factors* **7**, 155–162.

Broadbent, D. E., and Little, E. A. J. (1960). Effects of noise reduction in a work situation, *Occup. Psychol.* **34**, 133–140.

Bronzaft, A. L., and McCarthy, D. P. (1975). The effects of elevated train noise on reading ability, *Environ. Behav.* **7**, 517–527.

Bruel, P. V. (1976). Determination of noise levels, *in* "Man and Noise" (G. Rossi and M. Vigone, eds.). Minerva Medica, Turin.

Bursill, A. E. (1958). The restriction of peripheral vision during exposure to hot and humid conditions, *Quart. J Experiment. Psychol.* **10**, 113–129.

Cermak, L. S. (1972). "Human Memory: Research and Theory." Ronald, New York.

Cohen, A. (1974). Industrial noise and medical, absence, and accident record data on exposed workers, *in Proc. Int. Cong. Noise as a Public Health Problem* (W. D. Ward, ed.). U.S. Government Printing Office, Washington, D.C.

Cohen, A. (1976). The influence of a company hearing conservation program on extra-auditory problems in workers, *J. Safety Res.* **8**, 146–162.

Cohen, H. H., Conrad, D. W., O'Brien, J. F., and Pearson, R. G. (1973). Noise effects, arousal and information processing: Task difficulty and performance. Unpublished report, Department of Psychology and Industrial Engineering, North Carolina State Univ., Raleigh, North Carolina.

Cohen, S. (1978). Environmental load and the allocation of attention, *in* "Advances in Environmental Psychology," Vol. 1 (A. Baum, J. E. Singer and S. Valins, eds.). Erlbaum, Hillsdale, New Jersey.

Cohen, S. (1980). Aftereffects of stress on human performance and social behavior: A review of research and theory, *Psychol. Bull.* **88**, 82–108.

Cohen, S., Glass, D. C., and Singer, J. E. (1973). Apartment noise, auditory discrimination, and reading ability in children, *J. Experiment. Soc. Psychol.* **9**, 407–422.

Cohen, S., Evans, G. W., Krantz, D. S., and Stokols, D. (1980). Psychological, motivational, and cognitive effects of aircraft noise on children, *Am. Psychologist* **35**, 231–243.

Conrad, D. W. (1973). The effects of intermittent noise on human serial decoding performance and physiological response, *Ergonomics* **16**, 739–747.

Corso, J. F. (1952). The effects of noise on human behavior, *USAF WADC Tech. Rep.* No. 53-81.

Craik, F. I. M., and Lockhart, R. S. (1972). Levels of processing: A framework for memory research, *J. Verb. Learn. Verb. Behav.* **11**, 671–684.

Crook, M. A., and Langdon, F. A. (1974). The effects of aircraft noise in schools around London Airport, *J. Sound Vib.* **34**, 221–232.

Daee, S., and Wilding, J. M. (1977). Effects of high intensity white noise on short-term memory for position and sequence, *Brit. J. Psychol.* **68**, 335–349.

Davies, A. D. M., and Davies, D. R. (1975). The effects of noise and time of day upon age differences in performance at two checking tasks, *Ergonomics* **18**, 321–336.

Davies, D. R. (1968). Physiological and psychological effects of exposure to high intensity noise, *Appl. Acoust.* **1**, 215–233.

Davies, D. R. (1976). Noise and the autonomic nervous system, *in* "Man and Noise" (G. Rossi and M. Vigone, eds.). Minerva Medica, Turin.

Davies, D. R., and Gill, E. B. (1982). Noise, glucose and problem solving, *Bull. Brit. Psychol. Soc.* **36**, 78.

Davies, D. R., and Hockey, G. R. J. (1966). The effects of noise and doubling the signal frequency on individual differences in visual vigilance performance, *Brit. J. Psychol.* **57**, 381–389.

Davies, D. R., and Jones, D. M. (1975). The effects of noise and incentives upon retention in short-term memory, *Brit. J. Psychol.* **66**, 61–68.

Davies, D. R., and Jones, D. M. (1982). Hearing and noise, *in* "The Body at Work: Biological Ergonomics" (W. T. Singleton, ed). Cambridge Univ. Press, Cambridge.

Davies, D. R., and Parasuraman, R. (1982). "The Psychology of Vigilance." Academic Press, London.

Davies, D. R., Lang, L., and Shackleton, V. J. (1973). The effects of music and task difficulty on performance at a visual vigilance task, *Brit. J. Psychol.* **64**, 383–389.

Donnerstein, E., and Wilson, D. W. (1976). Effects of noise and perceived control on ongoing and subsequent aggressive behaviour, *J. Personality Soc. Psychol.* **34**, 774–781.

Dornic, S. (1975). Some studies on the retention of order information, *in* "Attention and Performance, V" (P. M. A. Rabbitt and S. Dornic, eds.). Academic Press, New York.

Duffy, E. (1957). The psychological significance of the concept of "arousal" or "activation," *Psychologic. Rev.* **64**, 265–275.

Duffy, E. (1962). "Activation and Behavior." Wiley, New York.

Easterbrook, J. A. (1959). The effect of emotion on cue utilization and the organization of behavior, *Psychologic. Rev.* **66**, 183–201.

Eschenbrenner, J. A. (1971). Effects of intermittent noise on the performance of a complex psychomotor task, *Human Factors* **13**, 59–63.

Eysenck, M. W. (1975). Effects of noise, activation level, and response dominance on retrieval from semantic memory, *J. Experiment. Psychol.* **104**, 143–148.

Eysenck, M. W. (1982). "Attention and Arousal." Springer-Verlag, Heidelberg.

Eysenck, M. W., and Folkard, S. (1980). Personality, time of day and caffeine: Some theoretical and experimental problems in Revelle *et al.*, *J. Experiment. Psychol.: Gen.* **109**, 32–41.

Finkelman, J. M., and Glass, D. C. (1970). Reappraisal of the relationship between noise and human performance by means of a subsidiary task measure, *J. Appl. Psychol.* **54**, 211–213.

Finkelman, J. M., Zeitlin, L. R., Filippi, J. A., and Friend, M. A. (1977). Noise and performance, *J. Appl. Psychol.* **62**, 713–718.

Finkelman, J. M., Zeitlin, L. P., Rosnoff, R. A., Friend, M. A., and Brown, L. S. (1979). Conjoint effect of physical stress and noise stress on information processing performance and cardiac response, *Human Factors* **21**, 1–6.

Fisher, S. (1972). A "distraction effect" of noise bursts, *Perception* **1**, 223–236.

Fisher, S. (1973). The "distraction effect" and information processing complexity, *Perception* **2**, 78–79.

Fleishman, E. A. (1975). Toward a taxonomy of human performance, *Am. Psychologist* **30**, 1127–1149.

Forster, P. M. (1978). Attentional selectivity: A rejoinder to Hockey, *Brit. J. Psychol.* **69**, 505–506.

Forster, P. M., and Grierson, A. T. (1978). Noise and attentional selectivity: A reproducible phenomenon, *Brit. J. Psychol.* **69**, 489–498.

Fowler, C. J. H., and Wilding, J. (1979). Differential effects of noise and incentives on learning, *Brit. J. Psychol.* **70**, 149–154.

Frankenhaeuser, M., and Lundberg, U. (1974). Immediate and delayed effects of noise on performance and arousal, *Biol. Psychol.* **2**, 127–133.

Gale, A. (1977). Some EEG correlates of sustained attention, *in* "Vigilance: Theory, Operational Performance, and Physiological Correlates" (R. R. Mackie, ed.). Plenum, New York.

Gawron, V. J. (1982). Performance effects of noise intensity, psychological set and task type and complexity, *Human Factors* **24**, 225–243.

Glass, D. C., and Singer, J. E. (1972). "Urban Stress: Experiments on Noise and Social Stressors." Academic Press, New York.

Goolkasian, P., and Edwards, D. C. (1977). The effect of loud noise on the psychological refactory period. *Bull. Psychon. Soc.* **9**, 139–141.

Graeven, D. B. (1975). Necessity, control and predictability of noise annoyance, *J. Soc. Psychol.* **95**, 86–90.

Grether, W. (1971). "Effects on Human Performance of Combined Environmental Stresses." U.S. Air Force, Wright Patterson Air Force Base, Dayton, Ohio, Aerospace Medical Research Laboratory Technical Report No. 68-70.

Grether, W. F., Harris, C. S., Mohr, G. C., Nixon, C. W., Ohlbaum, M., Sommer, H. C., Thaler, V. H., and Veghte, J. H. (1971). Effects of combined heat, noise and vibration stress on human performance and physiological functions, *Aerospace Med.* **42**, 1092–1097.

Grimaldi, J. V. (1958). Sensori-motor performance under varying noise conditions, *Ergonomics* **2**, 34–43.

Hack, J. M., Robinson, H. W., and Lathrop, R. G. (1965). Auditory distraction and compensatory tracking, *Perceptual Motor Skills* **20**, 228–230.

Hamilton, P. (1969). Selective attention in multisource monitoring tasks, *J. Experiment. Psychol.* **82**, 34–37.

Hamilton, P., and Copeman, A. (1970). The effect of alcohol and noise on components of a tracking and monitoring task, *Brit. J. Psychol.* **61**, 149–156.

Hamilton, P., Hockey, G. R. J., and Quinn, J. G. (1972). Information selection, arousal and memory, *Brit. J. Psychol.* **63**, 181–189.

Hamilton, P., Hockey, G. R. J., and Rejman, M. (1977). The place of the concept of activation in human information processing theory: An integrative approach, *in* "Attention and Performance, VI" (S. Dornic, ed.). Erlbaum, Princeton, New Jersey.

Harris, C. S. (1968). The effects of high intensity noise on human performance. *U.S. Army Medical Research Laboratory Report,* No. AMRL-TR-67-119.

Harris, C. S., and Sommer, H. C. (1971). "Combined Effects of Noise and Vibration on Mental Performance." U.S. Government Research and Development Report No. AD-731-146.

Hartley, L. R. (1973). Effect of prior noise or prior performance on serial reaction, *J. Experiment. Psychol.* **101**, 255–261.

Hartley, L. R. (1974). Performance during continuous and intermittent noise, and wearing ear protection, *J. Experiment. Psychol.* **102**, 512–516.

Hartley, L. R. (1981a). Noise, attentional selectivity, serial reactions and the need for experimental power, *Brit. J. Psychol.* **72**, 101–107.

Hartley, L. R. (1981b). Noise does not impair by masking: A reply to Poulton's "Composite Model for Human Performance in Continuous Noise," *Psychologic. Rev.* **88**, 86–89.

Hartley, L. R., and Adams, R. G. (1974). Effect of noise on the Stroop test, *J. Experiment. Psychol.* **102**, 62–66.

Hartley, L. R., and Carpenter, A. (1974). Comparison of performance with headphone and free-field noise, *J. Experiment. Psychol.* **103**, 377–380.

Hartley, L. R., Couper-Smartt, J., and Henry, T. (1977). Behavioural antagonism between chlorpromazine and noise in man, *Psychopharmacologia* **55**, 97–102.

Haveman, J. E., and Farley, F. H. (1969). "Arousal and Retention in Paired Associate, Serial and Free Learning." University of Wisconsin, Center for Cognitive Learning Technical Report, No. 91, Madison, Wisconsin.

Hiroto, D. S. (1974). Locus of control and learned helplessness, *J. Experiment. Psychol.* **102**, 187–193.

Hockey, G. R. J. (1970a). Effect of loud noise on attentional selectivity, *Quart. J. Experiment. Psychol.* **22**, 28–36.

Hockey, G. R. J. (1970b). Signal probability and signal location as possible bases for increased selectivity in noise, *Quart. J. Experiment. Psychol.* **22**, 37–42.

Hockey, G. R. J. (1970c). Changes in attention allocation in a multi-component task under loss of sleep, *Brit. J. Psychol.* **61**, 473–480.

Hockey, G. R. J. (1973). Changes in information selection patterns in multisource monitoring as a function of induced arousal shifts, *J. Experiment. Psychol.* **101**, 35–42.

Hockey, G. R. J. (1978). Attentional selectivity and the problems of replication: A reply to Forster and Grierson, *Brit. J. Psychol.* **69**, 499–504.

Hockey, G. R. J. (1979). Stress and the cognitive components of skilled performance, *in* "Human Stress and Cognition: An Information-Processing Approach" (V. Hamilton and D. M. Warburton, eds.). Wiley, Chichester.

Hockey, G. R. J. (1984). Varieties of attentional state: V. Effects of environment, *in* "Varieties of Attention" (R. Parasuraman and D. R. Davies, eds.). Academic Press, New York.

Hockey, G. R. J., and Hamilton, P. (1970). Arousal and information selection in short-term memory, *Nature* **226**, 866–867.

Hockey, G. R. J., and Hamilton, P. (1983). The cognitive patterning of stress states, *in* "Stress and Fatigue in Human Performance" (G. R. J. Hockey, ed.). Wiley, Chichester.

Hockey, G. R. J., McClean, N. P., and Hamilton, P. (1981). State changes and the temporal patterning of component resources, *in* "Attention and Performance," (J. Long and A. D. Baddeley, eds.), Vol. 9. Erlbaum, Hillsdale, New Jersey.

Houston, B. K. (1968). Inhibition and the facilitating effect of noise on interference tasks, *Perceptual Motor Skills* **27**, 947–950.

Houston, B. K. (1969). Noise, task difficulty and Stroop color-word performance, *J. Experiment. Psychol.* **82**, 403–404.

Houston, B. K., and Jones, T. M. (1967). Distraction and Stroop color-word performance, *J. Experiment. Psychol.* **74**, 54–56.

Jensen, A. R. (1965). Scoring the Stroop test, *Acta Psychologica* **24**, 398–408.

Jerison, H. J. (1959). Effects of noise on human performance, *J. Appl. Psychol.* **43**, 96–101.

Jerison, H. J., and Smith, A. K. (1955). "Effect of Acoustic Noise on Time Judgement." WADC U.S. Air Force Report 55-358. Wright-Patterson Air Force Base, Ohio.

Jerison, H. J., and Wallis, R. A. (1957). "Experiments on Vigilance: III. Performance on a Simple Vigilance Task in Noise and Quiet." U.S. Air Force, Wright Air Development Center Technical Regulation 57-318. Wright-Patterson Air Force Base, Ohio.

Jones, D. M. (1979). Stress and memory, *in* "Applied Problems in Memory" (M. M. Gruneberg and P. E. Morris, eds.). Academic Press, London.

Jones, D. M. (1983). Loud noise and levels of control: A study of serial reaction. Paper presented at the Fourth International Congress, Noise as a Public Health Problem, Turin, June 1983.

Jones, D. M. (1984). Performance effects, *in* "Noise and Society" (D. M. Jones and A. J. Chapman, eds.). Wiley, Chichester.

Jones, D. M., and Broadbent, D. E. (1979). Side-effects of interference with speech by noise, *Ergonomics* **22**, 1073–1081.

Jones, D. M., and Davies, D. R. (1984). Individual and group differences in the response to noise, *in* "Noise and Society" (D. M. Jones and A. J. Chapman, eds.). Wiley, Chichester.

Jones, D. M., Auburn, T. C., and Chapman, A. J. (1982). Perceived control in continuous loud noise, *Curr. Psychologic. Res.* **2**, 111–122.

Jones, D. M., Smith, A. P., and Broadbent, D. E. (1979). Effects of moderate intensity noise on the Bakan vigilance task, *J. Applied Psychol.* **64**, 627–634.

Jones, D. M., Thomas, J. R., and Harding, A. (1982). Recognition memory for prose items in noise, *Curr. Psychologic. Res.* **2**, 33–44.

Kahneman, D. (1973). "Attention and Effort." Prentice-Hall, Englewood Cliffs, New Jersey.

Kallman, W. M., and Isaac, W. (1977). Altering arousal in humans by varying ambient sensory conditions, *Perceptual Motor Skills* **49**, 19–22.

Kerr, W. A. (1950). Accident proneness of factory departments, *J. Appl. Psychol.* **34**, 167–170.

Key, K. F., and Payne, M. C. (1981). Effects of noise frequency on performance and annoyance for women and men, *Perceptual Motor Skills* **52**, 435–441.

Kryter, K. D. (1950). The effects of noise on man, *J. Speech Hearing Disorders, Monograph Supplement* 1, 1–95.

Kryter, K. D. (1970). "The Effects of Noise on Man." Academic Press, New York.

Kurz, D. P. (1964). Effects of three kinds of stressors on human learning and performance, *Psycholog. Rep.* **14**, 161–162.

Leonard, J. A. (1959). "Five-Choice Serial Reaction Apparatus." Medical Research Council Applied Psychology Unit Report 326/59.

Loeb, M., and Jones, P. D. (1978). Noise exposure, monitoring and tracking performance as a function of signal bias and task priority, *Ergonomics* **21**, 265–277.

Loeb, M., and Richmond, G. (1956). "The Influence of Intense Noise on the Performance of a Precise Fatiguing Task." U.S. Army Medical Research Laboratory Report 268. Fort Knox, Kentucky.

McGrath, J. J. (1963). Irrelevant stimulation and vigilance performance, *in* "Vigilance: A Symposium" (D. N. Buckner and J. J. McGrath, eds.). McGraw-Hill, New York.

McKennell, A. C. (1963). "Aircraft Noise Annoyance around London (Heathrow) Airport." Central Office of Information, London.

McLean, P. D. (1969). Induced arousal and time of recall as determinants of paired-associate recall, *Brit. J. Psychol.* **60**, 57–62.

Malmo, R. B. (1959). Activation: A neuropsychological dimension, *Psychologic. Rev.* **66**, 367–386.

Malmo, R. B., and Surwillo, W. W. (1960). Sleep deprivation: Changes in performance and physiological indicants of activation, *Psychologic. Monographs* **72** (502).

May, D. N., and Rice, C. G. (1971). Effects of startle due to pistol shot on control precision performance, *J. Sound Vib.* **15**, 197–202.

Mech, E. V. (1953). Factors affecting routine performance under noise: I. The influence of 'set', *J. Psychol.* **35**, 283–298.

Millar, K. (1979). Noise and the 'rehearsal-masking' hypothesis, *Brit. J. Psychol.* **70**, 565–577.

Millar, K. (1980). Word recognition in loud noise, *Acta Psychologica* **43**, 225–237.

Mohindra, N., and Wilding, J. M. (1982). Noise effects on rehearsal rate in short-term serial order memory, *Quart. J. Experiment. Psychol.* **35**, 171–182.

Moran, S. L. V., and Loeb, M. (1977). Annoyance and behavioral aftereffects following interfering and noninterfering aircraft noise, *J. Appl. Psychol.* **62**, 719–726.

Mullin, J., and Corcoran, D. W. J. (1977). Interaction of task amplitude with circadian variation in auditory vigilance performance, *Ergonomics* **20**, 193–200.

O'Malley, J. J., and Poplawski, A. (1971). Noise induced arousal and breadth of attention, *Perceptual Motor Skills* **33**, 887–890.

Park, J. F., and Payne, M. C. (1964). Effects of noise level and difficulty of task in performing division, *J. Appl. Psychol.* **47**, 367–368.

Pepler, R. D. (1959). Warmth and lack of sleep: Accuracy or activity reduced, *J. Comp. Physiologic. Psychol.* **52**, 446–450.

Percival, L., and Loeb. M. (1980). Influence of noise characteristics on behavioral after effects, *Human Factors* **22**, 341–352.

Plutchik, R. (1961). Effect of high intensity intermittent sound on compensatory tracking and mirror tracing, *Perceptual Motor Skills* **12**, 187–194.

Poulton, E. C. (1965). On increasing the sensitivity of measures of performance, *Ergonomics* **8**, 69–76.

Poulton, E. C. (1966). Engineering psychology, *Ann. Rev. Psychol.* **17**, 177–200.

Poulton, E. C. (1970). "Environment and Human Efficiency." Charles C. Thomas, Springfield, Illinois.

Poulton, E. C. (1976). Continuous noise interferes with work by masking auditory feedback and inner speech, *Appl. Ergonomics* **7**, 79–84.

Poulton, E. C. (1977a). Arousing stresses increase vigilance, *in* "Vigilance: Theory, Operational Performance and Physiological Correlates" (R. R. Mackie, ed.). Plenum, New York.

Poulton, E. C. (1977b). Continuous noise masks auditory feedback and inner speech, *Psychologic. Bull.* **84**, 977–1001.

Poulton, E. C. (1978). A new look at the effects of noise: A rejoinder, *Psychologic. Bull.* **85**, 1068–1075.

Poulton, E. C. (1979). Composite model for human performance in continuous noise, *Psychologic. Rev.* **86**, 361–375.

Poulton, E. C. (1981). Not so! Rejoinder to Hartley on masking by continuous noise, *Psychologic. Rev.* **88**, 90–92.

Poulton, E. C., and Edwards, R. S. (1979). Asymmetric transfer in within-subject experiments on stress interactions, *Ergonomics* **22**, 945–962.

Poulton, E. C., and Freeman, P. R. (1966). Unwanted asymmetrical transfer effects with balanced experimental designs, *Psychologic. Bull.* **66**, 1–8.

Rotton, J., Olszewski, D., Charleton, M., and Soler, E. (1978). Loud speech, conglomerate noise and behavioral aftereffects, *J. Appl. Psychol.* **63**, 360–365.

Salamé, P., and Wittersheim, G. (1978). Selective noise disturbance of the information input in short-term memory, *Quart. J. Experiment. Psychol.* **30**, 693–704.

Samuel, W. M. S. (1964). Noise and the shifting of attention, *Quart. J. Experiment. Psychol.* **16**, 264–267.

Schwartz, S. (1974). Arousal and recall: Effects of noise on two retrieval strategies, *J. Experiment. Psychol.* **72**, 299–316.

Schwartz, S. (1975). The effects of arousal on recall, recognition and organisation of memory. Unpublished manuscript cited by Smith and Broadbent (1981).

Seligman, M. E. P. (1975). "Helplessness: On Depression, Development and Death." Freeman, San Francisco.

Sherrod, D. R., and Downs, R. (1974). Environmental determinants of altruism: The effects of stimulus overload and perceived control on helping, *J. Experiment. Soc. Psychol.* **10**, 468–479.

Simpson, G. C., Cox, T., and Rothschild, D. R. (1974). The effects of noise stress on blood glucose level and skilled performance, *Ergonomics* **17**, 481–487.

Sloboda, W. (1969). The disturbance effect of white noise on human short-term memory during learning, *Psychonomic Sci.* **14**, 82–83.

Sloboda, W., and Smith, E. E. (1968). Disruption effects in human STM: Some negative findings, *Perceptual Motor Skills* **27**, 575–582.

Smith, A. P. (1982). The effects of noise and task priority on recall of order and location, *Acta Psychologica* **51**, 245–251.

Smith, A. P., and Broadbent, D. E. (1980). Effects of noise on performance on embedded figures tasks, *J. Appl. Psychol.* **65**, 246–248.

Smith, A. P., and Broadbent, D. E. (1981). Noise and levels of processing, *Acta Psychologica* **47**, 129–142.

Smith, A. P., Jones, D. M., and Broadbent, D. E. (1981). The effects of noise on recall of categorised lists, *Brit. J. Psychol.* **72**, 299–316.

Smith, K. R. (1951). Intermittent loud noise and mental performance, *Science* **114**, 132–133.

Smith, M. C. (1967). Theories of the psychological refractory period, *Psychologic. Bull.* **67**, 202–213.

Stevens, S. S. (1972). Stability of human performance under intense noise, *J. Sound Vib.* **21**, 35–56.

Stroop, J. R. (1935). Studies of interference in serial verbal reactions, *J. Experiment. Psychol.* **18**, 643–662.

Swets, J. A. (1977). Signal detection theory applied to vigilance, *in* "Vigilance: Theory, Operational Performance and Physiological Correlates" (R. R. Mackie, ed.). Plenum, New York.

Teichner, W. H., Arees, E., and Reilly, R. (1963). Noise and human performance: A psychophysiological approach, *Ergonomics* **6**, 83–97.

Theologus, G. C., Wheaton, G. R., and Fleishman, E. A. (1974). Effects of intermittent moderate intensity noise stress on human performance, *J. Appl. Psychol.* **59**, 539–547.

Tolin, P., and Fisher, P. G. (1974). Sex differences and effects of irrelevant auditory stimulation on performance of a visual task, *Perceptual Motor Skills* **39**, 1255–1262.

TRACOR (1971). "Community Reaction to Airport Noise", Vol. 1. NASA CR1176L-N71-29023.

Treisman, A. M. (1969). Strategies and models of selective attention, *Psychologic. Rev.* **76**, 282–299.

Tulving, E. (1972). Episodic and semantic memory, *in* "Organization of Memory" (E. Tulving and W. Donaldson, eds.). Academic Press, New York.

Tulving, E. (1979). Relation between encoding specificity and levels of processing, *in* "Levels of Processing in Human Memory" (L. S. Cermak and F. I. M. Craik, eds.). Erlbaum, Hillsdale, New Jersey.

Viteles, M. S., and Smith, K. R. (1946). An experimental investigation of the effect of change in atmospheric conditions, and noise upon performance, *Trans. Am. Soc. Heating Ventil. Eng.* **52**, 167–182.

von Wright, J., and Nurmi, L. (1979). Effects of white noise and irrelevant information on speeded classification: A developmental study, *Acta Psychologica* **43**, 157–166.

von Wright, J., and Vauras, M. (1980). Interactive effects of noise and neuroticism on recall from semantic memory, *Scand. J. Psychol.* **21**, 97–101.

Ward, L. M., and Suedfeld, R. (1973). Human response to highway noise, *Environment. Res.* **6**, 306–326.

Warner, H. D. (1969). Effects of intermittent noise on human target detection, *Human Factors* **11**, 245–250.

Warner, H. D., and Heimstra, N. W. (1971). Effects of intermittent noise on visual search tasks of varying complexity, *Perceptual Motor Skills* **32**, 219–226.

Warner, H. D., and Heimstra, N. W. (1972). Effects of noise intensity on visual target-detection performance, *Human Factors* **14**, 181–185.

Warner, H. D., and Heimstra, N. W. (1973). Target detection performance as a function of noise intensity and task difficulty, *Perceptual Motor Skills* **36**, 439–442.

Wechsler, D. (1958). "The Measurement and Appraisal of Adult Intelligence." Williams and Wilkins, Baltimore.

Weinstein, A., and MacKenzie, R. S. (1966). Manual performance and arousal, *Perceptual Motor Skills* **22**, 498.

Weinstein, C. S., and Weinstein, N. D. (1979). Noise and reading performance in open space school, *J. Educ. Res.* **72**, 210–213.

Weinstein, N. D. (1974). Effect of noise on intellectual performance, *J. Appl. Psychol.* **59**, 548–554.

Weinstein, N. D. (1977). Noise and intellectual performance: A confirmation and extension, *J. Appl. Psychol.* **62**, 104–107.

Weston, H. C., and Adams, S. (1932). "The Effect of Noise on the Performance of Weavers." Industrial Health Research Board Report 65, Part II. His Majesty's Stationery Office, London.

Weston, H. C., and Adams, S. (1935). "The Performance of Weavers under Varying Conditions of Noise." Industrial Health Research Board Report 70. His Majesty's Stationery Office, London.

Wilbanks, W. A., Webb, W. B., and Tolhurst, G. C. (1956). "A Study of Intellectual Activity in a Noise Environment." U.S. Navy School of Aviation Medicine Project NM001 104100 Report 1.

Wilding, J., and Mohindra, N. (1980). Effects of subvocal suppression, articulating aloud and noise on sequence recall, *Brit. J. Psychol.* **71**, 247–262.

Wilding, J., Mohindra, N., and Breen-Lewis, K. (1982). Noise effects in free recall with different orienting tasks, *Brit. J. Psychol.* **73**, 479–486.

Wilkinson, R. T. (1961). Interaction of lack of sleep with knowledge of results, repeated testing and individual differences, *J. Experiment. Psychol.* **62**, 263–271.

Wilkinson, R. T. (1962). Muscle tension during mental work under sleep deprivation, *J. Experiment. Psychol.* **64**, 565–571.

Wilkinson, R. T. (1963). Interaction of noise with knowledge of results and sleep deprivation, *J. Experiment. Psychol.* **66**, 439–442.

Wilkinson, R. T. (1965). Sleep deprivation, *in* "The Psychology of Human Survival" (O. G. Edholm and A. L. Bacharach, eds.). Academic Press, London.

Wilkinson, R. T. (1969). Some factors influencing the effects of environmental stressors upon performance, *Psychologic. Bull.* **72**, 260–272.

Woodhead, M. M. (1958). Effects of bursts of loud noise on a continuous visual task, *Brit. J. Indust. Med.* **15**, 120–125.

Woodhead, M. M. (1964). Searching a visual display in intermittent noise, *J. Sound Vib.* **1**, 157–161.

Yerkes, R. M., and Dodson, J. D. (1908). The relation of strength of stimulus to rapidity of habit formation, *J. Comp. Neurol. Psychol.* **18**, 459–482.

6

Noise Annoyance

F. J. LANGDON

Hemel Hempstead
Hertfordshire
England

I. GENERAL CONSIDERATIONS

Excessive and unwanted sound, in other words, noise, is an almost ubiquitous form of nuisance affecting more people more widely than perhaps any other disamenity. Nuisance is caused by noise in the home from neighbours and externally from transportation and in schools, hospitals, offices and workplaces generally. A curious and unique feature of noise which distinguishes it from other physical disamenities, such as overheating or discomfort glare in lighting, is the quite specific way in which people experience the undesirable effects. Although sound at very high pressure levels can produce physiological effects such as temporary shifts in hearing threshold and, with repeated exposure, hearing loss and although it can affect performance of certain tasks, over a range of quite moderate levels (say from 50 to 80 dB), its main effects are the distinctive and characteristic ones variously referred to as annoyance, disturbance, bother, intrusion—negative feelings or affects. All these terms—and psy-

143

chological studies have identified more than thirty English words and phrases with similar connotations—can be grouped under the general heading of nuisance, and this is what we shall be concerned with here.

The fact that this 'adverse subjective response'—to employ a term frequently used by psychologists—is referred to under so many different names and descriptions points to its capability for affecting people in a wide variety of ways. Some of these are purely attitudinal, as when someone says that 'the traffic noise here drives me up the wall'. Others are more closely related to various activities interfered with, such as reading, talking or watching television, though again it is largely the attitudinal aspect, the dissatisfaction occasioned by this interference rather than the degree of interference itself which people are most aware of. Further along the line of noise nuisance are the 'indirect' effects. Thus in warm weather, external noise will result in windows being closed with consequent overheating, poor ventilation and general discomfort. Or the garden cannot be used for sitting about in, and the dwelling cannot be used in the way it was designed to be (e.g., guests are entertained in the kitchen 'because it is too noisy in the front room').

The variety of these adverse effects and the kinds of responses they provoke is one reason for much of the controversy and seeming disagreement among those studying noise nuisance problems. It has seemed that no single attribute is capable of providing a term on which a scale of nuisance can be based. Although words such as 'annoyance' or 'bother' have been widely used, other researchers have felt them to be inadequate and have opted for terms such as 'intrusiveness' or 'disturbance'. Other workers have argued that it is impossible to know exactly what 'annoyance' means—at least when someone says he is 'highly annoyed'. It cannot be defined and our grasp of the meaning rests only on intuition and an appeal to our own experience. In consequence, objective and neutral terms have been employed such as 'acceptability–unacceptability' or 'dissatisfaction'. Some of these issues will be raised in the following section. For the present, the outstanding feature is this pervasive, specific and yet hard to define, one might say, protean character of noise and the resulting difficulties in trying to study its effects and arrive at practical norms for noise control.

II. SCALING AND EVALUATING ANNOYANCE

A. Scaling

Although I have begun by referring to noise nuisance, and to subjective adverse effects in particular, of which annoyance is but one descriptor, from now on the term annoyance will be used in this wider sense. This usage has been defined as 'a general feeling of displeasure or adversiveness towards a noise source believed to have a harmful effect upon a person's health and well-being' (Karolinska Institute, 1971). Borsky (1980), citing this definition, follows it up by the shrewd observation that it is relatively easy to ascertain whether a person has feelings of annoyance but that the measurement of the degree of annoyance presents many problems. One of these, which must be dealt with at the outset, is that of general scientific method.

As the response to noise in the form we are at present considering is largely attitudinal, one of feeling, we can hardly measure it in the way we would measure, for example, a judgment of loudness. It is not difficult to simulate a noise, such as that of road traffic, in a laboratory, but it is difficult to have a person simulate annoyance. To take part in an interesting experiment, and perhaps be paid for it, is an entirely different thing from having to put up with the noise all day at home. For this reason, studies of noise annoyance have tended to rely mainly on observation through social surveys rather than on controlled experiments. Annoyance has been studied in the laboratory under controlled conditions (Rice, 1977; Rylander *et al.*, 1977a; Flindell, 1979; Stephens and Powell, 1980) with some degree of success, although the results while comparable with those of field studies are not in general capable of being used directly for the establishment of control norms, for this self-same reason of the difficulty of genuinely simulating annoyance. Although social surveys, competently executed, draw considerable strength from the fact that they study 'the real thing', lack of control over experimental conditions has tended to limit the general theoretical conclusions we have been able to draw from them.

All surveys dealing with noise annoyance have involved the use of some kind of rating scale. Even a simple 'yes/no' question becomes a scale if it is applied to a sample large enough to enable the proportions of 'yes' to 'no' to be calculated. We can therefore pass from very simple assessments of the degree of annoyance to complex batteries, that is, collections of scales and questions inter-related in some way which all bear on the measurement of annoyance. The aim of these is to discriminate between different aspects of the noise and the way it affects people, and particularly at different levels of intensity.

One reason for using numerous questions and scales is the doubt as to whether single words such as 'bother' or 'annoyance' can elicit and record the appropriate response reflecting the variety of experience evoked by the noise. A second reason is that it is unwise to base an assessment on an informant's reply to a single question (McKennell, 1963; Guttman,

1944) and that the assessment produced by a group of tested questions will be more reliable. However, this assumes that there is some way of taking replies to each question into account, so that each influences the overall answer in its due proportion. The procedure resorted to for doing this is usually some form of factor analysis, a statistical procedure which measures the relative 'loading' of the component represented by each question on the dimension representing the overall concept of annoyance or dissatisfaction. More recently, opinion seems to have swung around in favour of single-question scale assessments (McKennell, 1970; Rice *et al.,* 1978). It can be argued that a respondent answering a single scale question on say, annoyance, carries out a kind of factor analysis in his own head, although the study of annoyance through factor analysis may still be invaluable, as will be seen later, if we want to know just what the respondent means by annoyance.

The construction of annoyance scales has been a fruitful source of disputation. Should they consist of verbal terms, such as 'not at all', 'a little', 'moderately', 'very much' (annoyed), or should they consist of a series of numbers lying between opposite semantic poles, say, 'definitely satisfied' to 'definitely dissatisfied'? And again, how many categories are best? Too few may distort the respondent's true feelings as he tries to approximate them to the nearest descriptor; too many may overtax his power of discrimination. One discussion and experimental study of these problems (McKelvie, 1978) suggests that so far as scale reliability is concerned, there is little to choose between verbal category and numerical scales, though the latter appear capable of producing higher statistical correlations with noise exposure [see also Griffiths *et al.,* (1980)].

The use of a numerical scale raises the question of just what the series of numbers lying between the poles actually means. What intensity of feeling is indicated by marking 'five' on a scale where 'one' represents 'definitely satisfied' and 'seven' corresponds to 'definitely dissatisfied'? One way to find out is to calibrate the scale. This can be done by finding out what answers to other questions, such as those concerned with interference with various activities, or beliefs about the adverse effects of the noise correspond with this score, and by extension with all the other values on the scale. Such a cross-tabulation is illustrated in Table I, which provides the basic data from which the intercorrelations between annoyance scale values and all the other information can be calculated to provide an overall calibration of the scale. Such a procedure, besides indicating the connotation of any form of adverse subjective response, tends to lessen the likelihood that over-reliance on replies to a single scale question will result in biassed or unreliable assessments.

One further problem must be overcome if we are to use an annoyance scale as a measuring instrument. We must be sure that the successive

TABLE I. Comparison of Overall Annoyance Responses with Aircraft on Noise with Selected Items (1975 Survey)[a]

		Annoyance scale		
		Low	Moderate	High
Annoyance—single question		$N = 261$	$N = 174$	$N = 856$
Low (0–4)		67%	20%	4%
Moderate (5–6)		18	36	12
High (7–9)		15	44	84
Annoyance with activity disturbance by aircraft noise				
TV or radio	Low	59	9	2
	Moderate	15	6	1
	High	26	85	97
Sleep	Low	98	84	30
	Moderate	1	5	2
	High	1	11	68
Rattle and vibrations	Low	93	75	28
	Moderate	5	12	4
	High	2	13	68
Rest and relaxation	Low	96	72	15
	Moderate	3	8	3
	High	1	20	83
Conversation	Low	65	16	2
	Moderate	13	8	1
	High	22	76	97
Startle	Low	43	9	1
	Moderate	24	10	2
	High	33	81	97
Volunteered things disliked				
Aircraft		17%	46%	57%
Direct question on degree disliked				
Aircraft noise	Low	42	11	3
	Moderate	20	16	5
	High	38	73	92
Poor neighbours	Low	97	95	93
	Moderate	1	1	4
	High	2	4	3
Felt like moving				
	Yes	28%	36%	49%
	No	72	64	51

[a] From Borsky (1980).

steps along the scale, whether these are numbers or verbal expressions, represent roughly equal intervals. If the scale is merely a series and no more, we shall be able to compute only rank-order correlations which indicate no more than that a succession of noise levels is related to a succession of responses. If we want to establish a quantitative relationship in the form of a regression equation, we must meet the statistical requirement of an interval or ratio scale. The literature dealing with these problems is very extensive, and the reader is referred to the Bibliography given at the end of this chapter.

For the present, the issue of annoyance measurement may be summarised by emphasizing that annoyance scales are developed by empirical testing to be reliable, in the sense of being both internally consistent and stable over time (Griffiths *et al.*, 1980; Bradley and Jonah, 1979) and valid; that is, of measuring what they purport to measure and not something else. Beyond this, they must be capable of being treated as roughly equal interval scales and of being related to other aspects of adverse subjective response and reported behaviour.

B. Components of Annoyance

When we ask what exactly noise annoyance is and what it is that makes us feel bothered by noise, we are really asking two questions. Or more precisely, our question has two aspects, that which relates to the source of the annoyance, the noise itself; and that which relates to ourselves and our internal psychological processes. The former concerns things such as the level of noise, its loudness, intermittency, duration, sound character and possible meaning. The latter covers the way we perceive the noise, its sensory and emotional effects, the way our peace of mind is disturbed or our activities interfered with. These aspects are not separate, for when we feel annoyed both are actively present and result in a feeling state that is to ourselves whole and indivisible and hardly capable of introspective analysis.

Nevertheless, in measuring and assessing the adverse effects of noise one of our first procedures is to take this general phenomenon apart and look at its main components. So far as concerns the sound itself, perhaps the most important characteristic is its loudness or perceived intensity. The degree of annoyance does not, of course, depend entirely on how loud a noise seems. An illustration of this is the effect of a dripping tap heard in a quiet house. Again, numerous studies (Scharf, 1977, 1980; Keighley, 1966) have shown that although judgements of annoyance and loudness are to some degree intercorrelated, the relationship is by no means perfect. And a further reason for this lack of complete correspondence lies in the character of the sound, that aspect of loudness which we may refer to as its *noisiness* (Berglund *et al.,* 1975, 1976; Kryter, 1968). Noisiness has been shown to relate more closely to judgements of loudness than to annoyance, though at very high sound levels all these judgments tend to coalesce. At lower levels, however, they are increasingly independent, as suggested by the example of the dripping tap.

In the case of transportation noise, in particular, that of aircraft and trains, characterised by the repetition of single events such as landings and take-offs and the passage of trains, the duration of each event has been claimed also to influence the extent of annoyance. Hence, two factors are simultaneously involved, the number of noise events and the duration of each. However, the precise effects of both still remain unclear and a subject of controversy. In a study of nuisance from rail noise, Aubree (1971a) took both into account and concluded that a better prediction of annoyance was thereby obtained. One the other hand, Fields and Walker (1980) in a similar but more recent survey failed to confirm this but obtained predictions just as good from a simple measure of dB(A) L_{eq} over a 24-hr period. In the case of aircraft noise, the U.K. statutory procedure for noise control takes into account the number of overflights exceeding 80 dB(A) over the whole time period (McKennell, 1963), as does a very similar procedure employed in France, likewise derived from the results of social surveys (Alexandre, 1970). Nevertheless, Rylander *et al.* (1980) have attempted to show in a series of studies that where numbers of overflights exceed 50 per day, increasing numbers of flights have little influence on the degree of annoyance, which can be predicted effectively from the maximum peak levels L_{max} of the noisiest aircraft; moreover, they used this formulation to re-interpret the results of the TRACOR (1976) studies, which showed an apparent and not easily explained decline in annoyance with increasing numbers of overflights. This inconsistency, an outcome of the use of a number-of-event-related formula, disappears when the noise data are treated in terms of Rylander's formula. Rylander has gone on to apply the same approach to the effects of city tramway noise (Rylander *et al.,* 1977b) by implication in agreement with Fields and Walker (1980) in ignoring numbers of noise events. From these various researches one can only conclude that although the occurrence of single events, and possibly their duration, may have some influence on the extent of annoyance, the precise significance of this in the measurement of noise effects remains far from clear.

In a different context, however, numbers of events have been shown to be important. Keighley (1966) studied annoyance caused by noise in offices where the overall noise level is effectively the product of numbers of discrete noises. In this case, each noise occurs at random and the peak level of each noise is also random. The greater the number of noise-events in a given period, the more they tend to overlap, for since their rise and decay times are not instantaneous they tend to fuse, forming the background level which increases with number of events. As the average level falls with decreasing numbers of events, annoyance increases since each discrete noise is perceived against a lower background level. For this reason, 'office quieting' through conventional treatments is often ineffective in reducing annoyance, which can be better dealt with by reducing noises at their sources. Thus carpeting the floor may be more effective than covering the ceiling with acoustic tiles.

An important component of annoyance, a particular aspect of noisiness, is the sound character of a noise. According to its spectral composition, a noise will appear more or less annoying, independent of its sound-pressure level. Two examples in which the spectral composition of the noise is taken into account are provided by the procedure for assessing the degree of community annoyance produced by industrial noise (Anonymous, 1963). This procedure, laid down in British Standard 4142 (British Standards Institution, 1967) requires variations in the form of weightings to be applied to the measured values of the noise obtained in decibels(A) according to time of week and time of day, and also according to spectral composition, allowing for the addition of 5-dB(A) to the measured value for the presence of identifiable pure tones, whines, screeches, etc. In the case of aircraft noise, allowance has also been made for the presence of high-frequency components. The unit of measurement for aircraft noise termed PNdB (perceived noise level in decibels), due originally to Kryter (1959), was based on a derived scale of subjective 'noisiness' (the 'noy' scale). In 1965 a further correction to the unit was made to take account of pure tone and duration effects and the corrected unit was termed EPNdB (effective perceived noise level in decibels) (Kryter and Pearsons, 1965; Kryter, 1968), and again in 1969 the suggestion was made to combine sound-pressure levels in bands below 355 Hz to take account of critical bands in auditory response (Kryter, 1969). All these refinements were introduced and tested through social surveys to enable measurement of noise to incorporate the sound character component of annoyance, though it has to be admitted that the claimed superiority of the corrected aircraft noise units have continued to be debated (Botsford, 1967, 1968).

Lastly, time of day may be mentioned as a component of annoyance to be taken into account. Time of day was one component considered in the Industrial Noise Rating method already mentioned. A further development of time-of-day weighting was that made by the recommendation of the U.S. Environmental Protection Agency (1974) for weighting measured or estimated community noise levels by 10-dB L_{eq} for the period 22.00 to

07.00 hours, termed a day–night level L_{dn} to replace the 24-hr L_{eq}, to allow for the greater degree of annoyance (and of sleep disturbance) caused by nighttime noise. However, a number of studies of annoyance from aircraft and road traffic noise have tended to cast doubt on the original proposals (Ollerhead, 1980; Langdon, 1976a). It has been suggested that the weighting is excessive and would require at most only 5 dB, possibly less, and that there may be relatively little interference with sleep after midnight, but that the period from 19.00 hours is the most important, together with that from 22.00 to 24.00 hours when people are trying to get to sleep (Langdon and Buller, 1977a). Much of the study of time-of-day effects has been connected with transportation noise, although in the course of surveys on annoyance caused by building and road construction noise (Baughan, 1980) it was found that operations extending beyond normal working hours into evenings and weekends provoked substantially more annoyance than was occasioned at ordinary times. Here, however, the results may not be evidence for a 'time-of-day' effect in the strict sense, as the increase in annoyance may have been due to the fact that more residents are at home during these leisure periods rather than any specifically time-related effects of noise. But this does not itself argue against a time-related weighting, since this is concerned with the effect, and not how it is caused.

Turning now to those components of annoyance which relate directly to the recipient of the noise, the 'internal' as opposed to the 'external' factors, perhaps the most important is the degree of personal sensitivity or alternatively the extent of noise tolerance. People vary greatly in the degree to which they are affected by noise. As will be seen when we come to consider the prediction of annoyance, individual differences in noise sensitivity are to a large degree responsible for our inability to account for more than a small part of the individual response to noise. For example, in the case of road traffic noise, it has often been found that such differences account for more variance than the noise itself, so that some individuals show high annoyance at 60 dB(A) while others remain unconcerned at 80 dB(A) (Langdon, 1976b).

Secondly, among the 'internal' components are a range of attitudinal factors which have been identified in the attempt to develop a structured model of annoyance, that is, a model which is operational in the sense that it defines annoyance, not by semantic connotation, but by identifying the differing mode of subjective response evoked by noise in the course of empirical studies. Social surveys have included questions bearing on the attitudes of respondents towards those believed responsible for the noise, whether they were thought to be concerned to limit or reduce the noise (Borsky, 1961), and beliefs as to the effectiveness of complaint and possible reasons for failing to complain despite feelings of annoyance (McKennell, 1963; Langdon, 1976a). The general outcome of this work has been to suggest that attitudes towards responsible authorities influence greatly the extent of annoyance. One attempt to measure such an effect directly may be cited. Jonsson and Sorensen (1967) interviewed residents on the perimeter of an airport, one-half of which had been exposed to a mailing shot of leaflets explaining the mode of operation of the airport and how the authorities acted to limit noise. Not surprisingly, these respondents produced significantly lower annoyance scores than did those in the area which had not been 'propagandised'. In a somewhat similar context related this time to the attitude of people towards the noise source itself, subjects in a laboratory experiment (Cederlof et al., 1963) were exposed to tape-recorded traffic noises while viewing projected transparencies of different road vehicles not related to the sounds being heard. Estimates of loudness and relative annoyance showed that the type of vehicle seen during the making of a judgement—though not related in any way to the sound—influenced the decision independently of noise level. The image of a motor cycle or sports car resulted in a higher loudness and annoyance score for the same sound level than did the image of a bus or delivery truck, the perceived social role directly influencing the annoyance occasioned by the noise.

In a series of studies, Kastka and his associates (Kastka, 1976, 1983; Kastka and Paulsen, 1979) have attempted to develop an annoyance model by considering not only the social and environmental influences, but also the nature of the internal response to noise. As already mentioned, people tend to express their feelings by a wide variety of words and phrases. Collections of these expressions obtained from a statistically adequate sample of exposed persons were subjected to factor analysis. From this, three main dimensions or areas of response were extracted. The first was a purely 'sensory' response to the intensity, frequency, duration and disturbing quality of the noise—in short, the external factors already considered. The second dimension, termed the 'somatic–emotional' component, centred on disturbance to well-being through anger, irritation, disturbance to relaxation and sleep. The third factor, termed the 'acoustic' component, related to the judged loudness of the sound and its capacity for interference with speech and hearing, watching television, reading, etc. This is in line with what has been said earlier, that the total subjective response termed annoyance is a complex one referring to different yet related areas of the subject's experience.

It is not, of course, suggested that the model developed by Kastka is, or is claimed to be, in any special sense 'correct' or definitive, for there may be other factors involved in the total response such as feelings of fear or insecurity, or indirect discomforts, such as that from overheating. There are also modifications introduced by the overall perceived quality of the environment (Aubree, 1971b; Langdon, 1976a). Nevertheless, it may be agreed that the annoyance response entails a psychological structure which may be modelled, even if in a simplified and incomplete way. On the other hand, while this model may carry us some way towards understanding and interpreting annoyance, when we attempt to use it for measuring and predicting the response for the practical purpose of noise control, we shall see that the constraints of mathematical and statistical treatments tend to limit its extended use.

III. MEASURING AND PREDICTING ANNOYANCE

A. General Problems

So far we have looked at the essential features of the subjective response to noise and mentioned some of the problems in scaling annoyance. But to measure annoyance in a practical way, to show how much noise is likely to occasion a given degree of annoyance and how much communities and individuals are affected by noise, we need reliable techniques of social survey capable of yielding useful predictions.

Not all surveys are aimed at predicting nuisance related to noise. We may simply want to know how much noise nuisance people are exposed to and how one location differs from another. For this purpose we need surveys of noise *impact* or *incidence,* such as the London Noise Survey (Parkin *et al.,* 1968), the U.K. National Survey of Road Traffic and the Environment (Morton-Williams *et al.,* 1978), the U.S. Nationwide Urban Noise Survey (Fidell, 1978) or the U.K. National Survey of Noise Nuisance from Neighbours (Langdon and Buller, 1977b). Some, though not all, of these studies were both physical and social noise surveys, but whether this was so or not, the object was in every case to obtain a picture of the current situation, its impact generally and the variations within the sample of people studied. Annoyance scaling must be consistent and reliable, and the sample studied must be adequately representative of the population at large if predictions are to be made from the survey sample. The essential question is therefore the assembly of a sample corresponding closely with the distribution of the regional or national population, and for this reason it may bear little relation to the distribution of noise sources.

If the cause of annoyance under study were, for example, road, rail or aircraft noise, there would be no reason for assuming such a correspondence since most of the population would not be exposed to this noise.

The U.K. survey (Morton-Williams *et al.*, 1978) found that only 6% of the sample was exposed to levels of road traffic noise likely to occasion serious annoyance. On the other hand, in the case of noise nuisance from neighbours, everyone except occupants of detached houses—about 15% of all housing—has in principle an equal chance of being annoyed by noise from the house or flat next door. Here noise nuisance does correspond with population distribution.

Hence to go beyond the study of incidence and attempt to relate annoyance quantitatively with noise level a survey must represent all levels of incidence. The U.K. traffic noise survey would have been rather inefficient for this purpose, giving a great deal of information about people not bothered by noise at all but very little about those seriously annoyed. Incidence or *synthetic* surveys therefore differ entirely from *analytic* ones concerned with making predictions about noise-related nuisance. The former have only to be statistically representative of the population. The latter make associative predictions; that is, they assert that a given level of annoyance is related to a particular level of noise. Indeed, they must do more, for they assert that this relationship extends over a range of noise levels and modes of response. What such surveys do not do, however, and this appears to have been generally overlooked until comparatively recently (Langdon and Griffiths, 1982), is to associate *changes* in annoyance with changes in noise levels. The subject matter of large-scale noise nuisance surveys is invariably a population exposed to stable and unchanging levels of noise. The information from such surveys helps to establish norms for acceptable environmental conditions but has little bearing on what would happen if noise levels were to change.

To express the relationship between noise and annoyance by means of a regression equation of the form $X = nY \pm C$, where X is annoyance, Y the noise level, and C a constant, it is necessary to measure or rate annoyance by means of equal interval scales. Assuming that this requirement has been met—and a great deal of controversy has raged about this (Hart, 1973; McKennell, 1965)—we can develop such an equation by the statistical procedure of fitting the generalisation to all the data by the method of least squares. This best-fitting line to all the observations of noise and annoyance corresponds to the highest correlation of probability and is usually expressed within the convention that 95% of all observations fall within one standard deviation from the regression line. One practical outcome of this is exemplified by community surveys, for which failure to achieve this level of accuracy limits very greatly the possibilities for noise reduction programmes, since with less accuracy noise levels would need to differ by more than 5 dB to guarantee a significantly different level of annoyance. To bring about such a difference is a costly and far from easy task.

The procedure outlined assumes that the relationship between noise and annoyance is one of the simplest, namely, a linear one. It may be that this relationship is more complicated, that annoyance increases by a square or cubic relation with noise. Nevertheless, in practice almost all the relationships derived through social surveys have been expressed by hypothetical linear relationships, largely because these are the simplest, requiring fewest assumptions and to be preferred on grounds of parsimony, particularly where no claim can be made to a real causal relationship.

The need to develop a structured model of annoyance, bringing in, with appropriate weightings, the components combined within the overall response, as suggested by Kastka, or to extend the physical measures beyond sound-pressure levels or even beyond acoustic quantities, will necessitate use of a multiple correlation and an associated multiple regression of the form $X = Y + Z + N \cdots \pm C$. The form is again purely linear, the further components being treated additively. Given this type of treatment, and in any case the nature and quality of the data rarely permit anything more sophisticated, the testing and application of a structured annoyance model which is in any way close to the complexities of the human response are likely to be well-nigh impossible. Statistical and mathematical procedures are by no means incapable of going beyond the simple forms referred to. But to go beyond them, the distribution of annoyance scores over the range, the dispersion of individual scores around each set of measured levels and their degree of reliability must conform to extremely rigorous criteria. As becomes evident when existing surveys are studied, this has so far hardly ever been the case.

B. Prediction of Annoyance for Communities

Noise annoyance does not arise in some vague or abstract fashion. It is the result of noise from road traffic or aircraft, from machines at work or from next-door neighbours. So most of what we know about predicting and limiting noise nuisance has been learnt through surveys dealing usually with one or other of these sources. Few studies which relate noise and annoyance have attempted to cover simultaneously the full range of environmental noise, partly because of differing patterns of incidence, and partly because many kinds of noise require a different approach to their measurement.

The type of problem which arises in predicting community annoyance is well illustrated through the example of road traffic noise, though much applies equally to other sources. A road traffic noise survey aimed at predicting annoyance at any given noise level—a prediction for the average or for the most typical person—involves interviewing enough people at enough locations so that the number of people at each location or site, and the number of sites, cover the range of experience from the quietest to the noisiest. Effective prediction requires that 95% of the data fall within one standard deviation from the prediction line. For community predictions the 'observations' are the points which represent the mean or median values of noise and annoyance at each site. Although these points must lie close to the line, the annoyance scores of the individuals they represent in the aggregate may be scattered over the entire scale. However, so long as these constitute a continuous, uni-modal distribution forming a roughly bell-shaped (Gaussian) curve, this is not of first importance. This may seem a rather irresponsible statement. How can we predict a community's response if it is as widespread and various as this? Surely, prediction would be better if there were less scatter? The answer to this is that, yes, it would be better, but not necessarily from the standpoint of community prediction. The really important question is whether the scatter is the result of poor measurement and inefficient scaling procedures, or whether it is a true representation of varied attitudes found in the community.

Let us assume that this is so. In this case we have only to remember that we are not trying to predict annoyance as a mere amusement, but for a practical purpose such as to establish acceptable limits to nuisance and to provide a basis for limiting noise. We are therefore interested primarily in those aspects of the environmental situation which we can manipulate, such as road widths, planning rules and volume and speed of traffic. If there are wide differences between individual annoyance scores at any given noise level, then, assuming the adequacy of our survey methods, they represent real differences which are facts of life and cannot be manipulated—unless we are to look forward to an environment where people are directed to live in this or that place according to their degree of noise sensitivity. Even if we trouble to record these differences, along with other demographic or economic variables, and take them into account, the principal gain will be in the coherence and scientific quality of our survey results. There will be no gain in community prediction since we cannot apply any of what we know in a predictive way without performing a kind of 'mini-survey' at every location to which we want to apply it.

The real point of concern about the wide dispersion of annoyance scores is that, as originally pointed out by Bryan and Tempest (1973), if community standards are established on the basis of the regression line representative of the average response, there is a danger that a proportion of the community will fail to derive any benefit since for them the standard will be so low as to permit noise levels they cannot tolerate. This is an important point; it might be met by setting the standard, not on the regression line, but some distance away from it, nearer to the average response of those more sensitive to noise. This necessarily assumes that the annoyance scores of the 'sensitive' people yield averages which run more or less parallel to those of the whole population. As will be seen later, however, the data from surveys which have used sensitivity tests to identify such individuals do not, unfortunately, appear to bear out this assumption.

The most important differences we need to consider in trying to establish noise control standards are those which may exist among different communities, such as urban or rural, or in different countries. There are also differences arising from the type of noise source, though these are of a different nature and will be considered separately. There is little strong evidence from national surveys of regional differences in dose-response relationships. The U.S. Urban Noise Survey (Fidell, 1978), one of the few studies which tried to cover most sources of urban noise, found no evidence of regional differences, at least for urban areas; and the U.K. survey of nuisance from train noise (Fields and Walker, 1980) arrived at similar conclusions; nor did the investigators find differences between urban and rural communities. On the other hand, a recent survey carried out in France (Vallet *et al.*, 1983) reported differences in annoyance from road traffic noise between urban and rural areas for residents living close to motorways, rural residents being more annoyed presumably because of lower background levels.

More striking are differences between different national communities. Table II shows results from three surveys of road traffic noise nuisance in

TABLE II. Results of Three Surveys Correlating Annoyance with Noise Level

Feature	Paris, 1971	London, 1972	Stockholm, 1968
Sample	693	1359	326
Noise range dB(A) L_{50}	53–75	57–77	43–67
Best results (Units)	$L_{50}\ r = 0.84\ (0.32)^a$	$r = 0.848\ (0.204)^a$	$r = 0.82$
	$L_{10}\ r = 0.84\ (0.32)^a$	$r = 0.848\ (0.205)^a$	$r = 0.88$
	$L_{eq}\ r = 0.84\ (0.32)^a$	$r = 0.84\ \ (0.203)^a$	$r = 0.96$
Noise level at 50th percentile in L_{50}	68	60	56 (disturbed only)
Suggested limit	Below 70 dB(A) L_{50}	61 dB(A) L_{50}	59 dB(A) L_{50}
Type of scale	Seven-point dissatisfaction plus interference with activities	Seven-point dissatisfaction	Ten-point index of bother and exposure
Type of traffic	Mainly free flow, some congestion	Free flow, no congestion	Free flow, little congestion

a Individual correlation.

Paris, London and Stockholm. Comparisons among different national surveys are likely to be more hazardous than those for regions within national surveys because different surveys are being compared, the placing of microphones and the ranges of exposure levels are likely to differ. Nevertheless, some comparison is possible. Using the median noise level L_{50}, the measure employed at that time by the Swedish and French surveys, and the mid-point of the annoyance scale—the level above which more than half the population studied were annoyed—as points of comparison, we find this level reached at about 68 dB(A) for Paris (Aubree, 1971b), 56 dB(A) for Stockholm (Fog and Jonsson, 1968) and 60 dB(A) for London [Langdon, 1976a (fieldwork, 1972)], suggesting some differences in noise tolerance, inhabitants of Stockholm and London showing more concern than those of Paris. A difference, again in the same direction, is suggested by a comparative study of noise annoyance in Stockholm and Ferrara (Jonsson et al., 1969). Although noise levels in Stockholm were between 5 and 8 dB(A) lower than in Ferrara, 51% of Stockholm respondents claimed to be annoyed by traffic noise as against 49% in the Italian city, the divergence increasing if comparison is confined to those claiming to be 'highly annoyed'. These results taken generally all suggest that noise annoyance is a social phenomenon which is influenced by, and reflects differences in, both life-style and psychological make-up. This last consideration is emphasized by results of a cross-cultural study (Thomas et al., 1983) covering noise annoyance in the United Kingdom, Germany and Japan. The investigators claim that while there were similarities in the concepts of 'noise' and 'annoyance' for respondents in all three countries, Japanese respondents differed considerably in their understanding of 'loudness'. There were also differences in the degree of acceptability of noise countermeasures and attitudes towards noise control.

In opposition to this way of looking at things, however, Poulton (1977), in a provocative discussion of the shortcomings of subjective studies, and in particular of social surveys, has claimed that the mid-points used as a basis for comparison are likely to differ, not for psychological or cultural reasons, but purely through statistical artefacts. Poulton argues that any such differences are 'range effects' that arise from differences in the ranges of noise levels encountered by different surveys. Although this contention may have some force when applied to laboratory experiments, in which case subjects make judgements over a range of stimuli, it would seem less applicable to social surveys, for here the respondent makes but a single rating—that related to his own circumstances—and does not experience a range of levels so far as making ratings is concerned. Nevertheless, as we do not know the actual range of environmental noise to which respondents are exposed, it is difficult to refute the claim that general noise experience might have influenced the respondent's answer.

Some attempts have been made to fill this gap in our knowledge. Maurin (1975) attempted to record in detail the daily noise dose for 35 subjects together with a diary of hour-by-hour activities. In similar fashion, Kono (Kono *et al.*, 1982) measured the total noise dose of 604 participants in a survey and concluded, as did Maurin, that the annoyance response is not explained entirely by noise exposure but depends on the general living environment. This, of course, brings us back to the conclusion of Jonsson (Jonsson *et al.*, 1969) already referred to, and which has also been stressed by Levy-Leboyer (1977, 1982; Levy-Leboyer and Moser, 1976), namely that the response to noise cannot be isolated and treated as a purely mechanical process but has to be evaluated as part of an overall environmental response. On the other hand, Schultz (1978) dispensed with these intellectual refinements and simply compared the results of 11 noise nuisance surveys from different countries, selected as using comparable methodologies, and concluded that there were no important differences in the dose–response relationships since all the data could be fitted to a single curve. Schultz's work has not been accepted without reservations, however. It is not easy to see from the published paper what adjustments have been made to intercept levels, and many of the scales used in the original surveys have been transformed to yield comparable forms of scoring. In particular, the author's assumptions have been strongly criticized by Kryter (1982), and more recently Griffiths (1983) has demonstrated that the procedure and methods used are statistically invalid.

Over all these attempts at comparison there hover reservations with respect to differences in method, both of noise measurement and annoyance rating, and of statistical treatment. There are also possible differences arising from the time at which surveys were conducted. For example, the traffic noise survey cited earlier (Langdon, 1976a) carried out in London during 1972 may be compared with a similar study using identical methodology, the fieldwork for which took place in 1967 (Griffiths and Langdon, 1968). Between the two surveys there is a difference of 4 dB(A) over the entire range of noise levels encountered at which annoyance scores were the same. From this it is difficult to avoid the conclusion that over this comparatively short period, but one in which the issue of noise nuisance as a matter of growing public concern culminated in the passage of the Land Compensation Act (Anonymous 1973), attitudes to noise hardened significantly.

Although it may be difficult to put numbers on, or to propose theoretical explanations for, some of these differences in the levels of noise tolerance in different communities, the general weight of evidence suggests that the differences are real. Apart from what has been said here, it has to be admitted also that differences in planning ordinances or legal limits to permissible noise are often the result not so much of survey findings as of differences in available resources or of public and political pressures on authorities (Alexandre, 1973).

C. Prediction of Individual Annoyance

In predicting annoyance for communities we do not directly concern ourselves with predicting the responses of individuals, at least, beyond the statistical analysis of scores in the interests of quality control. Here, the individual is a statistical artefact, the 'average' person or the average 'highly annoyed' person. To this extent, social surveys are capable of accurate prediction based on high levels of correlation between noise and annoyance ($r = 0.9$ for aircraft and approaching this for road traffic noise) and accounting for 70 to 80% of the response variance in terms of noise. This is accuracy of an order which enables us to forecast the effects arising from differences of less than 5 dB in noise levels, with the possibility of further improvement if the effects of environmental quality are taken into account.

We cannot do this for individuals. Even in the case of aircraft noise, where annoyance affects are most acute, we can obtain individual correlations of only about 0.6, accounting for less than 50%, and in the case of road traffic only about 10 to 12% of the response in terms of noise exposure. Even if we introduce intervening variables to take account of factors such as environmental quality or differences in noise sensitivity, we cannot add more than about a further 15 to 20% of explained variance. All these predictions of individual annoyance are, of course, just as significant statistically as the much higher levels of correlation obtained for community groups, for with the former the numbers involved are far greater, perhaps 2000 or more individuals as against 20 or 30 groups of people at particular locations.

Nevertheless, it is not uncommon to read in the literature of noise control that the generally low level of prediction for individual responses to noise demonstrates a shortcoming of social surveys, with the implied conclusion that they should not be relied upon for establishing community standards and noise controls. Quite apart from the fact that this criticism is wholly misconceived, it remains difficult to see what other scientifically preferable method might be used for this purpose.

If it is accepted that there are real individual differences in temperament and psychological makeup which produce different annoyance responses to noise, it follows that prediction of individual responses can never be precise. The prolonged study of an individual, incorporating psychological tests, would no doubt yield an accurate prediction of the likely response to a future noise experience. But such a 'prediction' would be of little practical use or scientific significance, nor would it be a very great technical achievement, being more like forecasting the winner of a race two metres from the winning post. What is required for a scientific account of noise annoyance is the ability to generalise from the response of one individual to that of another.

To do this we would need to take account of all the psychological, circumstantial and environmental differences among individuals. And we must distinguish among individual differences in response to noise and individual differences inherent to the respondent. Thus, in studying annoyance caused by road traffic noise to schoolteachers, it was found that far more variance was accounted for in individual response than had been the case with people at home (Sargent *et al*, 1980). This was simply because the teachers remained most of the time in one room and suffered a fairly constant level of exposure while performing a relatively unvarying task. People at home are, as against this, moving about the house and engaging in a variety of activities. Only part of the variation in response, therefore, may be truly said to relate to inherent individual differences in noise tolerance.

However, before attempting to detect the differences inherent to individuals which modify the response to noise, we need to examine carefully our measuring instruments, that is to say, our annoyance scales, for unless we can be sure of how reliable they are, in the sense of yielding the same answer on each occasion, we will be unable to apportion that part of the unexplained variance traceable to real individual differences and that due to errors in the annoyance scales (Langdon, 1978). Recent studies of scale reliability have shown that this topic is essential in trying to account for individual response to noise, and we shall therefore take a brief look at this question before dealing with individual annoyance prediction.

1. Consistency and Reliability of Annoyance Scales

It is usual to find measures of consistency such as Cronbach alpha (Guilford, 1954) in noise annoyance surveys performed by various methods, often through random sample subdivision and inter-comparison of responses. What such a procedure tells us is whether the annoyance scale is being used consistently, in the sense that all those taking part in the survey over the whole range of conditions tend to respond to the scale per se in the same way. Most of the studies referred to here cite values for consistency of $r > 0.95$, indicating that the scales used are very consistent.

TABLE III. Frequency Distribution of Difference
Phase 1 and 2[a]

Difference	±0	1	2	3	4	5	6
Percentage	37	36	16	7	2	1	1
Mean unsigned Difference 1.1					$n = 180$		

[a] From Griffiths and Delanzun (1977).

But a scale needs to be reliable and consistent in another equally important way, and this is the extent to which, given that circumstances remain unchanged, the same person will always give the same answer. This requirement, usually termed the measure of stability, can only be examined through a test–retest procedure involving at least two successive interviews. Table III, taken from a twice-repeated noise nuisance study (Griffiths and Delauzun, 1977), shows the distribution of the differences among answers to the seven-point dissatisfaction scale given during two interviews conducted at a two-month interval at four locations.

The coefficient of reliability computed from this survey was 0.6, indicating that the scale used was of only moderate reliability, yielding only 36% of reliable variance for individual responses. Further studies using both this and a four-point verbal scale of annoyance (Griffiths *et al.,* 1980; Langdon and Griffiths, 1982) have yielded similar values for interviews repeated four times in the course of a year, and comparable surveys in Canada (Bradley and Jonah, 1979; Hall and Taylor, 1980) have yielded not dissimilar results.

However, with all these surveys a considerable amount of time passed between successive interviews and it is possible that respondents might have in this time changed their attitudes towards the noise, even though exposure levels remained unchanged. This supposition is given credibility by data from a survey already cited (Langdon, 1976a) in which the same question was put to respondents at the beginning and the end of the interview. Here, 75% of respondents gave the same score on each occasion and only 9% differences greater than ±1 on the seven-point scale.

It may therefore be reasonably assumed that the actual reliability of annoyance scales is perhaps higher than the value of 0.6 given earlier. But assuming a good value, such as 0.75 (Bradley and Jonah, 1979), which gives about 65% of reliable variance, it is clear that a considerable part of the residual variance in prediction of individual annoyance is due, not to real differences between respondents, but to imperfections in the annoyance scale. These shortcomings have little effect on the prediction of community annoyance since the individual differences are random, just as

many individuals varying upward as downward in the repeated interviews, so that the errors cancel out. But what these studies do show is
that the effects of differences in noise sensitivity are not so great as was
originally assumed.

2. Individual Differences in Sensitivity to Noise

The main effect of differences in noise sensitivity is to increase the
dispersion of annoyance scores at all noise levels. But there is besides this
another curious effect which affects the accuracy of prediction. Implicit,
though nowhere explicitly stated, in the general notion of differences in
sensitivity discussed by Bryan and other researchers is the concept of a
distribution of scores at each noise level over the entire range. The most
sensitive are seen as occupying the upper, the least sensitive the lower
part of this distribution, with the average person somewhere in between.
As noise levels increase so do all the annoyance scores for each subpopulation according to their degree of sensitivity, and this constitutes the case
for special consideration being given to those with above-average sensitivity, when setting noise control standards.

The analysis of individual annoyance in the London traffic noise survey
(Langdon, 1976b), however, shows a rather different picture. As can be
seen from Fig. 1, those identified by a self-rating test as particularly
sensitive to noise show relatively little change in annoyance, remaining

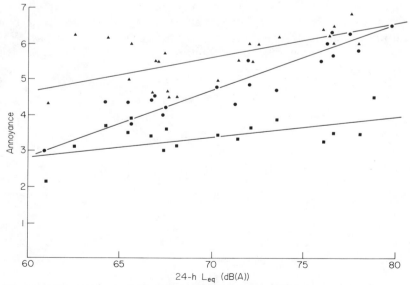

Fig. 1. Noise annoyance and sensitivity to noise. (▲, sensitive; ●, neutral; ■, nonsensi-
tive; $n = 1435$.)

annoyed even at relatively low levels of noise. The opposite appears to hold for those rating themselves not sensitive to noise.

It is therefore rather difficult to know what to prescribe to make those sensitive to noise more satisfied, since reducing noise levels by 5 or 10 dB will have very little effect.

A number of explanatory models for noise sensitivity have been put forward. A very simple mechanical model was outlined in the report cited, and this is illustrated in Fig. 2. The model assumes that differences in sensitivity correspond to differences in response thresholds. If the change in annoyance with noise level is represented by an ogive curve, in line with what is known of a graded biological response [see Sargent *et al.* (1980)], it will be seen that only the 'neutral' subpopulation attains a maximum dose–response relationship within the limits of the noise exposure range, the sensitives having attained the 'roll-off' at the top of the curve while the 'non-sensitives' do not begin theirs until the limits of the range are encountered.

Weinstein (1980) has put forward the idea that noise sensitivity is a particular instance of a more general personality trait. Thus the attitude termed by Weinstein 'criticalness', taken over from work by Frederiksen and Messick (1959), pervades the individual's approach to his whole environment. Such an attitude appears to be independent of beliefs about noise, such as its harmful effects, etc. This model is not incompatible with that already outlined and results from the survey referred to (Langdon 1976a,b) tend to accord with Weinstein's views.

On the other hand, more general attempts to link sensitivity to noise with fundamental personality traits have met with varied success. While such a link is not an unreasonable supposition, the results of experiments

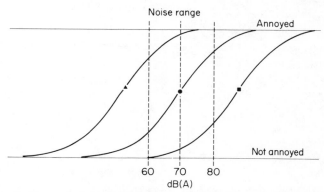

Fig. 2. Theoretical model of sensitivity to noise. (Key same as Fig. 1.)

designed to investigate the relation between noise sensitivity and specific traits such as neuroticism or introversion/extroversion seem difficult to interpret and not always in agreement. A number of workers claim to have demonstrated such a relation (Broadbent, 1972; Moreira and Bryan, 1972: Barbenza *et al.*, 1970). On the other hand, Griffiths and Delauzun (1977) failed to establish any relationship, employing the Eysenck Personality Inventory or the Gregory–Broadbent scale. Applying the Eysenck scale in simplified form to schoolchildren, Elliot (1971) found differences in noise tolerance for groups established on the basis of Eysenck scores. McLean and Tarnopolsky (1977) appear to accept a relationship between sensitivity and personality, although one of the difficulties in interpreting many of the findings in this area is the rather inadequate character of the statistical treatment and far from rigorous design of the laboratory experiments. It may be that differences in noise tolerance require more broadly based and less clinically derived criteria against which to be tested and that the traits embodied in the recognised clinical tests apply only to some of the quite large section of the population highly sensitive to noise.

Apart from differences in personality, there appear to be other reasons for individual differences in annoyance not related to noise sensitivity per se. Thus Borsky (1980) notes the effects of fear arising from belief in the possibility of an aircraft crashing. However, aside from the fear of 'misfeasance'—something going wrong—to use the term applied by Borsky and derived from his earliest surveys (Borsky, 1961), the general tendency of social surveys has been to discount any very large role for personal beliefs and attitudes, such as the possible effects of noise on health, in influencing the degree of annoyance. On the other hand, material factors have been shown to exert an influence. Aubree (1971b) found a significant difference in annoyance scores according to the amount of time spent at home, finding annoyance lower among people who spent more time there. This suggests that the value placed on peace, quiet and relaxation varied according to personal circumstances, quietness in the home being more highly valued by those who could only return to it from the outside on occasions. Along the same lines of thought, Levy-Leboyer (1977, 1982) has argued that noise annoyance cannot be viewed as a simple dose–response relationship but as a complex phenomenon in which the individual response is the product of a number of different sets of values placed upon environmental qualities which give meaning and significance to a person's existence. The degree to which these cluster around commonly shared attitudes within a culture governs the similarity of responses. Beyond these community norms there extend areas of particular value and significance for each individual, and these tend to produce deviations from the modal or group score.

Aside from the aspects considered by environmental psychology, there are economic factors at work. It has been shown (Langdon, 1978) that individuals with very low incomes tend to be less annoyed by noise and place a lower value on quiet, apparently because for them the greatest source of anxiety was that of maintaining a roof over their heads, a notion which is supported by results obtained by Rylander *et al.* (1972) and by Walters (1975), who, in surveying attempts to place an economic value on peace and quiet, concluded that freedom from noise is a luxury.

It may seem, looking back over the past 30 years or so devoted to studying the effects of noise, that a disproportionate amount of effort has been expended on the problem of individual differences in annoyance, largely because of a generally held belief in the importance of accurately predicting individual annoyance. In retrospect, this would appear to have been a mistake, for in fact the variance accounted for, both at community and individual levels, has increased systematically with greater sophistication of survey design and improved measuring techniques. And aside from this, it has come to be increasingly recognised that since individual differences are ineluctable, predictions which take account of them have little practical value. However, it is easy to be wise after the event, for while prediction is difficult, the exercise of hindsight does not require very great intellectual powers. The interest of the study of individual differences is in the development of a coherent and scientifically satisfying model of annoyance as one aspect of the effects of noise. Such a model is necessary as an intellectual support, for the levels of prediction accuracy now attained for identifiable groups within the community, which is what practitioners in noise control, planners and architects are concerned with, are now perfectly adequate for the purposes of administration and design.

IV. COMPARING DIFFERENT SOURCES OF NOISE ANNOYANCE

Reviewing the contemporary environment, Ward and Dubos (1972) remarked that "of all forms of pollution, noise is perhaps the most inescapable for the urban dweller. It pursues him into the privacy of his home, trails him in the street and quite often is the accompaniment of his labours." Accepting the truth of this ubiquitous exposure to noise, we might well ask, what is the worst, the most annoying and the most widespread in its effects? Here are a number of very different questions jumbled together. What kind of noise annoys most people and which annoys them most? And how can we compare these different things?

We can start by looking at the results of some broad-band incidence surveys. The BRS national survey of noise from neighbours (Langdon and Buller, 1977b) enquired about noise annoyance from all sources. The general picture for all types of urban dwelling—which did not vary greatly from one type to another—showed that more people, some 55% of the sample, heard road traffic than any other kind of noise, with aircraft noise, heard by 35%, coming a close second. But in terms of annoyance the picture looks quite different. Only about 5 to 6% were annoyed by traffic and even fewer, about 3.5%, by aircraft noise. Instead, the lead is taken by noise from neighbours heard through the party wall where dwellings are attached. Although only about 25% of the sample heard such noise, over 8% were annoyed by it.

The reason for these differences becomes apparent on reflection. Road traffic noise can be heard almost everywhere. But only in a few places where people live—as opposed to business and shopping centres—is it dense and near enough to be annoying. This is even more true for aircraft noise, which can be heard almost anywhere but only becomes annoying in the vicinity of major airports, of which there is not a very great number. As against this, almost everyone has neighbours—about 85% of the population—but if they are heard at all through the party wall the effect is likely to be annoying, because of presumed privacy, because the party wall is expected to give acoustic protection and because the experience is reciprocal—those who hear neighbours know that neighbours can hear them.

A not dissimilar picture is given by the results of the U.K. national traffic noise survey (Morton-Williams et al., 1978), which also gives first place to road traffic, heard by 89% of the sample, again only about 25% of the sample claiming to be bothered by it, though in this case traffic noise also heads the ranking of annoying noises. In this survey the proportion claiming to hear noises of all kinds is higher than that for the survey of neighbours' noise—although the questions often related to experience outside the home—but the ranking of noises heard is similar. A point-by-point comparison of the two surveys is hardly possible, partly because the traffic noise survey did not classify neighbours' noise separately (only the category 'people' was used) and partly because the sampling frames differed.

The U.S. nationwide urban noise survey (Fidell, 1978) also classifies the most widespread noises, although here the listing is only for annoyance, the number hearing but not annoyed not being recorded. Direct comparison is again difficult, for in this case the sampling frame used was 28 sites in 8 urban centres. Moreover, the listing does not particularise neighbours, children or railways as noise sources. Nevertheless, as with the U.K. surveys, transportation noise is well to the fore, with road traffic again the lead. While there is no doubt that this finding, general to national noise surveys, is correct, it is worth remembering that the types of noise reported heard or causing annoyance are to some extent the outcome of sample design, the questionnaire and the way results are presented.

Consequently, from these surveys we learn nothing about noise annoyance at work, in schools or in hospitals. Few urban noise surveys would have enquired about them, and in any case, the relative sparseness of schools and hospitals as compared to homes ensures that data related to such places would form only a tiny proportion of the whole. A comparative study of broad-band noise surveys does not seem to lead to answering the question posed at the outset for the comparison is beset by all kinds of methodological difficulties.

The reason why road traffic noise is the most widespread form of noise nuisance in the industrialised countries is fairly obvious. But it is nevertheless not the source of the most acute nuisance. Relatively few people experience intolerable noise in their homes from road traffic. As a proportion of the total population it is likely that even fewer are subjected to a similar degree of annoyance from aircraft noise. But this does not alter the fact that the highest levels of noise from aircraft are even less acceptable than those of road traffic and are made even more annoying by intermittency, often with very low noise levels between flyovers.

Perhaps the least recognised in importance is the annoyance caused by noise from neighbours. Although a well-researched topic (Chapman, 1948; Gray et al., 1958; Bitter and van Weeren, 1955; Bitter, 1968, Langdon et al., 1981, 1983), there has been a curious lack of public concern in this area, despite the high proportion of residents shown by surveys to suffer this kind of annoyance. It is, it must be admitted, not a 'fashionable' topic like aircraft or road traffic noise and has never developed into a campaign area. Perhaps this is because it is something that occurs in private and so the source is known and usually identified, and it may to some degree be taken for granted as a fact of life. Certainly, the respondents to the BRS surveys (Langdon et al., 1981, 1983) appeared to accept a certain amount of noise from next door as inevitable and reasonable. This is why when asked what they did if neighbours made too much noise, the greatest proportion said they tried to make less noise themselves.

Most surveys of noise nuisance have concentrated on a particular type of noise, or on a particular location, home, school or office. But in real life we are often exposed to many different sources, and some work has been concerned with comparing the relative 'annoyingness' of different noise sources and, beyond this, with considering the effects of noise coming from more than one source. More will be said on the comparison among various sources of noise when discussing transportation noise in Chapter 8. Here, our main concern is with some of the theoretical and methodological problems involved in trying to make such comparisons and to integrate the general noise environment.

The ability to compare the effects of noise from different sources requires, in the first instance, a suitable unit with which to measure the noise. For this purpose the equivalent or equal energy noise level L_{eq} has generally been used in Europe and the day–night noise level L_{dn} in the United States. Using L_{eq} as a comparator, the results of annoyance surveys dealing with road traffic, railway and aircraft noise have been brought together. A glance at Fig. 3, taken from a study of Ollerhead (1978), shows the absence of any universal dose–response relationship for annoyance from the three sources.

Ollerhead deemed the three surveys from which these data are taken to be comparable (Langdon, 1976a; Fields and Walker, 1980; Bottom, 1971).

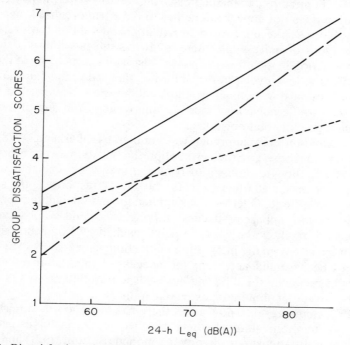

Fig. 3. Dissatisfaction with different noise sources. ——, road traffic (Langdon, 1978); ——, aircraft and road traffic (Bottom, 1971); ------, trains (Fields and Walker, 1980). [From Ollerhead (1980).]

At low noise levels, however, road traffic generates more annoyance than railways or aircraft. But as noise levels increase, annoyance from both road traffic and aircraft increases more sharply than does that from railways. The different dose–response curves make it clear that the whole of the data from these different sources could only be integrated by using a unit such as L_{eq} with the aid of weighting coefficients.

Given this sort of problem, to pass from comparison of different noise sources to estimating the annoyance produced by different, simultaneously heard noises is even more difficult. This has not discouraged study of the problem, however, since its solution would result in considerable simplification for urban noise control. Robinson (1971, 1972) was the first to propose a general solution by means of a unit, the noise pollution level L_{NP}, derived from fitting dose–response curves from surveys of annoyance caused by road traffic and aircraft noise, taking account not only of the sound pressure level but the degree to which that level fluctuated, to a greater extent than did L_{eq}. This unit would not necessarily yield better predictions than do existing units but was aimed at being applicable, in principle, to mixed noise sources. However, in practice this expectation does not appear altogether to have been realised, again presumably because the dose–response relationships differ too greatly for different types of noise while there is always the possibility of second-order interaction in the effects of noises heard simultaneously at different levels. Thus at low levels, traffic noise may help to reduce aircraft noise annoyance by raising the background level between flyovers. But at high levels, road traffic noise may increase annoyance from aircraft noise by imposing a greater total noise load.

A somewhat different approach was adopted by Ollerhead (1978) in the form of a model based on the assumption that all sources are proportionally constant (though the level of response may vary) and that where a single source masks all other sources, the overall response is the same as for that source alone. Ollerhead has applied the model to data from different surveys with apparent success, although it should be observed that this does not really go beyond the point reached by Robinson, since his claims were based on the integration of existing, single-source surveys. It seems to be assumed here that the successful integration of separate surveys to reconcile the differing dose–response relationships is equivalent to a survey of simultaneous mixed sources.

So far, to our present knowledge, there has been no full empirical test of this assumption, although there has been a partial investigation of mixed noise situations, one already mentioned (Bottom, 1971) and a more ambitious project employing a rather different approach by Rohrmann (1978) and Rohrmann *et al.* (1980). Rohrmann's surveys, while incorporating a general theoretical model of annoyance, do not include a formal mathematical treatment to integrate the contributions from different noise sources. Instead, city areas are selected, exposed to different simultaneous sources in different proportions, and at varying levels. In this way, eight city areas exposed to noise from road traffic, aircraft, railways and industry were used to generate 19 'clusters' representing different possible combinations of sources and levels. From these may be made a highly generalised prediction of nuisance, not perhaps very precise but allowing for considerable variation in the character, provenance and combination of the different sounds. Interesting and potentially useful as this approach may be, it is nonetheless not a general theoretical solution of the kind envisaged by Robinson. Basically, the problem comes down to our ability to integrate, not the sound sources, but their effects.

No general solution to the problem of mixed noise seems possible at present, mainly because of the primitive state of our ideas about annoyance and the restricted range of our acoustical measures. And the complexity of the problem is emphasised by Rohrmann's study, which found that while, as might be expected, traffic noise caused more annoyance than any other form of noise, the effects of different noises were different, aircraft and rail noise producing more interference with activities and communication. On the other hand, a more encouraging note is sounded by his general conclusion that the noise level appeared to have more effect overall than did the type of noise. In this situation, and given these kinds of findings, it may be that this empirical approach is the most promising for overall noise nuisance control.

BIBLIOGRAPHY

Alexandre, A. (ed.), Barde, J.-Ph., Lamure, C., and Langdon, F. J. (1975). "Road Traffic Noise." Applied Science, London.
Anonymous (1971). "Urban Traffic Noise: Strategy for Improved Environment." OECD, Paris.
Anonymous (1975). "Environment and Quality of Life: Damage and Annoyance Caused by Noise." EUR 5398. Commission of the European Community Directorate General, Luxembourg.
Burns, W. F. (1968). "Noise and Man." Murray, London.
Chalupnik, J. D. (ed.) (1977). "Transportation Noises." Ann Arbor Science, Inc., Ann Arbor, Michigan.
International Commission for Biological Effects of Noise (1980). "Noise as a Public Health Problem." ASHA Report 10. ICBEN, Freiburg. Am. Speech Language Hearing Assoc., Rockville, Maryland.
Kryter, K. D. (1970). "The Effects of Noise on Man." Academic Press, New York.
Oppenheim, H. N. (1966). "Questionnaire Design and Attitude Measurement." Heinemann, London.
Torgerson, W. S. (1958). "Theory and Methods of Scaling." Wiley, New York [Chapman and Hall, London].

REFERENCES

Alexandre, A. (1970). "Prévision de la Gêne due au Bruit Autour des Aeroports et Perspectives sur le Moyen d'y Remedier." Univ. of Paris, Paris.

Alexandre, A. (1973). "Le Temps du Bruit." Flammarion, Paris.

Anonymous (1963). "Noise: Final Report" [of the Wilson Committee]. Cmnd. 2056. Her Majesty's Stationery Office, London.

Anonymous (1973). Land Compensation Act. Her Majesty's Stationery Office, London.

Aubree, D. (1971a). "Étude Acoustique et Sociologique . . . sur le Bruit de Train." CSTB, Paris.

Aubree, D. (1971b). "Étude de la Gêne due au Trafic Automobile Urbain." CSTB, Paris.

Barbenza, C. M. de, Bryan, M. E., and Tempest, W. (1970). Individual loudness functions, J. Sound Vib. **11**, 399–410.

Baughan, C. J. (1980). "Nuisance from Road Construction: A Study at the A 31 Poulner Lane Division, Ringwood." Transport and Road Research Laboratory Supplementary Report 562. Transport and Road Research Laboratory, Crowthorne, Berkshire.

Berglund, B., Berglund, U., and Lindvall, T. (1975). Scaling loudness, noisiness and annoyance of community noises, J. Acoust. Soc. Am. **57**, 930–934.

Berglund, B., Berglund, U., and Lindvall, T. (1976). Scaling loudness, noisiness and annoyance of community noises, J. Acoust. Soc. Am. **60**, 1119–1125.

Bitter, C. (1968). "Noise Nuisance in Dwellings: A Survey in Broad Lines and in Detail." CIB Symposium on Noise, Commission W45 Proceedings. CSTB, Paris.

Bitter, C., and van Weeren, P. (1955). "Sound Insulation in Blocks of Dwellings." Research Institute for Public Health TNO Report 24. Delft, Holland.

Borsky, P. N. (1961). "Community Reactions to Air Force Noise." WADD Technical Report 6-689. Wright Air Development Center, Ohio.

Borsky, P. N. (1980). Review of community response to noise, in "Noise as a Public Health Problem". ASHA Report 10. ICBEN, Freiburg. Am. Speech Language Hearing Assoc., Rockville, Maryland.

Botsford, J. H. (1967). A simple method for identifying acceptable noise exposures, J. Acoust. Soc. Am. **42**, 810–819.

Botsford, J. H. (1968). The weighting game [Proceedings of 75th Meeting of the Acoustical Society of America, May 1968], J. Acoust. Soc. Am. **44**, 381.

Bottom. C. G. (1971). A social survey into annoyance caused by the interaction of aircraft noise and traffic noise, J. Sound Vib. **19**, 473–476.

Bradley, J. S., and Jonah, B. A. (1979). The effect of selected site variables on human response to traffic noise, Part I, J. Sound Vib. **66**, 589–604.

British Standards Institution (1967). "Method of Rating Industrial Noise Affecting Mixed Residential and Industrial Areas." British Standard 4142. British Standards Institution, London.

Broadbent, D. E. (1972). Individual differences in annoyance by noise, Sound **6**, 56–61.

Bryan, M. E., and Tempest, W. (1973). Are our noise laws adequate? Appl. Acoust. **6**, 219–232.

Cederlof, R., Jonsson, E., and Kajland, A. (1963). Annoyance reactions to noise from motor vehicles: An experimental study, Acustica **13**, 270–279.

Chapman, D. (1948). "A Survey of Noise in British Homes." National Building Studies, Technical Paper 2. Her Majesty's Stationery Office, London.

Elliot, C. D. (1971). Noise tolerance and extraversion in children, Brit. J. Psychol. **62**, 375–380.

Environmental Protection Agency (U.S.) (1974). "Information on Levels of Environmental

Noise Requisite to Protect Public Health and Welfare with an Adequate Margin of Safety," Appendix A. Government Printing Office, Washington, D.C.

Fidell, S. (1978). The Nationwide Urban Noise Survey, *J. Acoust. Soc. Am.* **64**, 198–206.

Fields, J. M., and Walker, J. G. (1980). "Reactions to Railway Noise: A Survey Near Railway Lines in Great Britain." ISVR Technical Report 102. ISVR, Southampton, England.

Flindell, I. H. (1979). A combined laboratory and field study of traffic noise, *in Proc. Inst. Acoust., Spring Meeting, Southampton*. Inst. Sound Vibration Res., Univ. of Southampton, Southampton.

Fog, H., and Jonsson, E. (1968). "Traffic Noise in Residential Areas." Report 36E National Building Research Institute. National Building Research Institute, Stockholm.

Frederiksen, N., and Messick, S. (1959). Response set as a measure of personality, *Educ. Psychol. Measurement* **19**, 137–157.

Gray, P. G., Cartwright, A., and Parkin, P. H. (1958). "Noise in Three Groups of Flats with Different Floor Insulations." National Building Studies, Research Paper 27. Her Majesty's Stationery Office, London.

Griffiths, I. D. (1983). Review of community response to noise, *in Proc. Int. Conf. Int. Comm. Biolog. Effects Noise, 4th, Torino*. Cent. Ric. Stud. Amplifon, Milan, Italy.

Griffiths, I. D., and Delauzun, F. R. (1977). Individual differences in sensitivity to traffic noise: An empirical study, *J. Sound Vib.* **55**, 93–107.

Griffiths, I. D., and Langdon, F. J. (1968). Subjective response to road traffic noise, *J. Sound Vib.* **8**, 16–32.

Griffiths, I. D., Langdon, F. J., and Swan, M. A. (1980). Subjective effects of traffic noise exposure: Reliability and seasonal effects, *J. Sound Vib.* **71**, 227–240.

Guilford, J. P. (1954). "Psychometric Methods." McGraw-Hill, London.

Guttman, L. (1944). A basis for scaling quantitative data, *Amer. Social Rev.* **9**, 139–150.

Hall, F. L., and Taylor, S. M. (1980). The reliability of social survey data on noise effects, *in Proc 1980 Meeting Acoust. Soc. Am.* Department of Geography, McMaster Univ., Hamilton, Ontario.

Hart, P. E. (1973). Population densities and optimal aircraft flight paths, *Regional Studies* **7**, 137–151.

Jonsson, E., and Sorensen, S. (1967). On the relationship between annoyance reactions to external environmental factors and the attitude to the source of annoyance, *Nord. Hyg. T.* **48**, 35–45.

Jonsson, E., Kajland, A., Pacagnella, B., and Sorensen, S. (1969). Annoyance reactions to traffic noise in Italy and Sweden, *Arch. Environ. Health* **19**, 692–699.

Karolinska Institute (1971). "Report of Fourth Karolinska Institute Symposium on Environmental Health: Measurement of Annoyance due to Exposure to Environmental Factors." Karolinska Institute, Stockholm.

Kastka, J. (1976). Untersuchungen zur Belastigungswirkung der Umweltbedingungen Verkehrslarm und Industriegeruche, *in* "Umweltpsychologie" (G. Kaminski, ed.). Klett, Stuttgart.

Kastka, J. (1983). An empirical model of noise annoyance reaction, *Proc. ICA* **11**; *GALF Rev. Acous. (Paris)*.

Kastka, J., and Paulsen, R. (1979). "Untersuchung uber die Subjektive und Objektive Wirksamkeit von Schallschutzeinrichten und ihre Nebenwirkungen auf die Anlieger." Institut für Hygiene, Univ. of Dusseldorf, Dusseldorf.

Keighley, E. C. (1966). The determination of acceptability criteria for office noise, *J. Sound Vib.* **4**, 73–87.

Kono, S., Sone, T., and Nimura, T. (1982). Personal reaction to daily noise exposure, *Noise Control Eng.* **19,** 4–16.

Kryter, K. D. (1959). Scaling human reactions to the sound from aircraft, *J. Acoust. Soc. Am.* **31,** 1415–1429.

Kryter, K. D. (1968). Concepts of perceived noisiness, their implementation and application, *J. Acoust. Soc. Am.* **43,** 344–361.

Kryter, K. D. (1969). "Possible Modifications to Procedures for the Calculation of Perceived Noisiness." Final Report NASA Contract NASl-6855. Stanford Research Institute, Stanford, California.

Kryter, K. D. (1982). Community annoyance from aircraft and ground vehicle noise, *J. Acoust. Soc. Am.* **72,** 1222–1242.

Kryter, K. D., and Pearsons, K. S. (1965). Judged noisiness of a band of random noise containing an audible pure tone, *J. Acoust. Soc. Am.* **38,** 106–112.

Langdon, F. J. (1976a). Noise nuisance caused by road traffic in residential areas: Part I, *J. Sound Vib.* **47,** 243–263.

Langdon, F. J. (1976b). Noise nuisance caused by road traffic in residential areas: Part III, *J. Sound Vib.* **49,** 241–256.

Langdon, F. J. (1978). Monetary evaluation of nuisance from road traffic noise: An exploratory study, *Planning A* **10,** 1015–1034.

Langdon, F. J., and Buller, I. B. (1977a). Road traffic noise and disturbance to sleep, *J. Sound Vib.* **50,** 13–28.

Langdon, F. J., and Buller, I. B. (1977b). Party-wall insulation and noise from neighbours, *J. Sound Vib.* **55,** 495–507.

Langdon, F. J., and Griffiths, I. D. (1982). Subjective effects of traffic noise exposure, Part II, *J. Sound Vib.* **83,** 171–180.

Langdon, F. J., Buller, I. B., and Scholes, W. E. (1981). Noise from neighbours and the sound insulation of party walls in houses, *J. Sound Vib.* **79,** 205–228.

Langdon, F. J., Buller, I. B., and Scholes, W. E. (1983). Noise from neighbours and the sound insulation of party floors and walls in flats, *J. Sound Vib.* **88,** 243–270.

Levy-Leboyer, C. (1977). "Étude Psychologique du Cadre de Vie." Editions du CNRS, Paris.

Levy-Leboyer, C. (1982). "Psychology and Environment." Sage, London [Engl. transl.].

Levy-Leboyer, C., and Moser, G. (1976). Que signifient les gênes exprimées? Enquête sur les bruits dans les logements, *Sondages* **38,** 7–22.

McKelvie, S. J. (1978). Graphic rating scales: How many categories? *Brit. J. Psychol.* **69,** 185–202.

McKennell, A. C. (1963). "Aircraft Noise Annoyance around Heathrow Airport." SS 337. Her Majesty's Stationery Office, London.

McKennell, A. C. (1965). Correlational analysis of social survey data, *Soc. Rev.* **13,** 157–181.

McKennell, A. C. (1970). *In* "Transportation Noises" (J. D. Chalupnik, ed.), 228–244. Univ. of Washington, Seattle.

McLean, E. K., and Tarnopolsky, A. (1977). Noise discomfort and mental health, *Psychol. Med.* **7,** 19–62.

Maurin, M. (1975). Qualité de l'environnement physique, *in* "Methodes pour Construire des Indicateurs de Nuisance due au Moyen de Transport," Chapter 6. IRT, CERNE, Bron, France.

Moreira, M. N., and Bryan, M. E. (1972). Noise annoyance susceptibility, *J. Sound Vib.* **21,** 449–462.

Morton-Williams, J., Hedges, B., and Fernando, E. (1978). "Road Traffic and the Environment." Social and Community Planning Research, London.

Ollerhead, J. B. (1978): Predicting public reaction to noise from mixed sources, *Internoise* 78, 579–584.

Ollerhead, J. B. (1980). Assessment of community noise exposure to account for time-of-day and multiple source effects, *in* "Noise as a Public Health Problem" ASHA Report 10, ICBEN, Freiburg. Am. Speech Language Hearing Assoc., Rockville, Maryland.

Parkin, P. H., Purkis, H. J., Stephenson, R. J., and Schlaffenerg, B. (1968). "The London Noise Survey." Her Majesty's Stationery Office, London.

Poulton, W. C. (1977). Quantitative subjective assessments are always biased, sometimes completely misleading, *Brit. J. Psychol.* 68, 409–425.

Rice, C. G. (1977). Investigation of the trade-off effects of aircraft noise and number, *J. Sound Vib.* 52, 325–344.

Rice, C. G., Large, J. B., Fields, J. M., and Walker, G. J. (1978). "Noise Research Criteria." *Noise Advisory Council Seminar and Exhibition, 1978, Darlington.* Traffic Engineering Control, London.

Robinson, D. W. (1971). Towards a unified system of noise assessment, *J. Sound Vib.* 14, 279–298.

Robinson, D. W. (1972). "An Essay in the Comparison of Environmental Noise Measures and Prospects for a Unified System." National Physical Laboratory, Acoustics Report Ac 59. NPL. Teddington. Middlesex.

Rohrmann, B. (1978). Design and preliminary results of an interdisciplinary field study on urban noise, *J. Sound Vib.* 59, 111–113.

Rohrmann, B., Finke, H. O., and Guski, R. (1980). Analysis of reactions to different environmental noise sources in residential areas, *in* "Noise as a Public Health Problem." ASHA Report 10. ICBEN, Freiburg. Am. Speech Language Hearing Assoc., Rockville, Maryland.

Rylander, R., Sorensen, S., and Kajlund, A. (1972). Annoyance reactions from aircraft noise exposure, *J. Sound Vib.* 24, 419–444.

Rylander, R., Sjostedt, E., and Bjorkman, M. (1977a). Laboratory studies of traffic noise annoyance, *J. Sound Vib.* 52, 415–421.

Rylander, R., Bjorkman, M., and Ahrlin, U. (1977b). Tramway noise in city traffic, *J. Sound Vib.* 51, 353–358.

Rylander, R., Bjorkman, M., and Ahrlin, U. (1980). Aircraft noise annoyance contours: Importance of overflight frequency and noise level, *J. Sound Vib.* 69, 583–595.

Sargent, J. W., Gidman, M. I., Humphreys, M. A., and Utley, W. A. (1980). The disturbance caused to school teachers by noise, *J. Sound Vib.* 70, 557–572.

Scharf, B. (1977). "Comparison of Various Methods for Predicting the Loudness and Acceptability of Noise." EPA 550/9-77-101. Government Printing Office, Washington, D.C.

Scharf, B. (1980). How best to predict human response to noise on the basis of acoustic variables, *in* "Noise as a Public Health Problem." ASHA Report 10. ICBEN, Freiburg. Am. Speech Language Hearing Association, Rockville, Maryland.

Schultz, T. J. (1978). A synthesis of social surveys on noise annoyance, *J. Acoust. Soc. Am.* 64, 377–405.

Stephens, D. G., and Powell, C. A. (1980). Laboratory and community studies of aircraft and noise effects, *in* "Noise as a Public Health Problem." ASHA Report 10. ICBEN, Freiburg. Am. Speech Language Hearing Association, Rockville, Maryland.

Thomas, J. R., Namba, S., Schick, S., and Kuwano, S. (1983). A cross-cultural study of noise annoyance: A comparison between Britain, Germany and Japan, *in Proc. Int. Conf. Int. Comm. Biolog. Effects Noise, 4th, Torino,* Abstracts, p. 160. Cent. Ric. Stud. Amplifon, Milan, Italy.

TRACOR (Connor, W. K., and Patterson, H. P.) (1976). "Analysis of the Effect of Number of Aircraft Operations on Community Annoyance." NASA Report CR-2741-N-76-30181 B 73 1-53. Langley, Virginia.

Vallet, M., Carrere, C., and Lacoste, P. (1983). La gêne due au bruit des liaisons interurbaines en rase compagne, *ICA 11 Proc.* **7,** 407. *GALF Rev. Acous. (Paris).*

Walters, A. A. (1975). "Noise and Prices." Oxford Univ. Press, Oxford.

Ward, B., and Dubos, R. (1972). "Only One Earth." Andre Deutsch, London.

Weinstein, N. D. (1980). Individual differences in critical tendency and noise annoyance, *J. Sound Vib.* **68,** 241–248.

PART III

7

Noise in Industry

W. TEMPEST

Department of Electronic and Electrical Engineering
University of Salford
Salford
England

I. INTRODUCTION

The problem of industrial noise seems to be as old as industry itself. Ramazzini (1713), who has been called the "Father of Occupational Medicine", noted that deafness was an occupational disease of millers and coppersmiths. In the next century Thackrah (1831) reported deafness amongst ship's carpenters, frizzers (who worked up the nap on cloth) and shear grinders. In the same period Fosbroke (1830–1831) observed that deafness is caused by explosion (of cannons) and by continued noise (such as that to which blacksmiths were exposed). Quantitative observations were made by Barr (1886), who reported that only 9% of the boilermakers (riveters, caulkers, platers and "holders-on") whom he studied in the Glasgow shipyards had normal hearing. By comparison, 79% of postmen and 46% of ironmoulders (foundry men) could hear normally.

The few references mentioned indicate that it has long been known, among medical men at least, that in certain trades and professions the

179

workers became permanently hard of hearing and that this is due to the intense noise levels involved. The first mention by the Factory Inspectorate (Anonymous, 1908) of the fact that deafness was caused by certain occupations cited boilermaking, the hammering of metal sheets and cylinders, the use of pneumatic tools, the beetling of cloth, engine driving and firing of guns. By the 1930s there must have been some knowledge by the general public of the effects of excessive noise. Walter Greenwood (1933) in his classic novel of working-class life in Salford, "Love on the Dole", wrote that in a local forge, the thump of the forging hammer made one giddy and that the din of the rivetting shop was insufferable. He goes on to say that every man was stone deaf after a spell of six months there! "Cassell's Modern Encyclopedia" (Hammerton, 1935) mentions under the heading "deafness" that "occupational deafness accounts for many cases of nerve deafness".

Despite the increasingly widespread knowledge of the hazards of noise [see, for example, Tempest and Bryan (1981)], it was not until 1963 that the first major initiative in the United Kingdom was taken by the Ministry of Labour (1963) to publish a guide to employers entitled "Noise and the Worker". This pamphlet gave advice on how to recognize and deal with a noise problem, setting out the essential features of a hearing conservation programme. However, it was only at the very end of the 1960s that two reports were published, providing for the first time satisfactory data on the quantitative relationship between noise exposure and hearing loss (Passchier-Vermeer, 1968; Burns and Robinson, 1970). This relationship is discussed in greater detail in Chapter 3, but the main conclusion to be drawn from both sets of data is that long-term exposure to levels below 80 dB(A) is not likely to injure hearing, whilst exposures to levels of 90 dB(A) and higher carry a significant risk. Due to the wide range of individual susceptibility to noise damage, it is not possible to postulate a single decibel level at which damage begins. It is only possible to demonstrate that at levels in excess of 80 dB(A) the percentage of persons at risk increases steadily with increasing noise levels. The choice of a level for legislative purposes continues to be disputed. Broadly speaking, trade union opinion would tend to support lower figures [say, 80 dB(A)], while, in many instances, management would prefer the higher [around 90 dB(A)]. The decision is essentially a political one, and we must weigh the costs of protection against the degree of risk involved.

II. THE INCIDENCE OF INDUSTRIAL NOISE

Noise is almost certainly the most widespread hazard in the modern industrial environment. A pilot survey carried out in 1971, by what was then the U.K. Factory Inspectorate (Anonymous, 1974), resulted in an estimate that of 6.4 million workers in manufacturing industry subject to the Factories Act, some 590,000 were exposed to noise of 90 dB(A) or more for 6 hr per day and that a further 570,000 were exposed for at least some of the time. In fact the total number of workers in the United Kingdom subjected to some degree of risk is larger than these figures would suggest, since the survey was limited to the occupations within the Inspectorate's purview. There are many other workers exposed to noise in transport, building, mining, etc., who were excluded. Furthermore, exposure to noise at levels between 80 and 90 dB(A) involves some degree of risk.

The incidence of high noise levels spreads across a wide spectrum of occupations, implying that progress towards the conservation of hearing must be aimed at the whole of industry. Table I gives a few examples of

TABLE I. Examples of Measured Noise Levels and Maximum Permitted Daily Exposures

Noise source	Noise level, dB(A) (L_{eq})	Maximum safe daily exposure[a]
Disco music	90	8 hr
Paper making	90	8 hr
Four wheel drive vehicle cab	95	2.5 hr
Weaving shed (Rapier looms)	95	2.5 hr
Flour mill	96	2 hr
Earth-mover cab	97	1.5 hr
Drop forging	98	1.25 hr
Weaving shed (Dobcross looms)	100	48 min
Pop group (performers)	100	48 min
Power station turbine hall	100+	<48 min
Newspaper presses	100+	<48 min
Steel fabrication shop	103	24 min
Diesel locomotive cab	104	20 min
Glass bottle making	106	12 min
Steel rolling mill	110	5 min
Planing machine (wood)	110	5 min
Circular saw (wood)	Up to 115	1.5 min
Stainless steel polishing	115	1.5 min
Rivetting	117	1 min
Fettling (cast iron)	118	46 sec
Diesel engine testing	120	30 sec
Chain saw	Up to 120	30 sec
Shot blasting	120	30 sec
Rock drilling	120	30 sec
Propellor chipping	132	2 sec

[a] Without hearing protection. Data from Department of Employment (1972).

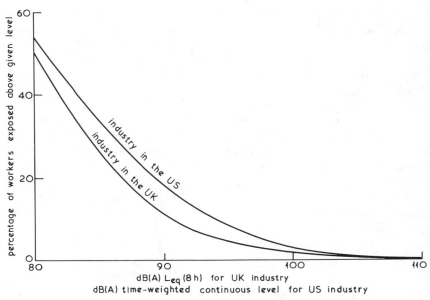

Fig. 1. Distribution of exposure to noise levels in excess of 80 dB(A) among workers in the United Kingdom and the United States. [Data from Health and Safety Executive (1982).]

occupational activities, together with the noise levels which they typically produce. The activities listed in the table give some indication of the wide variety of noisy employments and indicate clearly that no particular industry has a monopoly on this hazard. It seems likely that the single most widespread noisy "activity" is hand hammering. The way in which the exposure to noise levels in excess of 80 dB(A) is distributed amongst production workers is illustrated in Fig. 1. This figure shows the percentage of workers in the United Kingdom and in the United States who are exposed above a given level as a function of the daily equivalent continuous noise level [L_{eq} (8 hr)]. The data are obtained from the consultative document "Protection of Hearing at Work" (Health and Safety Executive, 1982).

III. HEARING CONSERVATION PROGRAMMES

The conservation of hearing by means of the control of the noise levels at the workers' ear is complex, requiring the co-operation of engineers, medical staff, management, workforce and trade unions if it is to be successfully accomplished. Table II sets out in brief form the requirements of a hearing conservation programme, which are considered in this section.

A. The Noise Survey

The technical aspects of noise measurement are dealt with in more detail elsewhere (Chapter 1). It is appropriate here to consider the particular aspects relevant to hearing conservation. Noise level measurements for hearing conservation are nowadays normally made in decibels on the A scale [decibels(A)]. When the noise sources are static and operate continuously, which is often the case in factories with fixed machinery, a sound level meter can be used to "map" the noise levels and to record them on a plan of the area involved. Where noise levels are variable, it is necessary to use some form of averaging meter, and in Western Europe, including Britain, this usually takes the form of an integrating meter to measure decibels(A), L_{eq}; this is the continuous, steady sound level which would feed as much (A-weighted) energy into the listener's ear as would the varying sound. An integrating meter can be used for noises with a cyclic or with a random variation in level, and also for impact noises such as hammering. With an integrating meter it is possible to obtain a measurement of L_{eq} by measuring one cycle of the noisy process involved. There are some cases, however, where the cycle of work is very long or the work pattern is rather variable (e.g., service or maintenance work),

TABLE II. Essentials of Hearing Conservation Programmes

Operation	Function
Noise survey	Identification of hazardous areas/occupations
Noise control	Reduction of noise at source
	Enclosure of noise source/operator to reduce noise to safe levels
	Use of sound absorbers
Hearing protection	Where noise control is not possible, provision, fitting and maintenance of ear plugs/muffs for personnel at risk, together with their education in the hazards of noise
Industrial audiometry	Monitoring of the effectiveness of hearing protection
	Pre-employment and serial audiometry to identify noise-sensitive workers
Organization of hearing conservation programmes	Co-ordination of work of medical, safety and occupational hygiene staff involved
	Education of management and workforce
	Referral and redeployment of workers with hearing loss/damage
Legal aspects	Statute and common law, legal liability, likely legislation

and it is necessary to measure noise over a long period. To deal with this situation noise dosemeters (or dosimeters) have been developed, which can fit a pocket with a microphone mounted on the meter or on a jacket lapel or helmet. The dosemeter is an integrating meter designed to be worn for a full working day, or shift, and to indicate the total noise dose received during the working period. Dosemeters are available to measure L_{eq}, as used in Europe, or to make the slightly different measurements required by U.S. practice.

The aim of the noise survey is to identify all the areas and/or occupations which comprise a hazard to hearing, and an essential part of this process is an adequately detailed and accurate record of all these recorded levels.

B. Noise Control

Noise control is the first and the most fundamental step in the hearing conservation programme. The basic approach is to reduce the noise at its source, which may pose problems in the areas of mechanical, aeronautical or hydraulic engineering. If the noise cannot be reduced at source, then some form of external noise reduction is needed involving enclosure of the source, or the recipient, often combined with other measures to absorb sound. These topics are dealt with in greater detail in Chapter 11.

C. Hearing Protection

The use of personal hearing protectors is in many cases the only alternative if noise control is impossible or impracticable. There are clearly some noisy processes which do not lend themselves at all to noise control techniques. Examples are the use of hand tools, such as chippers, grinders and whizzers in the rectification of metal castings. In such situations the second-best approach of personal hearing protection must be applied. Because personal protection is second best for many reasons, it needs to be used in conjunction with a scheme of regular audiometry in order to monitor the effectiveness of the protection.

Basically two types of protection are available, those which are inserted into the ear canal (i.e., plugs) and those which cover the entire external ear by means of a sealed cup (i.e., muffs). The choice between plugs and muffs depends on both the noise level and its frequency spectrum, and on certain other factors of the working environment. The correct procedure for determining whether a particular type of protector is suitable for a given noise environment is set out in Appendix 4 in the "Code of Practice" (Department of Employment, 1972). This method requires the attenuation of the plug or muff to be known as a function of frequency, and there is a standard measurement procedure (British Standards Institution, 1983) by which this should be measured.

Generally, plugs provide rather less sound attenuation than muffs and are useful in noise levels up to 100–105 dB(A). Muffs are normally considered satisfactory up to 110–115 dB(A).

A wide variety of plugs is available at the present time. Historically, the first reasonably comfortable plugs with satisfactory attenuation characteristics were probably the VR51A type used by some of the armed forces in the Second World War. This type of plug, moulded in a rubber or suitable soft plastic, is manufactured in five sizes and needs to be professionally fitted by a qualified person, who will determine the appropriate size and advise the user on the fitting and care of the plugs. Possibly because of the degree of discomfort caused by conventional plugs, a number of alternative insert protection devices have been developed, and these have tended to supercede the traditional ear plug. One of the most widely accepted is glass down, a form of extremely fine glass fibre which, although resembling cotton wool in appearance, has much greater sound-attenuating properties. This material was originally supplied in the form of a roll or piece, of which the user detached a portion, rolled it into a conical plug and inserted it into his ear. More recent developments include a pre-formed glass-down plug and a pre-formed glass-down plug in a thin plastic sheath. These developments were designed primarily to improve ease of insertion, and all three types, used properly, are satisfactory. The glass-down plugs are cheap enough to be disposable and seem to be widely accepted by workers in many industries. There are very occasional complaints of minor irritation of the ear canal, and it is also found in rare cases that fibres of glass down can become embedded in the natural wax in the ear canal and may need to be removed medically.

A more recent development in ear plugs is the foam-plastic plug, which has the property of expanding slowly after compression. This type of plug is cylindrical and can be rolled between the fingers to a small diameter; it expands after insertion to fill the ear canal. It is a relatively comfortable plug and, due to its excellent fit, provides better sound attenuation than most other plugs. The foam plug is, at the present time perhaps, not fully disposable, but it is washable and can be satisfactorily re-used many times. As a non-disposable plug it is not considered satisfactory in "dirty hands" occupations, where it might carry dirt or irritants into the ear canal. Further types of plug include a curable plastic material from which individual plugs can be moulded to fit individual ear canals and pliable, one-sized plugs which have a soft interior material covered by a smooth outer layer.

All plugs are subject to some extent to the problem that they cause a degree of discomfort, and from the point of view of management, who are trying to ensure that the protection is worn regularly, it is impossible to tell by observation whether plugs are being worn properly or not.

Ear muffs consist of an outer shell, normally of a rigid plastic, with a foam lining to help absorb sound and some form of foam or liquid-filled seal around the edges of the cup. The design is relatively standardised, although there are differences in detail among manufacturers, particularly in the seals. To work properly the seal must be in firm contact with the head right around the ear, and this means that the headband must exert adequate pressure. By comparison with plugs, muffs tend to provide greater attenuation, especially in the frequencies above 1 kHz. As with plugs, various objections can be reasonably raised against wearing muffs; these include discomfort due to the pressure of the seals on the head, increased perspiration under the muff and the fact that the muffs get in the way in confined spaces.

All forms of personal protection give rise to sensations of strangeness and isolation and frequently give the impression that communication is very difficult. The feelings of strangeness and isolation should disappear with usage, and there is evidence (Martin *et al.,* 1976) that the communication problem is not serious. It has been found that, in general, speech against a background of noise is heard as well by the protected as by the unprotected ear. The real problem is that, when speaker and listener are both wearing protection, the speaker tends to speak less loudly and thus cannot be heard above the background noise. Clearly this problem can be surmounted with practice in communication when protection is being worn.

Hearing protection does not seem to be a significant cause of disorders of the ears, but it can serve to aggravate any existing conditions which may cause discharge from the ears. It would seem appropriate to refer all cases with active ear disease for the opinion of an ENT consultant, regarding the feasibility of using hearing protection.

The difficulties and discomforts associated with the use of personal protection mean that it is essential that the workforce have adequate instruction on its use. This must include advice on how to use and maintain the device, and there must be checks to ensure that it is being worn correctly and that it is not damaged.

D. Industrial Audiometry

The provision of hearing protection needs to be supported by a programme of regular monitoring audiometry (Bryan and Tempest, 1980).

Anyone involved in the settlement of common law claims for occupational hearing loss soon discovers that if firms were engaged in the regular screening of employee's hearing, many of the claims would never even be put forward. Far too often a firm carries out a noise survey, brings in noise control measures (as far as is practicable) and provides personal hearing protection, but makes no attempt to measure the effectiveness of this hearing conservation programme. Subsequently hearing losses are found, and the firm and its insurers are put to great expense in defending themselves. On the other hand, the firm with a programme of regular audiometry will have available records of employees' hearing and will be able to refute any claims in those cases where the hearing has not deteriorated during employment.

The provision of a successful defence against spurious claims is only one benefit of a hearing conservation programme; a further and perhaps more important aspect is that it enables the medical staff to direct their attention to any employee whose hearing may be deteriorating, so that steps may be taken immediately to prevent the problem reaching significant proportions. It has been argued that industrial audiometry is neither necessary nor worthwhile since adequate noise control or personal protection should prevent noise-induced hearing loss from arising. However, in the real world of industry, noise control devices break down or may be removed; hearing protectors may be lost, damaged or simply not worn; and in practice, industrial audiometry provides an essential safeguard against these failures of the conservation programme. A further important aspect of audiometry is that it makes it possible to pick out the occasional, highly sensitive individual, who, despite the protective measures taken, is still at risk.

Industrial audiometry begins with the determination of a base-line audiogram before employment. This first audiogram then provides evidence of the initial state of the employee's hearing and can be compared with subsequent test results in order to monitor any deterioration. At the same time as the pre-employment audiogram, it is usual to administer a questionnaire including questions about prior noise exposure during employment and elsewhere, and any illnesses, injuries or other factors which may have affected hearing (Bryan and Tempest, 1980). During employment audiometric testing is usually repeated at annual intervals on all employees exposed to noise above 85 dB(A). In this case the relevant noise level is that to which the employee would be exposed *without* any hearing protection; i.e., it is not assumed that the hearing protection is worn. At the same time as the annual test, it is usual to administer a very brief questionnaire covering any factors which might have affected hearing since the last test. Those workers whose hearing shows a measurable deterioration since the last test are referred to the medical officer, who must decide whether it is necessary to recommend redeployment to an area of lesser noise level or simply attempt to improve the effectiveness and utilization of hearing protection; or if appropriate, he may refer the worker to a medical consultant for a second opinion. The document "Audiometry in Industry" (Health and Safety Executive, 1978) includes a scheme for the categorization of audiograms of industrial workers into groups, with recommendations such as "no action", "warn about hazard to hearing", "refer to medical officer".

Industrial audiometry is normally carried out by the method of pure-tone air-conduction testing in the frequency range 500–8000 Hz. This may be carried out by using either a manual or self-recording audiometer. If employed correctly, both methods should give identical results of similar accuracy (Bryan and Tempest, 1980). The type of audiometer chosen is determined mainly by the number of workers to be tested per year. A manual audiometer is cheaper and generally quite satisfactory for 200 to 300 tests per year, but above that number it is usually worthwhile to use a self-recording machine in order to reduce the workload of what is essentially a rather tedious, repetitive task on the medical staff. Whichever type of audiometer is used, it is essential that it be calibrated regularly and that the ambient noise of the test facility be satisfactorily low. Guidance on these topics is given in "Industrial Audiometry" (Bryan and Tempest, 1980) and in "Audiometry in Industry" (Health and Safety Executive, 1978). In many instances, the noise levels in industrial medical centres are excessive, and it is necessary to carry out testing in an acoustic booth.

It is also important that the workers should not be exposed to any high noise levels in the period immediately prior to the test. A typical recommendation is that put forward in the proposed EEC directive (EEC, 1982) that the subject should not have been exposed to a level exceeding 80 dB(A) during the 14 hr preceding the audiometric examination. In some cases this requirement can be met by wearing hearing protection while at work prior to the hearing test, but this condition can be difficult to enforce with certainty.

In addition to adequate calibration procedures and control of environmental noise, it is essential that the medical staff that carries out the screening be adequately trained. Training in industrial audiometry should include practical experience in obtaining audiograms and lectures on the care and maintenance of equipment, the assessment of audiograms and hearing conservation in general. Several institutions run courses, and the British Society of Audiology [see Bryan and Tempest (1980)] has prepared a syllabus to which such courses generally conform.

E. Organisation of a Hearing Conservation Programme

Since the operation of a hearing conservation programme has little *immediate* effect on the health and welfare of employees, or on the profitability of the organisation, special effort is often needed to ensure that the programme functions satisfactorily. Hearing conservation programmes are usually initiated by medical or safety departments, and in the first place must be "sold" to management, basically on the grounds already outlined—that hearing conservation promotes employee health and welfare because it reduces the risk of deafness and is thus a legal and moral responsibility of management; also it may be pointed out that by reducing the risk of compensation claims, it may ultimately save money. While the Health and Safety at Work Act (1974) does not impose specific limits on noise exposure, it does impose a general duty on employers (and employees) to maintain a safe place to work, and the Factory Inspectorate interprets this as including the requirement to reduce noise as far as practicable and to use hearing protection where noise control is not possible. Once management has been persuaded of the need for the programme, the next step is to involve the health and safety representatives/trade unions and to persuade them that hearing conservation is in the interest of the workforce and thus that it merits their co-operation. The third stage, with the help of management and unions, is to fully explain the programme to the workforce. In practice, this process of explaining hearing conservation is likely to require the use of lectures, visual aids, tape and slide programmes, films, etc., to educate management, unions and the workforce. In order to be effective a continuing process of worker education will be needed to get hearing conservation accepted and to ensure that the programme does not lapse after an initial surge of interest.

The implementation of the programme must be taken seriously by all concerned; for instance it is important that members of management always wear hearing protection when entering a noisy area. The argument that the boss does not need ear muffs because he only spends ten minutes a day there may be scientifically valid, but it is difficult to explain convincingly to the shop floor workers. The practical aspects of personal hearing protection are usually dealt with by the medical department, who must be able and willing to instruct workers on the use of plugs and muffs and on how they should be maintained. There should also be regular checks to see that protection is being worn correctly and is not damaged. While the majority of plugs used nowadays are disposable, or semi-disposable, muffs also have a limited life, usually of three months to a year, and budgets must allow for the cost of repairs and replacements. Audiometric testing provides an excellent opportunity for the nurse or doctor to explain the function and importance of wearing protection regularly, and when this is combined with a discussion of the individual's hearing, it may carry much more weight than any formal presentation. It provides an opportunity to explain the risks run in not wearing protection and to put over the idea that the protection needs to be worn continuously in noisy environments. Removing the protection, even for a short period per day, may let in sufficient noise to cause significant damage. In order to gain acceptance of hearing protection it is often helpful to offer a choice of hearing protectors, making two or three types available, if possible, including both plugs and muffs. The type chosen will depend on the particular condition of work; for example, plugs are usually not permitted in food packaging areas due to the risk of contamination of the product. Conversely, muffs may be unsatisfactory for some workers who need to operate in very confined spaces.

Audiometric testing will reveal all cases where, despite the use of noise control measures and/or personal hearing protection, hearing is deteriorating. These cases require investigation and possibly further action. If the problem is a failure to wear protection, then a warning must be given, together with an explanation of the risks involved. If the hearing deterioration persists, due either to a persistent failure to wear protection or to an unusual susceptibility to noise damage (sometimes it is not possible to tell which), then it may be necessary to advise management that a transfer to another job is necessary if serious hearing damage is to be avoided. In all such cases it is essential that the audiometry be as accurate as possible, and it is particularly important that the results should not be affected by temporary threshold shift due to noise exposure prior to the test. It is usually necessary to carry out a repeat test immediately after a weekend or other break, or even to suspend the employee from work for a few days prior to the test before any decision is made as to redeployment.

Obviously, the redeployment of an employee, often to a lower paid job, poses problems for the worker and for management. In some cases firms take the view that redeployment due to an injury sustained at work will not result in a loss of pay.

A further question often raised in connection with a hearing conservation programme is does the greater interest in hearing conservation lead to an increase in claims for compensation? This is a hard question to answer in general, since experiences vary from firm to firm. It has first to be borne in mind that many trade unions are well aware of the hazards of noise and may thus help their members to bring forward claims without any stimulation from the increased interest in hearing conservation. In such cases, the programme may tend to improve relations between management and workforce and possibly to reduce claims. In practice, claims often seem to arise from employees who feel no loyalty to their employers, i.e., those who have been, or are about to be, made redundant and who are looking for any possible financial support in a difficult situation.

Overall it might be fair to say that a hearing conservation programme is just one factor among several which can influence compensation claims and that the strong moral obligation on an employer to minimize the effects of a known hazard ought to over-ride any fears about possible claims.

IV. PROGRESS IN HEARING CONSERVATION

In the 1960s the main U.K. legislation pertaining to employment was the Factories Acts (1937–1961). This legislation, requiring employers in general terms to maintain safe working conditions, did not make specific reference to noise, and no court decision had defined the general terms as including noise. This did not preclude the Factory Inspectorate from taking such action as it could to encourage hearing conservation and to recommend it where appropriate, but its powers were limited. The pamphlet "Noise and the Worker" referred to previously was sent to all major employers in the United Kingdom in 1963 (Ministry of Labour, 1963).

The most important single incident in the promotion of hearing conservation was probably the successful court action for occupational deafness taken by Mr. Berry in 1971 against his employers (*Berry* v. *Stone Manganese Marine Ltd.,* 1972). Mr. Berry's award of £1250 drew attention to the existence of noise-induced hearing loss as a condition which could merit compensation; this judgement caused increased activity by employers and their insurers. In 1972 the "Code of Practice for Reducing the Exposure of Employed Persons to Noise" (Department of Employment, 1972) was published.

The "Code of Practice", for the first time, put forward a definite standard for the maximum permissible noise exposure level of workers [90 dB(A)] and included advice on noise measurement procedures, the recording of noise level data, methods of controlling noise exposure and hearing protection. Overall it comprised a valuable guide to hearing conservation, but it was an advisory code, its provisions were not mandatory, and it did not make any recommendations regarding audiometry.

In 1974 the Factories Acts were superceded by the Health and Safety at Work Act, which, like the Factories Acts, did not make specific regulations pertaining to noise. Provision was made in the Health and Safety at Work legislation to have "approved" codes of practice, which could be used in criminal proceedings as evidence that a statutory duty had been contravened. The "Code of Practice for Reducing the Exposure of Employed Persons to Noise (Department of Employment, 1972) was not an "approved" code.

Two years later a document called "Framing Noise Legislation" (Health and Safety Executive, 1976) put forward a positive recommendation for legislation to limit noise exposure to the 90-dB(A) level of the "Code of Practice". The proposals suggested, however, that a requirement to conduct audiometric tests should not be included in legislation, but should remain voluntary.

Two years after the publication of "Framing Noise Legislation", the long-expected document on audiometry appeared as a "discussion document" entitled "Audiometry in Industry" (Health and Safety Executive, 1978). While offering valuable advice on the conduct of audiometric tests, the document did not recommend a single specific level of noise exposure above which audiometry was needed.

By comparison the United States (U.S. Department of Labor, 1974) proposed that a noise limit of 90 dB(A) be coupled with regular testing of all workers exposed to 85 dB(A) or above.

The next step was the appearance of the "consultative document" entitled "Protection of Hearing at Work" (Health and Safety Executive, 1982), which put forward proposals on hearing conservation based on the recommendation of the 1972 "Code of Practice", and, in the area of audiometry, it followed the suggestions of "Audiometry in Industry" (1978). Also put forward were proposals to impose duties on designers, etc., of articles to "ensure, as far as is reasonably practicable, that the noise produced by their products is not injurious to hearing". Manufacturers would be required to "ensure the provision of information on the noise produced by products likely to cause persons to be exposed to noise above 90 dB(A) L_{eq} (8 hour)".

Soon after the publication of "Protection of Hearing at Work", the European Economic Community put forward its views (EEC, 1982), which appeared in the form of a "proposal for a Council Directive on the protection of workers from the risks related to exposure to chemical, physical and biological agents at work: noise".

The EEC proposals roughly follow the pattern of the "Code of Practice" in the areas of noise measurement, noise control and the use of personal hearing protection, but depart from the U.K. recommendations in two important respects. The first is the proposal that the maximum permissible noise level be fixed at 85 dB(A) [with a provision for a 'running-in' period of five years, during which 90 dB(A) would be allowed]. The second significant departure is in the recommendation on hearing tests; the proposal being that all workers who require hearing protection [i.e., all those exposed at levels in excess of 85 dB(A) *without* protection] shall have periodic audiometry at intervals of no more than three years. This proposal follows the U.S. practice of requiring audiometry for all at risk.

V. CONCLUSIONS

Progress in hearing conservation in the United Kingdom is now about twenty years old; it began with the publication of "Noise and the Worker" in 1963 and was greatly helped by the 1972 "Code of Practice". The successful actions for compensation at common law have also had a powerful influence. In the past decade progress has been rather modest, possibly a reflection on the prevailing economic climate. It now seems likely that the initiative taken by the EEC will prevail and ultimately lead to the general adoption of measures to conserve hearing throughout the community.

REFERENCES

Anonymous (1908). "Annual Report: Factories and Workshops", p. 206. Her Majesty's Stationery Office, London.
Anonymous (1974). "Annual Report of the Chief Inspector of Factories." Her Majesty's Stationery Office, London.
Barr, T. (1886). Enquiry into the effects of loud sounds upon the hearing of boilermakers and others who work amid noisy surroundings, *Proc. Glasgow Phil. Soc.* **17**, 223.
Berry v. *Stone Manganese Marine Ltd.* (1972). *Knights Industrial Rep.* **12**, 13–35.
British Standards Institution (1983). BS 5108:1983, "Measurement of Attenuation of Hearing Protectors at Threshold". British Standards Institution, London.
Bryan, M. E., and Tempest, W. (1980). "Industrial Audiometry." Bryan and Tempest, Noise Consultants, Lytham, St. Annes, England.
Burns, W., and Robinson, D. W. (1970). "Hearing and Noise in Industry." Her Majesty's Stationery Office, London.
Department of Employment (1972). "Code of Practice for Reducing the Exposure of Employed Persons to Noise." Her Majesty's Stationery Office, London.
EEC (1982). Proposal for a Council Directive on the protection of workers from the risks related to exposure to chemical, physical and biological agents at work: noise, *Offic. J. Europ. Commun.* **5.11.82** (P.C.289), 1–6.
Factories Acts (1937–1961). The Factories Act, 1937, Sec. 26(11) [as amended by the Factories Act 1959, and by the Factories Act, 1961, Sec. 29(1)]. Her Majesty's Stationery Office, London.
Fosbroke, J. (1830–1831). Practical observations on the pathology and treatment of deafness, *Lancet* **1**, 645.
Greenwood, W. (1933). "Love on the Dole." Jonathon Cape, London.
Hammerton, J. A., ed. (1935). "Cassell's Modern Encyclopedia", p 321. Cassell, London.

Health and Safety at Work Act (1974). Part I. Her Majesty's Stationery Office, London.

Health and Safety Executive (1976). "Framing Noise Legislation." Her Majesty's Stationery Office, London.

Health and Safety Executive (1978). "Audiometry in Industry" (discussion document). Her Majesty's Stationery Office, London.

Health and Safety Executive (1982). "Protection of Hearing at Work" (consultative document). Her Majesty's Stationery Office, London.

Martin, A. M., Howell, K., and Lower, M. C. (1976). Hearing protection and communication in noise, *in* "Disorders of Auditory Function II" (S. D. G. Stephens, ed.), pp. 47–61. Academic Press, London.

Ministry of Labour. (1963). "Noise and the Worker." Her Majesty's Stationery Office, London.

Passchier-Vermeer, W. (1968). "Hearing Loss due to Exposure to Steady State Broad-Band Noise." Report 35, Institute for Public Health Engineering, TNO, Netherlands.

Ramazzini, B. (1713). "De Morbis Artificium." Padua. [1964 edition entitled "Diseases of Workers", Hafner, New York.]

Tempest, W., and Bryan, M. E. (1981). Industrial hearing loss: Compensation in the United Kingdom, *in* "Audiology and Audiological Medicine" (H. Beagley, ed.), pp. 846–860. Oxford Univ. Press, Oxford.

Thackrah, C. T. (1831). "The Effects of Arts, Trades and Professions on Health and Longevity." [1957 edition, Livingstone, London.]

U.S. Department of Labor (1974). "Occupational Safety and Health Standard", Vol. 39, No. 125, Part 2. Government Printing Office, Washington, D.C.

8

Noise Arising from Transportation

F. J. LANGDON
Hemel Hempstead
Hertfordshire
England

I. ROAD TRAFFIC NOISE

Without doubt, road traffic is the most widespread and the most serious source of noise nuisance in the United Kingdom, as in most industrialised countries. By the early 1960s public pressure for noise control began to make itself felt and was most clearly expressed through the work of the Wilson Committee (Anonymous, 1963), the report of which laid down suggested standards of amenity and indicated the lines to be followed in dealing with the problem. A present re-reading of the report, compiled as it was at a time when there was comparatively little systematic research for the committee to draw upon, compels respect for the good sense and judgement of the committee, both as regards the standards of amenity to be aimed at and their perception of the nature of the problem.

The rapid increase in road traffic, particularly in towns and even more in large conurbations, ended a period of relative quiet in the streets which had probably endured since the disappearance of horse-drawn vehicles running on iron-tired wheels. Since the relationship between traffic volume and noise level is an exponential one, when traffic volume reaches about 1500 vph (vehicles per hour), the noise level thereafter rises but slowly to a maximum at about 2000 vph. As was quite early observed (Stevenson and Vulkan, 1967), this fact does not really limit noise nuisance since the main effects of further traffic growth are noise spreading

195

further into the evening and night, noise spreading into more streets and the proliferation of one-way systems through "traffic management" resulting in two noisy streets in place of one.

To illustrate the situation a few figures may be helpful. In the late 1940s and early 1950s, fewer than a quarter of the people living in London were bothered by traffic noise. By 1965 three-quarters of those questioned in a social survey (McKennell and Hunt, 1966) reported disturbance from traffic noise, over one-third claiming to be seriously disturbed. A study of overheating in school buildings (Langdon and Loudon, 1970) showed that between 40 and 50% of London teachers were so disturbed by traffic noise that windows had to be closed in summer weather. In the same year a working group set up by the Departments of Environment and Transport (Anonymous, 1970) estimated that over 30% of the U.K. population were exposed to the adverse effects of traffic noise and forecast that this figure would rise to 60% by 1980 on the basis of assumed growth in volume of traffic, a prediction which appears to have been borne out. Similar projections have been derived from Continental studies by the OECD (Anonymous, 1971a) and by various member countries of the EEC.

Attempts to quantify the problems of noise and related nuisance early encountered difficulties in measuring both traffic noise and its adverse effects. The noise from road traffic is more or less continuous though all the time varying in random fashion. It was quickly perceived that an "average" noise level could have little meaning, and the indices proposed were based on statistical accumulation of noise events over time. The one most generally adopted in the United Kingdom, decibels(A) L_{10}, with the level in decibels(A) exceeded for 10% of a given time period (24 hr, later modified to 06.00–24.00 hr), proved a very effective and robust measure, though not without its problems. For example, L_{10} represents largely the noise from single passing vehicles so that over a distance from point of emission to reception it attenuates at a different rate from that of the background noise of the traffic stream as a whole. Computation of noise levels is therefore rather complicated (Langdon and Scholes, 1968), particularly where estimates of possible noise level reductions are required.

The alternative unit, now almost universally adopted, is the equivalent noise level L_{eq}, effectively the mean of a logarithmically weighted summation of all noise levels at 5-dB intervals, which rests on the assumption that random variations in sound pressure level so summarised are energetically equivalent and by empirical validation equivalent in their degree of subjective effect to steady-state noise at that level. Numerous other indices, elaborations of the concepts embodied in L_{10} or L_{eq}, have tended to show that none yields results significantly superior to the simpler units (Langdon, 1976b; Hall and Taylor, 1977).

In order to estimate acceptable levels for controlling traffic noise the indices proposed must function effectively, not merely as noise measures, but also as objective correlates of adverse, mainly subjective, effects such as annoyance. The use of L_{10} and L_{eq} for this purpose has come about largely through social surveys designed to relate traffic noise to nuisance through regression models. From 1960 on, a large number of social surveys in Europe and America have built up a more or less consistent picture of traffic noise effects in terms of annoyance and interference with life, activity and relaxation through continuous improvement in scaling and survey techniques leading to effective prediction of nuisance and a general understanding of the problem.

Initially, the London survey of traffic noise carried out in 1965 (McKennell and Hunt, 1966) was unable to demonstrate a significant relationship between noise level and nuisance, mainly because of imprecision in relating measured locations to residents at the point of reception. In a similar way the study of nuisance from motorway noise near Paris (Lamure and Bacelon, 1967) produced only rather crude results with small predictive value, though it is interesting to note that the finding that at a level of 64-dB(A) L_{50} [68-dB(A) L_{eq}] some 50% of the survey population were bothered by noise is one which has been confirmed by almost all subsequent studies. Aubree (1971) extended this approach to cover a large number of sites in Paris, so bringing into view other types of traffic than that typified by motorways. While obtaining not dissimilar results to those already cited, Aubree found only a slow change in annoyance with increasing noise level, leading to difficulty in trying to predict the benefits likely to be obtained from noise control procedures, dispersion of annoyance scores being too great relative to differences in noise level. In the light of later analyses (Hall and Taylor, 1977; Fields, 1982) it seems that a likely reason for this was error in noise measurement which resulted in the noise-nuisance regression slope being flattened. A similar problem would seem to have emerged in the course of the survey of traffic noise in London (Griffiths and Langdon, 1968). Here, poor estimates of noise at low levels, probably the result of under-sampling and indicated by the very poor intercorrelations between L_{10} and L_{90}, resulted in only moderate correlation between L_{10} and adverse effect. A later survey using a larger number of sites (Langdon, 1976a) benefitted from advances in noise measurement technology and was able to produce effective and useful prediction of nuisance from either L_{10} or L_{eq} by using the same scaling methods as the earlier survey. In addition, the later survey brought in a far wider range of traffic conditions than the earlier studies, which had centred on the problems posed by noise from motorways. To this end, surveys had tended to select sites for study which carried freely flowing traffic. But more general urban traffic noise problems arose from disordered and highly congested as much as from free-flow conditions. Where non-free-flow conditions prevailed, it was found that existing acoustic indices performed less well and that to obtain useful predictions of nuisance it was necessary to apply measures of traffic flow and composition, either alone or in combination with acoustic measures (Langdon, 1976b; Rylander et al., 1976). The proportion of heavy, in effect diesel-engined, vehicles was found to be the predominant factor governing noise nuisance from congested traffic, a finding associated with the growing realisation that the heavy vehicle rather than the private car was largely responsible, disproportionately beyond its contribution to the measured noise level, for noise nuisance in these conditions (Watkins, 1981). This is to some extent a happy circumstance in view of the greater scope for noise reduction from heavy vehicles as compared with private cars. The contemporary private car has already achieved engine and exhaust levels low enough to make rolling noise from tires the major source [see Lamure (1975) and Priede (1971)].

Accurate and useful prediction of nuisance by social survey was fairly general by the 1970s for the United Kingdom and other European countries. It was also found possible to increase predictive accuracy by allowing for factors such as exposure within the dwelling and time spent at home (Aubree, 1971), the perceived quality of the environment (Langdon, 1976a) and, somewhat later, the effects of noise on the use of the dwelling (Lambert et al., 1983). There remained, nonetheless, a number of unsatisfactory features and serious misconceptions.

Thus prediction of individual annoyance remained at a low level so that even when allowance was made for individual differences in noise tolerance (Moreira and Bryan, 1972; Langdon, 1976a,b), less than 40% of the response variance was explained in terms of noise. Second, there remained serious errors in noise measurements. Not until the work of Hall and Taylor (1977) and later Fields (1982) was this fully explored. Fields showed that better results could be extracted from existing surveys (viz., Morton-Williams et al., 1978) by analysing the standard error term of the noise–nuisance equation. And third, a series of studies (Griffiths and Delauzun, 1977; Bradley and Jonah, 1979; Griffiths et al., 1980; Hall and Taylor, 1980) showed that there were quite large errors in annoyance measurement. These arose from the very moderate levels of reliability in scaling procedures. Reliability could be increased by improvements in scale construction and testing, which together with more accurate noise measurement and survey replication could raise the explained variance of the individual response to over 60%. These advances, together with the work of Kastka and his co-workers, aimed at developing a coherent noise nuisance model (Kastka and Buchta, 1977; Kastka and Paulsen, 1979), have indicated that a more or less comprehensive understanding of the effects of traffic noise capable of yielding reliable and accurate predictions now exists. This approach has been summarised in schematic form by Rohrmann (Rohrmann and Guski, 1981).

These are the main shortcomings which more recent noise nuisance studies have sought to overcome. So far as empirical results are concerned, later surveys show little variation from earlier ones. It would still seem that the upper limit of acceptability is agreed to lie around 65-dB(A) L_{eq}; that annoyance begins in the region of 55 L_{eq}; that a high proportion of heavy vehicles, particularly where traffic does not flow freely, can greatly increase noise nuisance, critically above 12 to 15%; and that disordered and congested traffic produces significantly more nuisance than free-flowing traffic at a similar noise level.

The identification and quantification of noise nuisance effects no longer occupies the centre of the stage. These tasks have yielded place to those associated with noise reduction. There are many ways of reducing noise nuisance, and various studies, many of them before-and-after exercises, have indicated that not all are equally effective and moreover that serious misconceptions exist with respect to norms of acceptability.

One way of reducing noise in an area is by diverting traffic. A number of by-pass schemes in the United Kingdom were studied by the Transport and Road Research Laboratory (Watkins, 1981). The results of a series of before-and-after social surveys tended to show that the reduction of annoyance was considerably greater than would have been expected given the amount of noise reduction. On the other hand, a similar series of before-and-after studies made by Kastka (Kastka and Ritterstedt, 1981) in Dusseldorf and Wuppertal, where noise barriers had been erected, indicated gains in amenity far less than might have been predicted from the reduction in noise levels; in some cases, annoyance remained unchanged or even increased. It would seem that these different ways of going about reducing noise had produced rather different results and indicated that accurate prediction of the benefits to be expected from noise control procedures has not yet been achieved.

A number of reasons for these contradictory outcomes has been put forward. It has been pointed out that in the second case, that of barriers, noise at source is not reduced. The effect of noise constancy, a perception of the true source level by "taking into account" the attenuation properties of the intervening barrier (Robinson et al., 1963) would tend to diminish any ameliorative effect dependent on a reduction of noise only at the point of reception. Apart from this, there is the simple acoustical fact that barriers, like acoustic double-glazed windows, do not attenuate sound equally throughout the spectrum, being most effective at the high frequencies and least at the low frequencies most characteristic of heavy vehicles [see Langdon and Griffiths (1982)].

There is a further complication here which with a little reflection brings us to the heart of these problems. It may well be asked how the expected noise nuisance reductions, overfulfilled in some cases and falling short of expectation in others, were forecast. So far as can be gathered, the U.K. predictions were based on regression equations from social surveys, the German predictions on a regression derived from the "before" noise values. What seems to have been overlooked in both cases is that such equations, derived from the study of long-term, unchanging conditions, are not applicable to environmental change. So a whole new area opens up and this is the establishment of guide norms from empirical studies of noise control programmes, including longitudinal studies recording changes in attitude as the new, quieter environment gradually becomes accepted as the normal, everyday one. The period of this process of social adaptation is not as yet known. In the case of the sites studied by the TRRL (Watkins, 1981), where noise levels decreased, very few changes in annoyance were found by further studies up to a year or so after the earlier surveys. Similar results were reported by Vallet (Vallet *et al.*, 1978) from before-and-after studies carried on over a two-year period. On the other hand, a different situation may exist where noise levels rise suddenly. A study by Lawson and Walters (1973) reported a sharp rise in annoyance immediately after the opening of a new motorway close to high-rise apartment blocks. After a year, however, a repeat study found that annoyance scores had fallen to the original level recorded prior to the opening of the motorway.

Lastly, there is the case of traffic noise reduction achieved by means of acoustic double glazing, the benefits of which in many places having only recently become available. This constitutes a topic for current and future research. Some researchers, for example, Rohrmann and Guski (1981), have already gone so far as to hazard that nuisance reduction by this method is likely to be less satisfactory in the long run for a variety of reasons; in particular, that such measures protect only the house interior, which is a severe reduction of amenity. Only reduction in noise levels at source, through planning and re-routing of traffic or by vehicle quietening, it is argued, can result in real improvement to amenity. At the same time it is fair to point out that, since the predicted levels of annoyance have been derived from stable-state surveys and cannot be used to forecast levels of "improvement" to be expected, it is difficult to see precisely what constitutes a satisfactory, as opposed to an unsatisfactory, outcome where measures such as acoustic double glazing are employed.

II. AIRCRAFT NOISE

While not as widespread a form of nuisance as that from road traffic, it is evident from a large number of community surveys that for people living in the vicinity of civil or military airports, aircraft noise causes far more annoyance and disturbance than does any other form of transportation noise.

As a main user of transport aircraft, the United States had a predominant interest in research aimed at controlling aircraft noise, and so the earliest work on the problem (Stevens *et al.*, 1955) was that carried out in the United States. Unlike road traffic noise, the study of which began later and thereby profited from being linked from the outset with social survey research, the development of aircraft noise rating scales went ahead solely on the basis of acoustical engineering technology associated with a number of purely hypothetical assumptions about the adverse effects of noise on the community and the likely responses of people suffering from it.

The outcome of this was the proliferation of a variety of rating procedures and indices such as the Composite Noise Rating (CNR) (Rosenblith and Stevens, 1953; Stevens *et al.*, 1955; Stevens and Pietrasanta, 1957; Galloway and Pietrasanta, 1964), Noise Exposure Forecast (NEF), (Bishop and Horonjeff, 1967) and L_{dn} (Anonymous, 1973a). Only with the poineering work of Borsky (1961) was there an attempt at direct empirical validation of rating scales through systematic studies of human response [see Galloway and Bishop (1970)]. Yet Borsky's work seems to have had little influence on noise index development in the United States, although providing the inspiration for noise surveys in the United Kingdom and elsewhere. Not until the "Seven Cities" study (TRACOR, 1970) was there a systematic comparison of the performance of the different indices.

By this time, the Noise and Number Index (NNI) (Anonymous, 1973b) had been adopted for use in major U.K. civil airports. The NNI, derived from the survey carried out at London (Heathrow) Airport (McKennell, 1963), predicts effectively a variety of adverse effects including annoyance, interference with activities, communication, etc., and has been confirmed by subsequent studies (Anonymous, 1971b; Waters, 1973). It has also been shown, however, that equally good predictions are given by L_{eq} (Ollerhead, personal communication, 1976), and over the past few years various criticisms have been directed against the NNI. These have been summarised (Brooker, 1983) and at the present time a major survey of U.K. airports is in progress, largely with the object of re-examining the NNI in relation to other indices, in particular, L_{eq} and L_{dn}.

The point to be stressed here, however, is that almost all these indices, and other similar rating systems developed in France, Germany, Scandinavia and Japan, are in essence very similar and intercorrelate highly with each other. They generally display elaborate refinements taking account of spectral frequency differences, duration, time of day, number of overflights, peak levels and many non-acoustical parameters, such as attitudes to, and beliefs about, the operation of the airport, none of which appear to make very great differences to their predictions of noise nuisance. On the other hand, the degree of correlation with human response, while adequate for community predictions, is universally low as regards individual responses. The earlier indices, such as the CNR and the NEF, although developed without benefit of social surveys, nevertheless appear to have functioned perfectly well for their designed purpose of environmental noise control. This is not a matter of lucky choices in devising the indices. The later studies which incorporated social surveys merely used these same noise units or suggested only slight adjustments to them. In fact, up to the present there is nothing to indicate that any of the characteristics of human response to noise which have been identified have influenced the design of noise indices. Whether psychological studies will in the future bring about the design of noise indices is an open question, but taking into account the small likelihood of commercial noise measurement equipment being available in order to measure aspects of sound other than those at present recorded, the answer may well be negative.

One reason why this is in any case likely to be a rather unprofitable line to follow is simply the enormous range of variability in individual response to noise. Thus McKennell (1963) noted that 15% of his sample living in quiet areas below 84 PNdB rated themselves as "very annoyed", while at the same time 32% of residents living in areas above peak levels of 104 PNdB were "not bothered" by aircraft noise. In consequence, only a small part, about 15%, of response variance can be accounted for by noise. The largest part of the residual variance, other than that attributable to scaling error, is the outcome of differences in sensitivity to noise. A recent careful study using the powerful technique of path analysis (Taylor, 1983) has shown that this is in fact the main determinant of response together with speech interference, aircraft noise level measured as L_{eq} only coming third in rank order of explanatory variables. Hence to attempt to discriminate between the effectiveness of indices such as the NNI in one or another of its various forms, L_{eq} (or aircraft noise L_{eq} with $k = 4$), the CNR or the NEF is likely to be unsuccessful.

Perusal of social surveys dealing with aircraft noise nuisance, and similarly with road traffic noise studies, makes it clear that the "improvements" in predictive accuracy for individual responses which many surveys demonstrate arise from the incorporation of additional variables, some attitudinal in relation to the noise source, others related to more general living and environmental conditions. Thus McKennell's original Heathrow survey (McKennell, 1963), the second Heathrow survey (Anonymous, 1971b) and the extremely detailed Munich Airport survey (Finke *et al.*, 1973) all introduced a considerable number of variables, some expanding the concept of annoyance and dealing with disturbance to relaxation and communication, some related to fears of accidents and to beliefs about the effects of noise on health, some related to attitudes towards the neighbourhood and towards those believed to be responsible for operation of the airport. Nevertheless, none of these can be truly incorporated into the scales which measure the individual response to noise as such, so that the direct noise–nuisance relationship remains as imprecise as ever. Even if all the variables which are likely to predispose individual sensitivity can be exposed and their effect upon it measured, for individual cases studied clinically, the brute fact of threshold differences remains. Borsky, reviewing aircraft noise nuisance surveys (Borsky, 1980), stresses the predominance of non-acoustic and psychosocial factors as indicated by numerous studies (Borsky, 1974; Bradley and Jonah, 1977; François, 1978) and concludes that all these taken together have far more influence on peoples' responses than the extent of the noise itself.

Despite the range of such differences, it has been possible to make fairly accurate predictions for communities. The well-established results of the two London (Heathrow) surveys, together with those for Gatwick, which showed only slight differences in annoyance above 44 NNI , have enabled guidelines for land use to be laid down covering the range from 35 NNI, at which annoyance begins to make itself felt, to 60 NNI and above. These are shown in Table I.

Similar zoning controls have been published by governments in other countries—although to present knowledge there are still no federal land zoning controls for U.S. airports—using two, three or sometimes four zones to restrict land use. Almost all the noise-rating systems employed in connection with such zoning schemes are based on measurements or estimates of peak levels, sometimes taking account of overflight duration variously defined and time of day—which may be divided only between day and night or sometimes separating evening from night, together with a

TABLE I. Recommended Criteria for Control of Development in Areas Affected by Aircraft Noise[a]

Level of aircraft noise to which site is, or is expected to be, exposed	60 NNI and above	50–59 NNI	40–49 NNI	35–39 NNI
Dwellings	Refuse	No major new developments. Infilling only with appropriate sound insulation		Permission not to be refused on noise grounds alone
Schools	Refuse	Most undesirable; when, exceptionally, it is necessary to give permission, e.g., for a replacement school, sound insulation should be required	Undesirable	Permission not to be refused on noise grounds alone
			Appropriate sound insulation to be required	
Hospitals	Refuse	Undesirable	Each case to be considered on its merits	Permission not to be refused on noise grounds alone
		Appropriate sound insulation to be required		
Offices	Undesirable	Permit	Permit but advise insulation on conference rooms, depending upon position, aspect, etc.	
	Full insulation required			
Factories, warehouses, etc.		Permit		
	It will be for the occupier to take necessary precautions in particular aspects of factory depending on processes and occupancy expected			

[a] From Department of Environment Circular 10/73, "Planning and Noise" (Anonymous, 1973c).

measurement of overflights—although where a cumulative total energy measure such as L_{eq} is employed, this becomes unnecessary since the energy term is, of course, influenced by numbers of overflights.

Against all these systems, a group of researchers in Sweden, led by Rylander (Rylander *et al.*, 1972, 1974, 1980) have attempted to show that above a more or less critical threshold (between 36 and 63 take-offs in 24 hr), the number of flights has no effect on nuisance, which is governed solely by the peak noise level of the noisiest aircraft. Rylander concedes that the peak value measure offers no advantages over equal-energy indices (or pseudo-equal-energy measures such as the NNI), but it does not follow from this that the proposed new measure is pointless. For example, where the noise contour is determined by using the NNI, a reduction in numbers of the noisiest aircraft has little effect on the contour. Using the peak noise measure, however, retrofit or relegation of noisy aircraft has a large and impressive effect in reshaping the contour (Rylander *et al.*, 1974).

It has, however, been noted that the greatest number of daily overflights in the survey data exhibited by Rylander was only 174 (the number of flights for the data used in later studies is given only as >50), a figure much lower than those for major international airports, and it has been suggested on the basis of laboratory studies (Rice, 1977) that above about 400 flights per day, the number again begins to exert an influence, a finding similar to those reported in a study of disturbance to television viewing (Langdon *et al.*, 1974). The peak noise index therefore remains a controversial and contentious issue; it has not been adopted by the Swedish authorities, who have established a unit FBN [*flyg buller nivä* (aircraft noise level)] which is closely related to L_{dn} and is at present the subject of a number of comparative surveys to be published shortly. In the course of reviewing the claims of the peak noise level index as advocated by Rylander, in comparison with those of existing measures, Borsky (1980) remarks that although it appears that L_{eq} tends to overestimate annoyance at some levels and underestimate it at others, on the whole it seems as capable of useful predictions as any of the more complex indices employed. Given that the time-related weighting of 10 dB for the night period, added to L_{eq} and transforming it to L_{dn}, is excessive (Borsky, 1976; Ollerhead, 1980) and could well be reduced to a value around 4 to 5 dB, it would seem that a measure of this modified type is capable of estimating adverse effects more simply and just as effectively as the more complex number-related indices.

It has been suggested that an adverse effect unique to aircraft noise is that on mental health, a study carried out as far back as 1968 appearing to reveal a significantly greater incidence of mental illness in high-noise-exposure zones (Abey-Wickrama *et al.*, 1969). Later, more carefully controlled studies (Tarnopolsky *et al.*, 1978; Tarnopolsky and Morton-Williams, 1980) have shown that this is not in fact the case. While psychiatric patients appear to be more vulnerable to noise, there is little correlation between frank psychiatric disorder and the proportion of people highly annoyed at any given level of noise, and these findings are in agreement with those of François, derived from the study of a thousand residents around Paris Orly Airport (François, 1975). On the other hand, it does seem that one effect of aircraft noise at high levels is to induce feelings of depression and malaise and to lower people's general standard of health and well-being (François, 1975).

III. RAILWAY NOISE

Rail noise is a far less widespread nuisance than that from road traffic. For example, Fields and Walker (1980) estimate that about 170,000 people in the United Kingdom are directly exposed to railway noise and that the number of people exposed to equivalent levels of road traffic noise is between 10 and 25 times as great. The U.K. road traffic noise survey (Morton-Williams *et al.*, 1978) suggests that about 2% of the U.K. population is bothered or disturbed by railway noise, a figure close to the estimate of 3% for the Dutch population (Peeters, 1981). Nevertheless, nuisance from train noise has attracted increased attention in the past few years, partly because of growing concern over environmental noise generally and partly because of the development of high-speed trains—and particularly high-speed freight trains—and the likelihood of their enhanced noise levels resulting in unacceptable conditions for people living close to railroad tracks.

As a number of researchers have observed (Fields and Walker, 1980; Schümer *et al.*, 1980), comparison between the potential of railways and road traffic for generating noise nuisance is not an easy matter, being complicated by a variety of factors of which the difference between the noise of trains and that from road vehicles is but one. Road traffic noise, particularly where the volume of traffic is sufficient to constitute a source of nuisance, tends to be varying but more or less continuous. Train noise, on the other hand, is intermittent and for many people living close to rail lines the ambient level between train passages may be relatively low. But, to a greater extent than road traffic, noise from fast freight trains may continue right through the night, and track maintenance can be a source of noise nuisance which has no direct counterpart in road traffic.

One question to which much research has been devoted is whether the same type of noise index can be used for rail as for road traffic. This is important, both as regards making direct comparisons and in relation to overall environmental noise control. Although the weight of evidence from various surveys has suggested that the equivalent energy measure L_{eq} over 24 hr yields the best predictions of annoyance—although this may depend on how annoyance is defined—this conclusion seems hedged about with many qualifications.

Thus Aubree (1973) found annoyance, measured by means of a composite scale incorporating factors such as interference with conversation and watching television, best predicted by L_{eq} over 24 hr, although other variables such as the duration of train pass-bys and the total volume of traffic also appeared to contribute. Peeters (1981) also found decibels(A) L_{eq} "without the addition of any special coefficient" an adequate predictor of annoyance, obtaining a correlation for individual annoyance scores of $r = 0.41$ and $r = 0.47$ for disturbance ($n = 671$), exactly the same levels of correlation obtained by incorporating data for numbers of pass-bys and distance from track. Aubree found that more important were variables such as the proportion of rooms in the dwelling facing the track, the perceived quality of the local environment and the attitude towards railways as a means of transport. Using all these variables together with decibels(A) L_{eq} as a noise measure, Aubree claimed to account for some 40% of individual response to physical measures ($r = 0.64$). On the other hand, when the other variables were neglected, noise alone accounted for only 10% of individual variance.

Noise nuisance appeared to increase markedly from 68 to 72 dB(A) L_{eq}, and on the basis of this inflexion of the regression line Aubree went so far as to suggest that 72 dB(A) be regarded as a maximum acceptable limit for train noise. This is some 5 dB(A) higher than suggested earlier by the same author (Aubree *et al.*, 1971) for road traffic and corresponds to his statement that rail noise needs to be 5 dB(A) higher than that from road traffic to produce the same level of annoyance.

These results are quite different from those obtained by Fields and Walker (1980) from a much larger survey using a truly representative random probability sample of 1453 residents at 403 measured trackside locations in the United Kingdom. While these authors also found decibels(A) L_{eq} 24 hr the best predictor of noise annoyance, they concede that there remains a significant effect due to the duration of train pass-by which the measure fails to account for. Again, they failed to confirm Aubree's finding that annoyance is influenced by attitudes toward environment or rail transport. More importantly, they found no evidence for any sharp inflexion in the regression slope between noise level and annoyance and therefore state that there is no self-evident limit to train noise nuisance—a limit, if imposed, needs to be a conscious political or social decision—although at a level exceeding 75 dB(A) L_{eq}, over 50% of people would be bothered or disturbed.

Two novel and important points emerge from this survey: first, that noise from track maintenance is a much more serious source of annoyance than actual passage of trains and, second, that the level of annoyance is greatly influenced by the type of train (passenger, freight) and traction (overhead electric, third rail, diesel). Thus freight trains produced more annoyance (at equivalent noise levels) than did passenger trains and diesel-drawn or third-rail electric trains more annoyance than overhead electric, a difference in the latter case equivalent to 10 dB(A) at levels above 55 dB(A) L_{eq}. The authors are apparently unable to explain this difference after having examined a number of possible explanations, including whether short-jointed or long, welded track was used.

In agreement with other researchers Fields and Walker also find rail noise much less annoying than that from either road traffic or aircraft. Thus they claim a difference of at least 10 dB(A) in favour of rail noise for equivalent levels of annoyance compared with road traffic. At the highest levels of exposure, using results from the U.K. traffic noise survey (Morton-Williams *et al.,* 1978) and the BRE London survey (Langdon, 1976a) for purpose of comparison, this difference increases to about 19 dB(A). Compared with aircraft noise, using the NNI as a noise index and the 1967 second Heathrow survey (Anonymous, 1971b) for comparison, rail noise is less annoying by a range of 13 to 30 NNI according to noise level. This wide range of increase in the difference is due to the sharper increase of nuisance from aircraft noise at the higher noise levels.

A similar survey carried out in West Germany (Schümer *et al.,* 1981) in which over 1000 residents at 14 sites alongside rail lines also concluded that rail noise was less annoying than that from road traffic, though in this case the difference claimed for equivalent annoyance levels was only some 4–5 dB(A) L_{eq}, far less than that found by Fields and Walker and smaller than the difference found in an earlier German survey (Heimerl and Holzmann, 1978). The authors also found a larger difference between the effects of rail and road noise in urban as opposed to rural areas, again in contrast to Fields and Walker, who found no regional or urban–rural differences.

Just as Fields and Walker had found that high-speed overhead electric passenger trains produced less annoyance than did other types, a Japanese study (Sone *et al.,* 1973) had already reached a similar conclusion when comparing noise nuisance produced by the high-speed Shinkansen lines (Tokaido and Sanyo lines) with that from conventional passenger networks. Again, decibels(A) L_{eq} was found to be the best predictor, the authors observing, as did Aubree, that equivalent energy level is influenced by the number and duration of train passages and requires no additional coefficient to allow for these. Although the Shinkansen line trains travel at a maximum speed of 210 km/hr, annoyance was significantly lower than that from conventional trains. Thus at 60 dB(A) L_{eq}, 80% of residents close to the ordinary lines were annoyed as against only 50% living near the Shinkansen lines, a difference estimated at 5 dB(A) over the range of annoyance scores. While rail noise was found to be as annoying as road noise at lower levels, at higher levels road noise was more disturbing. The authors concluded that part of the reason for the difference in annoyance produced by the high-speed and the conventional trains is that the latter use jointed rails. This suggests that the character of the noise has an influence since the difference in annoyance holds for the same noise levels measured in L_{eq}. However, it is interesting to note that in considering various hypothetical explanations for variations in annoyance within their survey, Fields and Walker examined the possibility that jointed rail systems may produce more annoyance than welded track and eventually rejected it. On the other hand, while a more recent study of the Shinkansen lines (Nimura et al., 1981) has confirmed the earlier difference of 5 dB(A) between the Shinkansen and conventional rail lines, the previous explanation in terms of jointed rails appears less credible, partly because these differences amount at most to only 3 dB(A) and partly because the differences in annoyance persist, although during the intervening period jointed rails are increasingly being replaced by long, welded sections.

In conclusion, it is apparent that there exists fair agreement among major surveys of annoyance from rail noise as regards the suitability of decibels(A) L_{eq} as a noise unit and the rejection of a peak level measure, while all seem to agree that rail noise is less annoying than that from road traffic. Beyond this, however, there are wide differences of opinion, as regards both the extent of the difference between road and rail noise and the possible reasons for it. A large number of possible explanations have been canvassed, ranging from differences in spectra and sound character to the long historical experience of railways as part of the environment, the regular time-tabled occurrence of train passages and favourable attitudes towards railways as a transport mode. Yet for one reason or another, each of these explanations fails to satisfy a critical test applied by one or other of the various researchers.

Thus Fields and Walker, unlike Aubree, do not find any special esteem for, pride in, or respect for, British Rail. Some 80% of the respondents used the rail system less than once a month and 50% no more than once a year. Although we do not know the exact comparison figures for the SNCF, the overall statistics of use suggest a much greater rate of travel than this. Fields and Walker also point out that the argument based on regular time-tabled pass-bys is upset by the passage of unscheduled freight trains at varying and unpredictable times. The overall effect of the various rail noise surveys is to create the impression that the explanations for the differences between rail and other transportation noises are manifold and complex and may, moreover, be different for different communities. Hence, for the present they can only be recorded as facts of life. Given the relatively modest contribution of train noise to noise nuisance overall, its comparatively low inherent capacity for producing annoyance and the small proportion of the total population affected by it, it is quite possible that our present lack of understanding will continue without much change, since the incentive for continued research is largely absent.

REFERENCES

Abey-Wickrama, I., a'Brook, M. F., Gattoni, F. E. G., and Herridge, C. F. (1969). Mental hospital admissions and aircraft noise, *Lancet* **2,** 1275–1277.
Anonymous (1963). "Final Report of the Committee on the Problem of Noise" (Wilson Committee). Cmd 2056. Her Majesty's Stationery Office, London.
Anonymous (1970). "Report of the Working Group on Research into Road Traffic Noise." TRRL Report LR 357. Transport and Road Research Laboratory, Crowthorne, Berkshire.
Anonymous (1971a). "Urban Traffic Noise." Organization for Economic Co-operation and Development, Paris.
Anonymous (1971b). "Second Survey of Aircraft Noise Annoyance around London (Heathrow) Airport." MIL Research Ltd. Her Majesty's Stationery Office, London.
Anonymous (1973a). "Public Health and Welfare Criteria for Noise." U.S. Environmental Protection Agency Report 550/9-73-002. Government Printing Office, Washington, D.C.
Anonymous (1973b). "Building and Buildings: The Noise Insulation Regulations." Statutory Instruments 1963–1973. Her Majesty's Stationery Office, London.
Anonymous (1973c). "Planning and Noise." Circular 10/73. Department Environment. Her Majesty's Stationery Office, London.
Aubree, D. (1971). "La Gêne due au Bruit de Circulatuon en Site Urbain." CSTB, Paris.
Aubree, D. (1973). "Enquête Acoustique et Sociologique Permettant de Definir une Échelle de la Gêne Eprouvée par l'Homme dans son Logement du Fait des Bruits de Train." CSTB, Paris.
Aubree, D., Auzou, S., and Rapin, J. M. (1971). "Étude de la Gêne due au Trafic Automobile Urbain." CSTB, Paris.
Bishop, D. E., and Horonjeff, R. D. (1967). "Procedures for Developing Noise Exposure Forecast Areas for Aircraft Flight Operations." FAA Technical Report FAA DS-67-10. Washington, D.C.
Borsky, P. N. (1961). "Community Reactions to Air Force Noise." U.S. Air Force WADD TR 60–68, AD 267 052. Wright Air Development Center, Ohio.
Borsky, P. N. (1974). "Annoyance and Acceptability Judgements of Noise Produced by Three Types of Aircraft by Residents Living Near JFK Airport." NTIS Report N75-17092.

Borsky, P. N. (1976). Sleep interference and annoyance by aircraft noise, *Sound Vib.* **10**(12), 18–21.

Borsky, P. N. (1980). Review of community response to noise, 1978, *in* Noise as a Public Health Problem," pp. 453–475. ASHA Report 10. ICBEN, Freiburg. Am. Speech Language Hearing Assoc., Rockville, Maryland.

Bradley, J. S. and Jonah, B. A. (1977). "A Field Study of Human Response to Traffic Noise." Report to Transport Canada. Univ. of Western Ontario, Ontario.

Bradley, J. S., and Jonah, B. A. (1979). The effects of site selected variables on human responses to traffic noise II, *J. Sound Vib.* **67**, 395–407.

Brooker, P. (1983). Public reaction to aircraft noise: Recent U.K. studies, *Internoise* **2**, 951–956.

Fields, J. M. (1982). The effect of errors in specifying noise environments on results from community response surveys, *Internoise* **2**, 609–612.

Fields, J. M., and Walker, J. G. (1980). "Reactions to Railway Noise: A Survey near Railway Lines in Great Britain," Vol I. Technical Report 102. ISVR, Univ. of Southampton, Southampton, England.

Finke, H. O., Guski, H., and Rohrmann, B. (1973). An interdisciplinary study of the effects of aircraft noise on man, *in Proc. Int. Congr. Noise as a Public Health Problem, Dubrovnik.* Environ. Prot. Agency, Washington, D.C.

François, J. (1975). "Les Repercussions du Bruit des Avions sur l'Equilibre des Riverains des Aeroports." Institut Francais d'Opinion Publique (IFOP), Paris.

François, J. (1978). Aircraft noise, annoyance, and personal characteristics, *in* "Noise as a Public Health Problem," pp. 594–599. ASHA Report 10. ICBEN, Freiburg.

Galloway, W. J., and Bishop, D. E. (1970). "Noise Exposure Forecasts." U.S. Federal Aviation Administration Report FAA-NO-70-9. Washington, D.C.

Galloway, W. J., and Pietrasanta, A. C. (1964). "Landuse Planning Relating to Aircraft Noise." U.S. Department of Defense Report AFM 86-5, TM 5-365, NAVDOCKS P-98.

Griffiths, I. D., and Delauzun, F. R. (1977). Individual differences in sensitivity to traffic noise: An empirical study, *J. Sound Vib.* **55**, 93–107.

Griffiths, I. D., and Langdon, F. J. (1968). Subjective response to road traffic noise, *J. Sound Vib.* **8**, 16–32.

Griffiths, I. D., Langdon, F. J., and Swan, M. A. (1980). Subjective effects of traffic noise exposure: Reliability and seasonal effects, *J. Sound Vib.* **71**, 227–240.

Hall, F. L., and Taylor, S. M. (1977). Predicting community response to road traffic noise, *J. Sound Vib.* **52**, 387–399.

Hall, F. L., and Taylor, S. M. (1980). The reliability of social survey data on noise effects, *in* Acoust. Soc. Am. Proc. Department of Geography, McMaster Univ., Hamilton, Ontario.

Heimerl, G., and Holzmann, E. (1978). "Ermittlung der Belästigung durch Verkehrslärm in Abhängigkeit von Verkehrsmittel und Verkehrsdichte in einem Blaungsgebiet (Strassem und Eisenbahnverkehr)." Universitat Stuttgart, Stuttgart.

Kastka, J., and Buchta, E. (1977). "An Analysis of Road Traffic Noise Annoyance Reactions by Survey Method." Institute of Acoustics, London Conference on Traffic Noise, London.

Kastka, J., and Paulsen, R. (1979). "Untersuchung über die Subjektive und Objektive Wirksamkeit von Schallschutzeinrichten und ihre Nebenwirkungen auf die Anlieger." Institut Für Hygiene, Universitat Dusseldorf, Dusseldorf.

Kastka, J., and Ritterstedt, U. (1981). Subjective and objective effects of noise barriers, *Internoise* **2**, 817–820.

Lambert, J., Simmonet, F., and Vallet, M. (1983). Patterns of behaviour in dwelling houses exposed to road traffic noise, *in Proc. Int. Conf. Int. Commission for Biological Effects of Noise, 4th, Torino.*

Lamure, C. (1975). *In* "Road Traffic Noise," (A. Alexandre, ed.), Chapter 4. Applied Science Publishers, London.

Lamure, C., and Baçelon, M. (1967). "La Gêne due au Bruit de la Circulation Automobile." Cahiers du CSTB 762, Paris.

Langdon, F. J. (1976a). Noise nuisance caused by road traffic in residential areas, *J. Sound Vib.* **47,** 243–263.

Langdon, F. J. (1976b). Noise nuisance caused by road traffic in residential areas II, *J. Sound Vib.* **47,** 265–282.

Langdon, F. J., and Griffiths, I. D. (1982). Subjective effects of traffic noise exposure II. Comparisons of noise indices, response scales, and the effects of changes in noise levels, *J. Sound Vib.* **83,** 171–180.

Langdon, F. J., and Loudon, A. G. (1970). Discomfort in schools from overheating in summer, *J. Inst. Heating Ventilation Eng.* **37,** 265–274.

Langdon, F. J., and Scholes, W. E. (1968). The traffic noise index: A method of controlling noise nuisance, *Arch. J.* **147,** 813–820.

Langdon, L. E., Gabriel, R. F., and Creamer, C. R. (1974). Judged acceptability of noise exposure during television viewing, *J. Acoust. Soc. Am.* **56,** 510–515.

Lawson, B. R., and Walters, D. (1973). The effects of a new motorway on an established residential area, *in Proc. Int. Conf. Env. Psychol.* University of Surrey, Surrey, England.

McKennell, A. C. (1963). "Aircraft Noise Annoyance around London (Heathrow) Airport." COI SS 337. Her Majesty's Stationery Office, London.

McKennell, A. C., and Hunt, E. A. (1966). "Noise Annoyance in Central London." SS 332. Her Majesty's Stationery Office, London.

Moreira, N. M., and Bryan, M. E. (1972). Noise annoyance susceptibility, *J. Sound Vib.* **21,** 449–462.

Morton-Williams, J., Hedges, B., and Fernando, E. (1978). "Road Traffic and the Environment." Social and Community Planning Research, London.

Nimura, T., Sone, T., and Kono, W. (1981). Evaluation of train/railway noise, *Internoise* **2,** 803–809.

Ollerhead, J. B. (1980). Accounting for time of day and mixed source effects in the assessment of community noise exposure, *in* "Noise as a Public Health Problem," pp. 556–561. ASHA Report 10. ICBEN, Freiburg. Am. Speech Language Hearing Assoc., Rockville, Maryland.

Peeters, A. L. (1981). Annoyance due to railway noise in residential areas, *Internoise* **2,** 821–826.

Priede, T. (1971). Origins of automotive vehicle noise, *J. Sound Vib.* **15,** 61–73.

Rice, C. G. (1977). Investigation of the trade-off effects of aircraft noise and number, *J. Sound Vib.* **52,** 325–344.

Robinson, D. W., Bowsher, J. M., and Copeland, W. C. (1963). On judging the noise from aircraft in flight, *Acustica* **13,** 324–336.

Rohrmann, B., and Guski, R. (1981). Zeitschrift für Umweltpolitik, *Frankfurt* **2,** 183–212. [Psychological aspects of environmental noise (in English).]

Rosenblith, W. A., and Stevens, K. N. (1953). "Handbook of Acoustic Noise Control." U.S. Air Force WADC TR-52-204. Wright Air Development Center, Ohio.

Rylander, R., Sorensen, S., and Kajland, A. (1972). Annoyance reactions from aircraft noise exposure, *J. Sound Vib.* **24,** 419–444.

Rylander, R., Sorensen, S., and Berglund, K. (1974). Reanalysis of aircraft noise annoyance data against the dB(A) peak concept, *J. Sound Vib.* **36**, 399–406.

Rylander, R., Sorensen, S., and Kajland, A. (1976). Traffic noise exposure and annoyance reactions, *J. Sound Vib.* **47**, 237–242.

Rylander, R., Bjorkman, M., Ahrlin, U., Sorensen, S., and Berglund, K. (1980). Aircraft noise annoyance contours: Importance of overflight frequency and noise level, *J. Sound Vib.* **69**, 583–596.

Schümer, R., Kasubek, W., Knall, V., and Schümer-Kohrs, W. (1981). Reactions to road and railway traffic noise in urban and rural areas, *Internoise* **2**, 827–830.

Sone, T., Kono, S., Nimura, K., Kaneyana, S., and Kumagai, M. (1973). Effects of high speed train noise on the community along a railway, *J. Acoust. Soc. Jpn.* **29**(4), 214–224.

Stevens, K. N., and Pietrasanta, A. C. (1957). "Procedures for Estimating Noise Exposure and Resulting Community Reactions from Air Base Operations." U.S. Air Force WADC TN 57-10. Wright Air Development Center, Ohio.

Stevens, K. N., Rosenblith, W. A., and Bolt, R. H. (1955). A community's reaction to noise: Can it be forecast? *Noise Control* **1**, 63–71.

Stevenson, R. J., and Vulkan, G. H. (1967). Urban planning against noise, *Offic. Architecture Planning,* 643–655 (May).

Taropolsky, A., and Morton-Williams, J. (1980). "Aircraft Noise Prevalence of Psychiatric Disorders." Social and Community Planning Research, London.

Tarnopolsky, A., Barker, S. M., Wiggins, R. D., and McLean, E. K. (1978). The effect of aircraft noise on the mental health of a community sample: A pilot study, *Psychol. Med.* **8**, 219–233.

Taylor, S. M. (1983). Path modelling of aircraft noise annoyance, *Internoise* **2**, 1207–1210.

TRACOR (1970). "Community Reaction to Airport Noise," Vol. 1. NASA CR-1761. Langley, Virginia.

Vallet, M., Maurin, M., Page, M. A., and Pachiaudi, G. (1978). Annoyance from and habituation to road traffic noise from urban expressways, *J. Sound Vib.* **66**, 459.

Waters, C. S. (1973). "Report to the Counties of Surrey, East and West Sussex on the Social Survey Carried Out around London (Gatwick) Airport." Loughborough Consultants Ltd. Loughborough Univ. of Technology, Loughborough, England.

Watkins, L. H. (1981). "Environmental Impact of Roads and Traffic." Applied Science Publishers, London.

9

Noise in Transportation

D. Williams

Department of Mechanical and Production Engineering
Stockport College of Technology
Stockport
England

I. INTRODUCTION

In the past two decades we have seen an alarming increase in the speed, power and, in some cases, sheer numbers of cars, goods vehicles, aircraft and railway locomotives. Although seemingly this increase is a measure of our technological progress, it has not been without its social cost. It has meant increased pollution and congestion and has no doubt added to the stress of our daily lives.

Accepted as one form of pollution and recognised as a hazard to health and safety, excessive noise is, fortunately, being given increasing attention by machine designers, for it is only at the design stage that truly effective noise reductions can be achieved. This is particularly so for the internal noise of enclosed transportation systems. The manufacturer, however, will have to be given strong commercial incentives to reduce noise, for in a highly competitive field the increased cost of noise reduction may make his goods less attractive. Customers and users of transport can bring about changes by demanding less noise and requiring noise levels to be written into specifications. Legislation would have an even

215

greater effect, but a regulation or law which cannot be enforced may be worse than none at all. Fuchs (1975) observes that regulations without effective enforcement only penalise those who honestly try to comply and the full benefit is lost to the general public. Legislation must be enforceable, and therefore it must be based on criteria commensurate with the technical possibilities.

The population of the Western world is highly mobile, spending a considerable part of each day's activities in some form of transportation. Because of the speed and nature of these systems, the traveller is exposed to levels of noise which he would find unacceptable outside transportation. There is evidence that, in some cases, people are travelling in noise levels which, after some time, not only produce fatigue, lack of concentration or vigilance, but may cause hearing loss. Drivers of commercial vehicles, agricultural tractors and off-highway earth-moving machinery are particularly vulnerable, whilst there is no doubt that noise in ships' engine rooms, aircraft interiors and hovercraft interiors deserves particular attention.

Hearing losses have been reported among airline stewards (Pinto, 1962; Kryter, 1970), commercial vehicle drivers (Nerbonne and Accardi, 1975) and agricultural tractor drivers (Helmsman, 1976).

The Health and Safety at Work Act (Anonymous, 1974) should have some effect on the working environment of those employed in transport vehicles; however, the act is vague and unspecific, needing the support of various codes of practice. An examination of the act shows that workers in transport (except air crews when in flight and ships' crews when at sea) are covered by the legislation. The exclusion of the crews in flight and

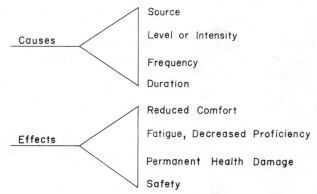

Fig. 1. Effects of noise on drivers and passengers of transport vehicles. [Based on ISO (1978).]

at sea may be the result of the difficulties of enforcement outside the United Kingdom.

Considerable evidence is available to show that currently accepted noise levels in road, rail, sea and air transportation vehicles are unsatisfactory for passenger comfort and optimum performance of those operating them. Figure 1 illustrates the effects of noise on travellers and particularly on operators and drivers.

II. MOTOR CARS

Motor cars represent the majority of all forms of transport and have, in general, been developed with comfort and performance in mind. The prospective buyer, however, has no indication of the noise levels he can expect in any particular car, and it is a serious criticism of car manufacturers that they do not provide a clear, concise and quantified measurement of the expected interior noise levels. The manufacturers and the motoring press restrict themselves to qualitative statements such as "quiet," "whisper" and "silence" and rarely quantify the levels of noise. The inevitable result is confusion for the motorist, who is often misled by the sales literature and test specifications.

One possible reason for this state of affairs is the lack of criteria for noise acceptability in road vehicles (Bryan, 1976; Bryan *et al.,* 1978). Another and possibly more important reason is that until recently (BSI, 1981) there existed no U.K. standard method for measuring the interior noise of motor vehicles. It should be noted that this standard sets out procedures for the measurement of decibels(A) octave- and one-third-octave band levels for the evaluation of hearing damage risk, speech interference, checking of compliance with noise specifications and orientation of a programme of measurements for studying noise reduction procedures. The specified procedure is *not* intended for the assessment of noise exposure for comfort and security such as fatigue and vigilance.

A study of the literature published shows that private cars have been developed to provide an interior noise environment of between 62 and 85 dB(A) when operating between 30 and 70 mph (miles per hour) (48–112 km/hr). Since the majority of cars fall within the middle of this range, continued exposure by long-distance driving, although contributing to fatigue, affecting vigilance and in turn influencing road safety, is unlikely to carry a deafness risk. As Saunders (1970) concluded, the noise in motor cars is not a serious problem and efforts aimed at improving the environment should be considered carefully against the desire of the motorist to

pay for perhaps marginal increases in comfort and the need of the manufacturer to keep the price of his goods competitive.

The personalised nature of the motor car makes it quite unlike other
forms of transport. The car and the way it is driven may reflect the
owner's personality. The sports car enthusiast, often an extrovert,
equates noise with power and speed; therefore quietness is not a buyable
commodity to him. In contrast, quietness is a luxury of the rich car
owner, who may be prepared to pay considerable sums in his search for
increased comfort. Often he is misled, for the truth is that no car is
"quiet" when travelling under power; they all produce noise, but it seems
that our level of tolerance to noise is higher when travelling in transport
than it is outside the transport environment (Bryan *et al.*, 1978).

One of the first attempts to measure subjective reaction to internal
noise of cars was made by Bristow (1952). He concluded that, provided
that the octave band spectrum of the noise was measured, it was possible
to place cars in order of subjective reaction to noisiness.

A brief review of the work done up to 1964 in predicting all the possible
sources in order to give efficient noise control at the design stage was
presented by Gladwell (1964), although he gave little mention of the subjective reaction. Some years later, Ford *et al.* (1970), using 6 cars and 12
subjects who were each required to judge 48 different situations, concluded that the subjective response to internal car noise could adequately
be measured by using the A-weighted network of a sound level meter,
although they also found good correlation between subjective assessment
and Zwicker phons, Stevens phons and decibels(B).

Laboratory-based experiments using the "dummy head" method proposed by the MIRA and applied by Philips and Christie (1974) showed
that it was possible to rank cars in order of annoyance. Using five vehicle
test methods, they found good agreement between in-vehicle and laboratory judgements based on a ranking rather than a rating technique.

There is considerable well-documented evidence of infrasound in transportation, although the results so far are far from unanimous in their
conclusions. Aspinall (1966) studied the noise in cars down to 10 Hz,
whilst Tempest and Bryan (1972), Hood and Leventhall (1971), Williams
and Tempest (1975a,b) and Broner (1978) considered measurements down
to 2 Hz. All their results confirm the presence of considerable energy in
the infrasonic region, and there is now some evidence that this very low
frequency, which does not show up on the decibels(A)-weighted meter,
does in some way influence the subjective assessment, ranking or rating
of car interior noise (Bryan *et al.*, 1978). The noise spectra of Fig. 2 show
the results obtained inside four passenger vehicles, indicating convergence at the low frequencies, although there is considerable difference in

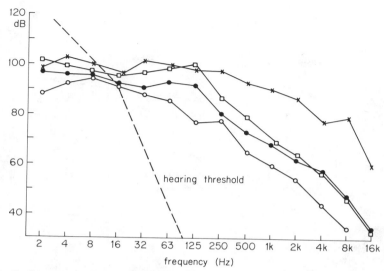

Fig. 2. Octave band noise levels inside four passenger vehicles, with all windows closed and at a sustained speed of 97 km/hr. Four-litre saloon (O—O), noise rated quiet/noticeable; 3-litre saloon (●—●), noise rated intrusive; 1-litre saloon (□—□), noise rated annoying; 2.25-litre vehicle (×—×), noise rated unbearable.

the sound pressure levels above the octave centred at 32 Hz. Figure 3 shows the ranges of interior noise levels measured in both private motor cars and heavy goods vehicles.

The large differences between the readings taken on the "linear" and decibels(A) scales of the sound level meter confirm the presence of substantial energy at low frequencies. The noise levels in the cars typically increase by about 6 dB(A) between 40 and 70 mph (64–112 km/hr). This means that at the higher speeds, most of them are over 70 dB(A) and have moved out of Bryan's (1976) "quiet" rating into the "noticeable" category. Of course, his criteria were tentative and more recent work has shown that vehicle noise criteria are more meaningful when applied to one vehicle group. Obviously, the sound pressure level, decibels(A) weighted, in the cab of a "quiet" lorry could be the same as that inside a "noisy" motor car. Each road vehicle type is sufficiently different to be divided into groups: cars, lorries, buses. There is also some indication that the subjective resolution of jurors using a "noisiness" rating for cars seems to be between 2 and 3 dB (i.e., it requires a change of about 2.5 dB to give a change of response of 1 unit) and that the subjective response is affected by the presence of low-frequency energy. In the study by Bryan *et al.* (1978) laboratory observers were asked to compare reproduced car noise

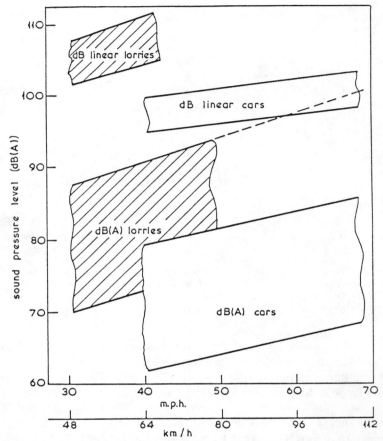

Fig. 3. Comparison of objective measurements of interior noise of 47 heavy goods vehicles and 68 private motor cars.

for a number of vehicles under two conditions. Under the first condition, the noise was reproduced without modification, but in the second the frequencies below 20 Hz were filtered out. Figure 4 shows that the unfiltered noises were rated higher (less acceptable) than filtered noises, which may suggest that future field studies may need to include the low frequencies if correlation with in-car response is to be obtained.

Attaining improved comfort in future cars implies the need for a substantial reduction of the noise levels in the passenger compartments (Jennequin, 1971). The many noise generators—engine, auxiliaries, aerodynamic noises, road noise, panel resonances and transmission noise—all act simultaneously, giving the interior a particular noise signature. The closeness of the passenger to the noise, in addition to the small volume of the driving compartment along with the motion of the car, means that the noise environment is quite unlike others one meets in everyday life.

By the mid-1970s the consensus view on the subjective evaluation of noise inside motor vehicles appeared to be that decibels(A) was the best of the units in widespread current use, but was not entirely satisfactory. The limitations in the value of decibels(A) in assessing sound with a large energy content at low frequencies have already been mentioned (Tempest, 1973). A study by Callow and Hedges (1979) has looked at the problem more broadly and has resulted in proposals for a modified decibels(A) unit referred to as the CRP (composite rating of preference).

The CRP is based on studies of the subjective response to vehicle interior noise, tape-recorded and reproduced stereophonically in the labo-

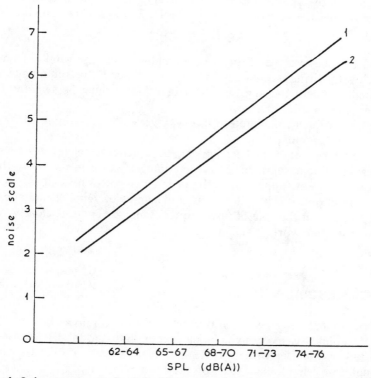

Fig. 4. Laboratory comparison of subject assessment of vehicle noise with and without low-frequency energy: (1) unfiltered and (2) high-pass filtered noise (20-H_z cutoff).

ratory. Using up to 40 subjects, the noise levels from five mid-market cars (each costing about £2500 in 1975) were assessed. A series of constant-speed conditions for each car was evaluated, and it was found that decibels(A) levels showed a high correlation with subjects' preferences ($r = 0.9$). A detailed study of the subjects' responses showed that these were influenced by subjectively dominant parts of the spectra, and these dominant regions were associated with two frequency ranges, one above 800 Hz and the other below the firing frequency of the engine.

From a multidimensional analysis of the subjects' preferences a formula for CRP was developed as follows:

$$CRP = \{[dB(A)]^2 - 1.5(HF)^2 + 0.5(SB)^2\}^{1/2};$$

when $dB(A)$ = measured sound level,

$$HF = dB(A) - SIL$$

where SIL is the speech interference level, which is here defined as the average of the octave band levels in the 1-, 2- and 4-kHz bands.

$$SB = LF - SIL,$$

where LF is the average of the octave band levels in the band of frequencies below the firing frequency of the engine. Since firing frequency varies with engine speed, the appropriate octave bands must be selected for a given engine speed.

The CRP values calculated in this way gave a correlation of 0.96 with subjects' judgements.

The development of the CRP appears to represent a useful step forward in the subjective assessment of vehicle interior noise in the sense that it has explored a new approach to the problem, with some success. As the authors point out, it is based on a limited sample of vehicles and does not take into account the role of low-frequency noise. There is clearly scope for much further work in this area.

III. COMMERCIAL VEHICLES

There is increasing evidence that the level of cab noise in some commercial vehicles is excessive, being higher than the 90-dB(A) criterion suggested by the Department of Employment's code of practice (1972). Although few reliable data exist on the road safety aspects of noise, it is generally thought that high levels of noise tend to cause fatigue and irritation to drivers. Furthermore, high levels of internal noise in vehicles could mask audible warning signals from other vehicles and from within the vehicle itself. It therefore seems important on the grounds of road safety to make every effort to reduce the level of noise in goods vehicle cabs.

Under current U. K. regulations, a solo driver may drive for up to 8 hr in one day; with a co-driver he can be travelling in the cab for a period of 11 hr. The Code of Practice (Department of Employment, 1972) suggests that for a daily exposure of 8 hr, the noise level should not exceed 90 dB(A), and for 11 hr the acceptable level is reduced to 88.6 dB(A). An examination of the relevant literature shows that the noise levels in many cabs exceed these criteria and may therefore be the cause of hearing loss.

Nerbonne and Accardi (1975) report indications of noise-induced hearing loss amongst a truck driver population. Their tests on 113 drivers showed good correlation between total number of years driving experience and hearing loss, particularly at 4 kHz. The hearing level of the right ears was generally higher than the left ears, and their results show a noticeable depression in hearing level at 0.25 to 0.5 kHz, possibly related to the exposure these men experienced to both auditory and vibratory stimuli with primary energy at extremely low frequencies, a criterion known to be prevalent in trucks.

In order to reduce the cabs' noise it is of paramount importance first to identify the noise sources and to measure the levels objectively. Our understanding of the nature and sources of commercial vehicle noise owes a great deal to the work of Priede (1967, 1971), Mills and Aspinall (1968), Waters (1974) and Williams and Tempest (1975a,b).

Priede's paper gives some maximum band spectra, which are similar to those found by Mills and Aspinall and further confirmed by Waters. Priede showed that one-third of his sample of 15 vehicle types exceeded the NR-85 [≈90-dB(A)] contour, whilst the highest levels reached NR 95 [≈100 dB(A)]. Waters concluded that the interior cab noise originated from the same source as the exterior noise, mainly airborne noise from the power unit, and suggested that any modifications to reduce exterior noise would have a similar effect on interior noise.

Using a 9-ton truck and the ISO test (ISO, 1964) for external noise, Waters measured reductions of 8 dB(A) from 87 to 79 dB(A) in a vehicle with modifications, primarily, to the power unit source. Apart from the engine, it was necessary to modify the exhaust, cooling fan and air intake to achieve the reductions. It seems from a practical view that the law of diminishing returns becomes a serious problem beyond a 7- to 9-dB(A) reduction (Berry, 1970).

Williams and Tempest (1975b) measured noise in 47 heavy goods vehicles. Their results are summarised in Fig. 5, which shows the frequency

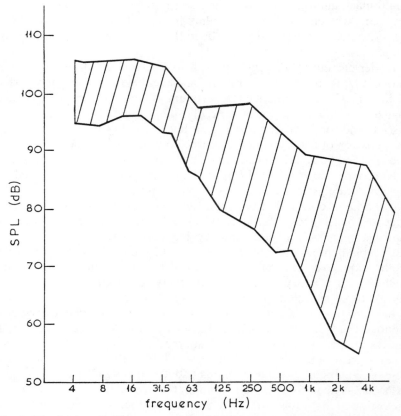

Fig. 5. Distribution of noise spectra obtained in 47 heavy goods vehicles, demonstrating convergence at the lower frequencies.

distribution of noise levels in the vehicles. Their measurements show large differences between the readings taken on the linear and decibels(A) scales of the sound level meter and indicate the presence of substantial noise energy at the low frequencies. This is confirmed by similar results published by Broner (1978) and Bryan (1976). Reports indicate that the high levels of infrasonic noise present in lorry cabs travelling at cruising speed may in some way affect driver vigilance (Bryan, 1976; Tempest, 1976).

The importance of noise was shown in three quite independent surveys conducted amongst lorry drivers, one in Sweden, one in North West England and the other randomly distributed at six locations around England. The Swedish survey rated noise second to driving position on a "comfort" rating. The results of the North West England survey showed that of the parameters vibration, noise, temperature, ventilation, seat comfort and visibility, noise level was considered second in order of importance to seat comfort (Williams and Tempest, 1975a).

The results of the other British survey showed that 28% suffered from ear troubles (Morris, 1976). Suspension and seating design were both considered more important than noise levels.

Manenica and Corlett (1973) observed that general comfort in a vehicle arises from a large number of different factors, each at a different level, these levels being randomly distributed within one vehicle. This is particularly true of commercial vehicles, especially tractors with unladen articulated trailers in which "cab nod" probably swamps all other discomfort effects.

It is obvious then that driver comfort cannot be considered in terms of noise levels only, but we should be aware that the noise in many existing lorry cabs does present a hazard to hearing, well-being and, ultimately, safety. It has often been argued that an uncomfortable driver is an unsafe one. He is, potentially, an unhealthy one, too.

What of the future? Government departments along with two U.K. manufacturers have worked on a project to produce a quiet lorry having interior cab noise below 80 dB(A), with an eventual target of no more than 75 dB(A) inside the cab at the operator's ear (Nelson, 1978; West, 1979). Developments in heavy commercial vehicle cabs reported by Lewis and Collins (1978) show that it is possible, even on production vehicles under conditions of maximum acceleration, to reduce interior noise levels down to 77.5 dB(A) in the vicinity of the driver's ear.

Waters indicates that a typical noise reduction package would result in an increase of the order of 0.5% on the running cost per ton mile. The government receives over £500 million each year from truck operators. If only a small percentage of this sum were redirected into vehicle noise research, we might even see, or at least hear, the 75-dB(A) lorry.

In the near future it is hoped that we will obtain a better understanding of infrasound in lorries, which will eventually lead us towards its reduction. European and American truck manufacturers claim to have overcome many of their noise problems; it is hoped that their achievements will be emulated in the United Kingdom.

IV. AGRICULTURAL TRACTORS

There is evidence that noise levels in tractors in the 1960s were between 90 and 105 dB(A) (Clarkson, 1972). Current U.K. legislation requires all farm tractors to be fitted with an approved safety cab. The cab is known as a "quiet cab", designed to restrict the noise level to a maximum of 90 dB(A), which, although high, is a move in the right direction. It will be interesting to see how the legislation will be enforced, particularly in the remote rural areas. There are currently in use many noisy tractors, and these will be even noisier when fitted with a cab, for the development of safety has often been made at the expense of noise. In view of the high noise levels encountered, there would seem to be a case for periodical audiometric testing of tractor drivers.

The measurement procedure for assessing noise in agricultural tractor safety cabs is detailed in the British Standard 4063 (BSI, 1973). It recommends the use of the decibels(A) scale and the fast response of the meter at the load corresponding to the maximum noise in the gear giving the nearest forward speed to 7.25 km/hr. There are also recommendations for the style of the report and the conditions of acceptance with regard to safety.

V. RAILWAY TRANSPORT

The Wilson Committee Report (1963) observed that the noise inside most British Railways coaches was uncomfortably high and compared unfavourably with that inside the coaches on some foreign railways. They concluded that, except in the case of some special Pullman trains and some London Transport Executive Underground lines, no real effort appeared to have been made to reduce noise inside passenger compartments. They made no attempt to quantify their findings nor did they suggest acceptable noise levels for railway coach interiors. Their remarks may have initiated the obvious research effort that has been applied to railway coach acoustics during the past decade, making high-speed travel tolerable or even pleasant, particularly on the inter-city electrified services. Vehicle designers are now conscious that the attraction of a transport service depends to a high degree on the comfort of passengers.

Eade and Stanworth (1976), in considering the criteria for acceptable internal noise levels, found that the ability to converse normally was of prime importance. They suggest that an ideal background noise in a railway vehicle would allow conversation in a normal voice over a distance of 3 m. This requires a maximum preferred speech interference level (PSIL) of 50 dB, which in turn implies a level of approximately 60 dB(A). It also seems that the desirability to have some privacy as a result of masking by the background noise is of prime importance to the railway passenger.

Railway transport passenger vehicles can be divided into those vehicles having their own self-contained power units, such as diesel and electric multiple units, and those with separate power locomotives. High internal noise levels are a serious problem among diesel multiple units caused by structure-borne and airborne noise from the main power unit.

Bryan (1976) considered the subjective acceptability of the noise inside a British Rail multiple-unit diesel train by measuring noise in a motorised coach and an unmotorised coach. He found a noise level of 82 dB(A) in the motorised unit, which was rated "very annoying", and a level of 78 dB(A) in the unmotorised coach, which was considered "slightly annoying". His measurements clearly highlight one of the problems in determining subjective ratings, for a 4-dB(A) difference between the coaches does not adequately reflect the differences in "annoyance" between the two coaches. The noise spectra for these coaches are shown in Fig. 6, which shows the presence of a large difference in the sound pressure levels below 63 Hz, which may account for the considerable difference in annoyance scaling.

The treatments to improve the noise insulation of modern stock have been more effective at high frequencies than at low frequencies, thus increasing the percentage of low-frequency noise in the internal sound field and so, possibly, creating a smaller subjective change in the comfort level than the decibels(A) measurements indicate.

For noise in road vehicles, it has been shown that decibels(A), decibels(B) and loudness level (LL) correlate well with subjective ratings of

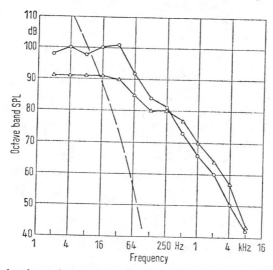

Fig. 6. Octave band sound pressure levels (dB re 20 μPa) in diesel multiple-unit passenger trains (DMU). Unmotored coach, 64 km/hr ($\triangle$—$\triangle$); motor DMU, 80 km/hr ($\bigcirc$—$\bigcirc$); hearing threshold (———). [From Tempest (1973).]

comfort and annoyance (Ford *et al.*, 1970; Bryan, 1976; Williams and Tempest, 1975a; Philips and Christie, 1974).

In aircraft, the preferred speech interference level (PSIL) is generally regarded as important. The alternative noise criteria (NCA) and noise ratings (NR) were used as a framework for earlier rail vehicle design criteria, both of which are based on the concepts of speech interference level and loudness level, typical values being NCA or NR 60 (Koffman, 1967; Dietrich, 1966). British Rail investigations have shown that two weighted SPL measurements, decibels(A) and decibels(B), used in conjunction can provide a good measure of subjective noise comfort (Eade and Hardy, 1977).

Some German requirements are detailed by Stüber (1975), who reports that in Germany the Office for Experiments of the German Federal Railway in Munich deals with, among other things, the requirement to increase the comfort for the passengers by having low noise levels in passenger coaches and low noise levels in the driving cabs in order to ensure the health and safety of the driver. The results show that in a range of trains, at differing speeds, the noise levels are within the range 51–74 dB(A) inside the coaches and between 68 and 85 dB(A) inside the driving cabs. It is suggested that in the interest of comfort and safety, the noise levels should not exceed, even above 200 km/hr, 80 dB(A) and, preferably, be nearer to 60 dB(A), which appears to be about the speech interference level. Eade and Stanworth (1976) quote levels between 57 and 82 dB(A), measured at speeds between 96 and 200 km/hr.

Noise spectra for a representative sample of British Rail are shown in Fig. 7, along with internal noise specifications for inner suburban stock and main line stock.

Underground railways tend to be particularly noisy. They are usually designed for a lower level of comfort than main-line stock, and the layout, involving numerous wide doors for rapid loading and unloading, makes it particularly difficult to adequately insulate the passengers from rail, wheel and motor noise. Data on noise levels in New York City Transit Authority rapid transit cars are quoted by Remington and Wittig (1979) as commonly exceeding 90 dB(A). In their paper, concerned mainly with the control of noise in existing rolling stock, they discuss possible noise criteria at some length. Due to the very high noise levels, the considerations are quite different from those pertaining in long-distance rail travel, and their main concerns are the risk of occupational deafness and the ease of communication among passengers. Their recommendation is for a maximum level of 82 dB(A), corresponding to a reduction of more than 10 dB(A) on the levels then current. They describe their 82-dB(A) level as meeting the U.S. Environmental Protection Agency (EPA) criteria for protection rela-

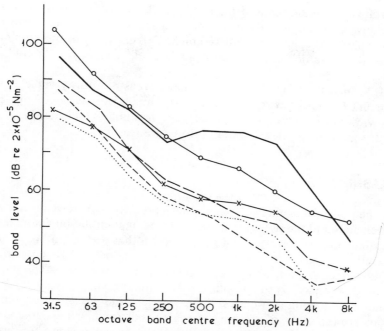

Fig. 7. Internal noise levels in trains. BR Mark I second class at 145 km/hr (———); BR Mark IID first class at 145 km/hr (— —); BR Mark III second class at 145 km/hr (------); SNCF 52t coach first class (compartmented) at 140 km/hr (·····); BR inner suburban stock specification [72 dB(A) and 90 dB(B)] (—O—). BR main-line stock specification [65 dB(A) and 73 dB(B)] (—×—). NB Stock at full speed on good-quality continuous welded rail in open country. [Data from Eade and Hardy (1977.]

tive to the risk of noise-induced hearing loss, reducing the average loudness in the vehicles by more than one-half and permitting adequate speech communication among persons seated together.

The literature relating to the internal noise of railway vehicles is unanimous in the view that the most obvious improvement has been due to welded rail construction, which has reduced considerably the transmission of train noise into the passenger compartment. Unfortunately, the noise problem has been compounded by the move away from compartmented stock, by higher speeds and by coaches of lighter construction. All have added to the problems facing acoustic designers.

Despite the clear gains that have been made, the drivers of the powerful earlier generation diesel and electric locomotives are subjected to high noise levels for long periods, although some improvements have been achieved by isolating the driving cabin from the engine room by acoustic curtains. Draughts caused mainly by lack of maintenance to door and window seals apparently cause more discomfort to drivers than noise, particularly at speeds around 100 mph. Drivers seem of the opinion that they would rather have a warm, noisy compartment than a cold, quiet one.

It seems, therefore, that there remain areas in railway vehicles which warrant further research effort. Clearly, much can be gained by seeking the views of the drivers and passengers, for, unless their needs are satisfied, research and development efforts are wasted.

VI. SHIPS

The powerful engines in today's ships produce high noise levels, which, when added to the noise from air conditioning, ventilation equipment and auxiliary plants for electrical power, mean that people on board are frequently subjected to excessive noise for long periods, even during the hours of relaxation and sleep (Hines, 1966). Since most of the noise aboard originates at the propulsion unit, the problem may be partly overcome by ensuring that crews' living quarters and passenger accommodations are remotely sited from the engine room and propulsion end. Even then, the problem of structure-borne noise arises, giving ship designers many problems.

The working environment in many ships, particularly that of engineering officers, is far from satisfactory. Clarkson (1972) quotes levels of 98 to 108 dB(A) as being typical of engine room noise and Williams (unpublished data) has found similar levels in the engine rooms of tugs, ferry vessels and harbour dredgers. The noise environment experienced daily by human operators for the local harbour trust and the navy was investigated by Broner (1978). The levels encountered were generally high, and in all cases he was able to identify the major noise sources. His results indicate the presence of considerably high levels of low-frequency sound, particularly in the engine rooms.

Gibbons et al. (1975) report on some work carried out on behalf of a shipping company which was primarily aimed at quantifying noise and vibration levels on a number of tankers. At the same time, as a secondary consideration, they studied urine samples in order to study 17-ketosteroid excretion and urinary volumes, both of which are believed to be sensitive to noise-induced stress (Sakamoto, 1959). Although their study was limited to only four ships, it was possible to obtain a simple statistical analysis which showed a significant decrease in the level of 17-ketosteroids and urinary volume whilst the ship's officers were serving at sea, compared to the values when the men were on leave. In general, subjects used in the experiment found it difficult to differentiate between annoyance due to noise or vibration when both were present unless one parameter was at an extreme level. Richards and Middleton (1976) found this to be particularly true when the noise and vibration had large low-frequency components. In view of this, Gibbons *et al.* (1975) proposed a tentative procedure for combining NR (noise rate) and VR (vibration rate) to give NVR, a compound rating for noise and vibration. They also used questionnaires which indicated that within a few days of the start of a voyage adaptation had occurred so that noise and vibration levels were not particularly annoying. This effect occurs to some extent in all forms of transport and has been observed by the author during subjective tests on road transport. It seems that it is the nature rather than the level of noise which disturbs or causes annoyance, and this fact suggests that a subjective scale based on annoyance is unlikely to be of more than academic interest. The main noise complaints came from engineering officers whose normal working environment was about 30 dB(A) higher than the quieter areas of the ship's accommodation. Their principal complaint was that noise interfered with communication, whilst some believed they had suffered some hearing loss.

Noise acceptability in ships was considered by Hoyland (1976), who compared the requirements of the regulations and recommendations existing in some European countries. He found that noise regulations, with varying degrees of enforcement, existed in some countries, with particular reference to ships. The acoustical environment in the working areas of the ship must be determined by the consideration of efficiency of work, the danger of deafness and thus health, and the need for limited aural and telephone communications (Richards and Middleton, 1976). The noise in the living areas, they concluded, must be determined by the need to communicate readily, to hear music and to be able to sleep. It was desirable to limit the maximum permissible sound levels at watch-keeping and manoeuvring positions which would allow better communications when running in poor visibility or when manoeuvring. Their results showed that levels as high as 115 dB(A) occurred in engine rooms, although the average noise on the platform level was nearer to 105 dB(A), differing very little with engine power and size of ship, nor for that matter with the speed of the engine unit. Although the engine room noises were high, the levels in the control rooms were 30 dB(A) less than in the main plant area. It would seem, therefore, that the Department of Employment's code of practice 90-dB(A) criterion would not be exceeded, provided that the engine room personnel spent the majority of their working hours in the control room.

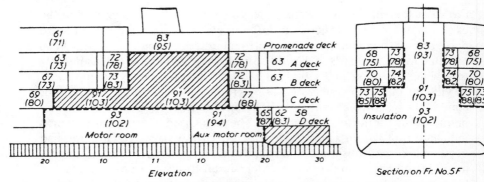

Fig. 8. Sound levels in an acoustically insulated motor ship (data in brackets are for an uninsulated ship in same location). [From Richards and Middleton (1976).]

Richards and Middleton suggest that 60 to 65 dB(A) are acceptable levels for crews' quarters, which is of course higher than levels in living and sleeping accommodations in shore-based quarters.

Figure 8 shows the benefits which may be achieved by careful consideration to acoustic insulation. These results indicate that reductions of 9 dB(A) are possible in the main engine room and 6 dB(A) on upper decks.

Measurements by Richards and Middleton (1976) on the sundeck of a passenger ship showed levels of 85 to 95 dB(A), making it unattractive to most passengers. This vessel also suffered from exhaust noise problems, which resulted in considerable low-frequency noise annoyance.

The problem of ensuring adequate privacy between adjacent cabins is becoming increasingly important in non-passenger ships such as tankers, which can undertake long voyages, sometimes with family units on board (Jones and Berry, 1978).

An examination of the existing literature leaves the reader in no doubt that ships are unnecessarily noisy, and the long-term aims of achieving desirable noise climates on board ship will be assisted by the introduction of more stringent specifications stimulated by regulations and codes of practice, operated under the general guidelines of the Health and Safety Act.

VII. HOVERCRAFT

The commercial hovercraft in the United Kingdom is seagoing, although the Road Traffic Act (Anonymous, 1962) brings them within the definition of motor vehicles.

The Wilson Committee (1963) recognised that the hovercraft was intrinsically noisy, although their comments in the main were directed at external noise, particularly around the terminal areas.

Macmillan (1975) reports that hovercraft are without doubt the noisiest form of surface transport but, since they are few in number and operate mostly away from built-up areas, the noise problem chiefly affects the passengers and crew. Inside the craft, transmission and engine noise are predominant and sound pressure around 83 dB(A) is typically constant at cruising speeds, with short peaks of 91 dB(A) on arrival and departure (Duerden, 1972). This may not be a serious problem, since most hovercraft journeys are of short duration, although the crews may be affected.

The hovercraft is a unique and comparatively new form of transport; therefore the literature relating to its internal environment is relatively sparse. It seems, however, that the designer is faced with solving other problems such as vibration and poor visibility from the passenger space as well as noise levels.

REFERENCES

Anonymous (1962). Road Traffic Act. Her Majesty's Stationery Office, London.

Anonymous (1974). Health and Safety at Work Act. Her Majesty's Stationery Office, London.

Aspinall, D. T. (1966). "An Empirical Investigation of Low Frequency Wind Noise in Motor Cars." MIRA Report No. 1966/2. Motor Industry Research Association, Nuneaton.

Berry, W. C. (1970). Practical means of implementing vehicle noise control II, *J. Sound Vib.* **13**, 459–460.

Bristow, J. R. (1952). Noise in private cars, *Inst. Mech. Eng., Auto. Div. Proc.* **1**(1), 9–22.

Broner, M. (1978). Low frequency and infrasonic noise in transportation, *Appl. Acoust.* **11**, 129–145.

Bryan, M. E. (1976). A tentative criterion for acceptable noise levels in passenger vehicles, *J. Sound Vib.* **48**, 525–535.

Bryan, M. E., Tempest, W., and Williams, D. (1978). Vehicle noise and the passenger, *Appl. Ergonom.* **9**, 151–154.

BSI (1973). British Standard 4063, "Method of Assessing Safety Cabs of Agricultural Tractors." British Standards Institution, London.

BSI (1981). British Standard 6086, "Measurement of Noise Inside Motor Vehicles." British Standards Institution, London.

Callow, G. C., and Hedges, R. (1979). "The Subjective Response of Occupants to the Noise inside Vehicles." MIRA Report 1979/1. Motor Industry Research Association, Nuneaton.

Clarkson, B. L. (1972). The social consequences of noise, *Inst. Mech. Eng. Proc.* **186**, 97–107.

Department of Employment (1972). "Code of Practice for Reducing the Exposure of Employed Persons to Noise." Her Majesty's Stationery Office, London.

Dietrich, C. W. (1966). "The M.B.T.A. South Shore Project: Passenger Noise and Vibration Criteria." Bolt, Berranek and Newman Inc. Report 1428.

Duerden, C. (1972). "Noise Abatement." Butterworths, London.

Eade, P. W., and Hardy, A. E. J. (1977). Railway vehicle internal noise, *J. Sound Vib.* **51**, 403–415.

Eade, P. W., and Stanworth, C. G. (1976). Acoustic aspects of railway vehicle design, *Inst. Mech. Eng. Proc.* **190**, 515–525.

Ford, R. D., Hughes, G. M., and Saunders, D. J. (1970). The measurement of noise inside cars, *Appl. Acoust.* **3**, 69–84.

Fuchs, G. L. (1975). Subjective evaluation of transport noise in Latin America, *J. Sound. Vib.* **43**, 387–394.

Gibbons, S. L., Lewis, A. B., and Lord, P. (1975). Noise and vibration on board ship, *J. Sound Vib.* **43**, 253–261.

Gladwell, G. M. L. (1964). A review of noise and vibration in motor cars, *J. Sound Vib.* **2**, 202–210.

Helmsman, A. (1976). Safe and sound, *Occup. Health,* 514–517 (November).

Hines, W. A. (1966). "Noise Control in Industry." Business Publications Ltd., London.

Hood, R. A., and Leventhall, H. G. (1971). Field measurement of infrasonic noise, *Acustica* **25**, 10–13.

Hoyland, A. (1976). The noise and vibration environment on board ship, *in Proc. Vibration Noise Conf.,* pp. 1–12. Institute of Marine Engineers, London.

ISO (1964). ISO R362. "Methods of Measurement of Noise Emitted by Vehicles." International Organization for Standardisation, Geneva.

ISO (1978). ISO 2631. "Guide to the Evaluation of Human Exposure to Whole Body Vibration." International Organization for Standardisation, Geneva.

Jennequin, G. (1971). Is the computation of noise levels inside a car feasible? *Inst. Mech. Eng.* **C108**, 132–137.

Jones, A. J., Berry, G. (1978). Noise control in ships, *Noise Control Vib. Isolation,* 371–375 (November/December).

Koffman, J. L. (1967). Design for comfort, *J. Inst. Locomotive Eng.* **57**, 428–478.

Kryter, K. D. (1970). "The Effects of Noise on Man." Academic Press, New York.

Lewis, R. P., and Collins, J. F. (1978). Recent developments in heavy commercial vehicle cab noise, *Noise Control Vib. Isolation,* 383 (November/December).

Macmillan, R. H. (1975). The control of noise from surface transport, *J. Sound Vib.* **43**, 173–187.

Manenica, I., and Corlett, E. N. (1973). A model of vehicle comfort and a method for its assessment, *Ergonomics* **16**, 849–854.

Mills, C. H. G., and Aspinall, D. T. (1968). Some aspects of commercial vehicle noise, *Appl. Acoust.* **1**, 47–66.

Morris, P. (1976). The trouble with trucks, *Occup. Health,* 285–289 (June).

Nelson, P. M. (1978). Controlling road traffic noise, *Noise Control Vib. Isolation,* 363–367 (November/December).

Nerbonne, M. A., and Accardi, A. E. (1975). Noise induced hearing loss in a truck driver population, *J. Auditory Res.* **15**, 119–122.

Philips, A. V., and Christie, D. (1974). Techniques of interior noise evaluation, *in Proc. Int. Congr. Acoustics, 8th, London.*

Pinto, R. M. N. (1962). Sex and acoustic trauma: Audiologic study of 199 Brazilian airline stewards and stewardesses (Varig), *Rev. Brasil Med. (R10)* **19**, 326–327.

Priede, T. (1967). Noise and vibration problems in commercial vehicles, *J. Sound Vib.* **5**, 129–156.

Priede, T. (1971). Origins of automotive vehicle noise, *J. Sound Vib.* **15**, 61–73.

Remington, P. J., and Wittig, L. E. (1979). Retrofit noise control of rapid transit cars, *J. Sound Vib.* **66**, 419–441.

Richards, E. J., and Middleton, A. (1976). Practical noise reduction in ships, *in Proc. Vibration Noise Conf.*, pp. 13–21. Institution of Marine Engineers, London.

Sakamoto, H. (1959). Endocrine dysfunction in noisy environment, *Mie Med. J.* **9**, 39–74.

Saunders, D. J. (1970). Human reaction to noise and vibration in motor cars, *Environ. Eng.*, 12–16 (May).

Stüber, C. (1975). Air and structure-borne noise of railways, *J. Sound Vib.* **43**, 281–289.

Tempest, W. (1973). Low frequency noise annoyance, *Acustica* **29**, 205–209.

Tempest, W., ed. (1976). "Infrasound and Low Frequency Vibration." Academic Press, London.

Tempest, W., and Bryan, M. E. (1972). Low frequency sound measurements in vehicles, *Appl. Acoust.* **5**, 133–139.

Waters, P. E. (1974). Commercial road vehicle noise, *J. Sound Vib.* **35**, 155–222.

West, J. (1979). The quiet heavy vehicle programme, *Automotive Eng.* **4**, 3 (June/July).

Williams, D., and Tempest, W. (1975a). Noise and the lorry driver. Paper read at the Institute of Acoustics Conference, Salford University, 23/24 January.

Williams, D., and Tempest, W. (1975b). Noise in heavy goods vehicles, *J. Sound Vib.* **43**, 97–107.

Wilson, A. (1963). "Noise—Final report." Her Majesty's Stationery Office, London.

10

Noise in the Home

G. M. JACKSON AND H. G. LEVENTHALL

Atkins Research and Development
Epsom, Surrey
England

I. INTRODUCTION

Noise levels in the home are governed by a multiplicity of factors, the majority of which are outside the control of the recipient. Due to socio-economic reasons, it may be impossible for him to choose exactly where he lives, whether it be in a house alongside a busy main road with peak levels of over 90 dB(A) or in a quiet country cottage with an ambient outside noise level of maybe 30 dB(A). Because of these extremes in the residential noise environment, a particular domestic appliance might well be described as "noisy" in a quiet environment but not even noticeable when used in a house adjacent to a main road. For this reason, the types of appliances measured and reported in this chapter have been chosen to cover a wide range of noise levels and so help to pinpoint those appliances or groups of appliances in most need of quietening.

THE NOISE HANDBOOK

II. DOMESTIC ROOM ACOUSTICS

Due to the different sound-absorbing qualities of individual rooms, the noise level produced by an appliance may well differ from one room to another. If the room is acoustically "dead", the noise level may be lower than in an acoustically "live" or "hard" room.

We might well question the use of measuring and comparing appliance noise levels if these levels depend upon the acoustics of the domestic room in which the appliance is used. In this situation, we need to look at the structure of the sound field produced by a noise source in a room. This sound field has two main parts—the "direct" and "reverberant" components. As the terms imply, the direct sound comes straight to the listening point from the noise source, whilst the reverberant sound is the combination of sound rays reaching the listener after reflection at the surfaces of the room. Some additional importance is connected with the sound reaching the listener after only one reflection. This first-reflected sound corresponds to the primary image in the nearest reflecting plane to the appliance, often the floor. The effects of room acoustics on the sound levels produced within a room can be calculated from a knowledge of the room constant R and the directivity factor Q of the noise source. The room constant depends on the sound-absorbing material in the room, whilst Q is a measure of the extent to which the source concentrates its radiated sound in one direction. The relevant theory, and some practical data, are presented in the Appendix at the end of this chapter. The main conclusions are that, in the case of kitchens, the noisiest room will produce a reverberant field some 3 dB higher than the average room, for the same appliance. In the case of living rooms, the difference is about 1.7 dB. These are calculated values, and additional experiments were conducted in a further 50 living rooms and 50 kitchens (Jackson and Leventhall, 1973) by using a standard noise source. These showed that the noisiest living rooms and kitchens were about 4 dB above the average, in contradiction to 1.7 dB for living rooms and 3 dB for kitchens, as predicted above. These discrepancies were shown to be due to different mounting conditions of the noise source. When placed on a partially absorbing floor, the power output of the noise source may be reduced with a resulting decrease in noise level. These results show the importance of the first reflected sound and illustrate the importance of considering both the appliance and its mounting as the total noise source (Allen and Potter, 1962). For these reasons, all the domestic appliance levels measured and reported here were conducted either *in situ* or in the simulated domestic environments developed for the purpose (Jackson and Leventhall, 1973).

III. DOMESTIC APPLIANCE NOISE LEVELS

Little detailed information has been published on the noise of household appliances (Bender, 1973; Jackson and Jeatt, 1974). Decibels(A) noise levels presented by Bender (1973) were all measured at a distance of 3 ft from the appliance. In this study we have generally been more concerned with average user locations.

For the purposes of comparing noise levels it is convenient to separate domestic appliances into different categories. This may be done in several ways, for example, by noise level. Due to the influence of the acoustic environment (whether kitchen or living room) upon the noise emitted by appliances, it was decided to catalogue them under the type of room they are ordinarily installed or used in. The demarcation lines between these groups are not rigid, but it was felt that this type of grouping would show those household areas with greatest average noise level. This approach might also be of use to the architect designing blocks of flats if he wishes to keep noise-sensitive areas, such as bedrooms, away from areas of inherently high noise level. We have divided the appliances tested into the following categories of use—kitchen appliances, living room/bedroom appliances, bathroom appliances and miscellaneous domestic appliances.

A. Kitchen Appliances

1. *Food Preparing Devices*

The appliances studied in this group were food mixers, liquidisers, whistling kettles, electric toasters, coffee percolators and gas cookers. Only the first three types of appliance are likely to produce any serious noise problem, but fortunately their period of operation is generally quite short, although the frequency of operation of whistling kettles could be quite high.

Figure 1 refers to food mixers. Most of the models tested had the three speed settings slow, medium and fast. Ten models with medium speed settings (some of them had only one speed) were tested. Seven of these models had, in addition, both slow and fast settings. The third octave band average noise levels and standard deviation of these levels are shown, as measured at a characteristic user position 1 m from the appliance. Over the frequency range 250 Hz to 12.5 kHz, the average third octave band sound pressure levels increased by approximately 6 dB when the speed was increased from the slow to medium settings, and a further 5 dB on increasing to the fast speed. These same increases are also shown in the average decibels(A) levels.

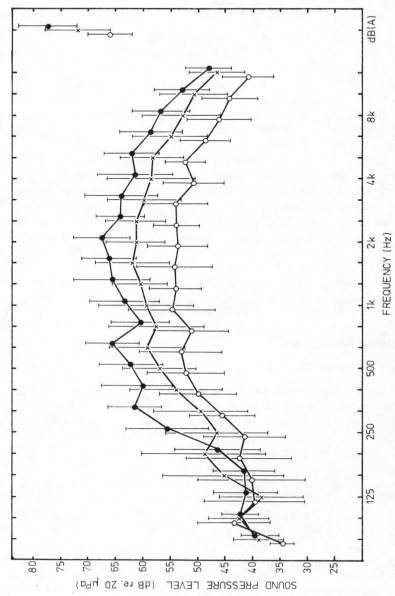

Fig. 1. Noise spectra of food mixers, average and standard deviation, at slow (○), medium (×) and fast (●) speed settings.

By way of comparison, liquidisers are shown in Fig. 2. We see the similarity for the fast and medium speeds between the spectra and noise levels of both food mixers and food mixers with liquidiser attachments. For the slow speed setting the spectra changed markedly above about 2 kHz, the food mixer with liquidiser attachment being about 8 dB quieter than when operating in the food mixing mode. We must, however, differentiate between food mixers with liquidiser attachments and purpose-built liquidisers. Three purpose-built liquidisers were tested. Average third octave band levels were calculated and shown in Fig. 2 with the spreads of level. It will be seen that the spread in levels produced by the food mixers with liquidiser attachments (also three models tested) are about twice those of the purpose-built models. Between 250 Hz and 12.5 kHz purpose-built models were, on average, 12 dB noisier than the food mixer attachments. Clearly there is scope for quietening these purpose-built liquidisers, as a level of about 89 dB(A) at a distance of 1 m is excessively noisy. Even the more sturdily constructed food mixer attachments are too noisy, with levels of about 78 dB(A).

All of the nine whistling kettles studied had characteristic double peaks in their third octave band noise spectra, as measured at a distance of 1.5 m. These peaks most often occurred in the 2.5-kHz and 5-kHz bands, although the highest peak measured was in the 6.3-kHz band, as shown in Fig. 3. This diagram shows the average level, standard deviation and spread of levels in third octave bands. It will be noticed that occasionally the lowest level in a band is more than the standard deviation away from the average; this is due to the skewed distribution of levels in these bands. Whereas it is desirable for whistling kettles to be heard, it is debatable whether or not they need to be as loud as 94 dB(A). The average level of about 80 dB(A) should be high enough to make them audible above most other household appliances, especially with their characteristic piercing noise. Whistling kettles are only likely to become a nuisance to neighbours if left on for any length of time or used at unsociably late hours of the night. By comparison, 10 electric kettles were measured while boiling; their average level was 50 dB(A) [compared with kettles on gas rings having an average level of 47 dB(A)]. Electric kettles with steam-sensing devices to turn them off automatically have great advantages, acoustically, over the whistling type.

The results of measurements on the quieter appliances in this group of kitchen aids are shown in Fig. 4. Five different models of electric toaster were analysed when they "popped up". These impulsive sounds were surprisingly noisy. At a measurement distance of 1.5 m, the average decibels(A) level was 78. Due to the frequency of operation and short duration of the noise, it is unlikely that these appliances will cause dis-

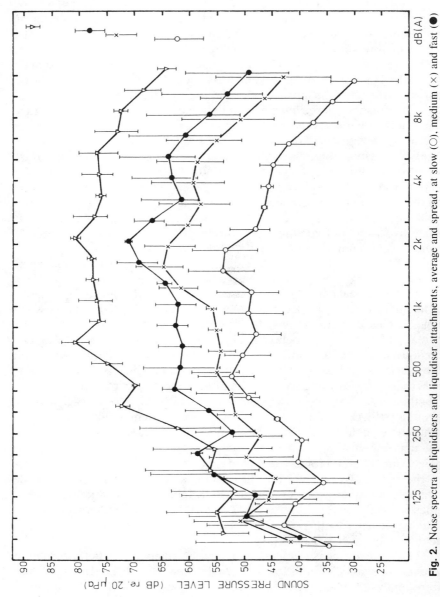

Fig. 2. Noise spectra of liquidisers and liquidiser attachments, average and spread, at slow (○), medium (×) and fast (●) speed settings on attachments; ▽, single-speed liquidisers.

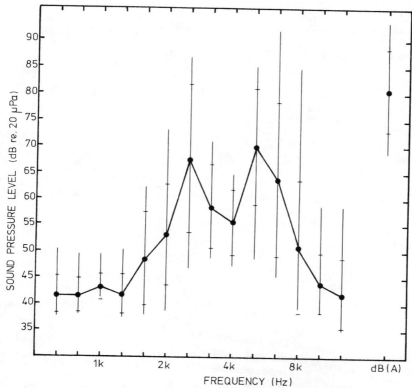

Fig. 3. Noise spectra of whistling kettles, average, standard deviation and spread.

tress to anyone. However, one of the toasters studied had a warning sound indicating that the toast was stuck, a particularly unpleasant rasping noise of 70 dB(A), which would certainly be annoying if allowed to continue for any length of time.

Comparing the purpose-built liquidisers of Fig. 2 with the coffee percolators and gas cookers shown in Fig. 4 shows the extremes in noise level associated with this group of kitchen appliances. Average third octave band levels and standard deviations of these levels are shown, measured at a distance of 1.5 m from 10 gas cookers (with two burners alight) and 10 coffee percolators. On average, the "bumping" of the percolators is 8 dB greater than the gas cookers. Neither of these appliances is likely to cause any noise annoyance and is only included for the sake of comparison. If a liquidiser or food mixer manufacturer managed to produce an appliance

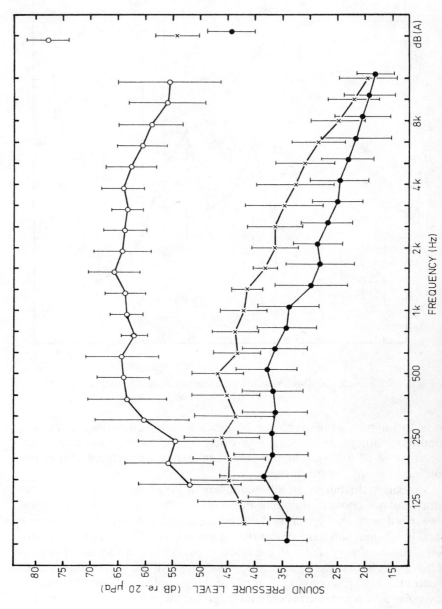

Fig. 4. Noise spectra, average and standard deviation, of electric toasters (○), coffee percolators (×) and gas cookers (●).

as quiet as a gas stove, it would be a meritorious achievement, but unfortunately one that is probably technically impossible.

2. Cleaning Devices

Figures 5, 6 and 7 refer to this group of appliances. Nine washing machines were studied, each having approximately the same washing load. Figure 5 shows the third octave band average noise levels and standard deviations of these levels for both the washing and spinning modes. Noise levels were measured at a distance of 1.5 m from the appliance at operator ear height. Most of the appliances had fairly broadband spectra, with levels decreasing at a rate of about 5.5 dB per octave, above about 500 Hz.

A number of different washing actions were studied: an oscillating four-finned pillar in the base of the tub, a rotating-drum tumbling action and a many-bladed rotating disc in the side of the tub. A number of models employing each of these water-agitating methods were studied. No particular washing action was consistently quieter than the others, although on average, the rotating drum tumbling action was about 7 dB(A) quieter than the others.

All the washing machines tested also had a drying mode. In the case of the tumble action, the same horizontally mounted rotating drum was used for both washing and drying, whilst for the other types of machine a second tub was provided, rotating about its vertical axis. There was, on average, no difference in noise level between these two spinning actions.

Spin dryers are compared in Fig. 6 with a single example of a hot-air tumble drier, a domestic version of those found in many commercial launderettes. Only three models of spin drier were available for test, so that only the averages and spreads in noise levels at 1.5-m distance are shown. Above 500 Hz the hot-air tumble drier was about 7 dB quieter than the average of the spin driers, with an overall level 10 dB(A) lower than the average of the spin driers. Comparing Fig. 5 with Fig. 6, we see a marked similarity between the spectrum shapes and levels of the spin driers and drying modes of the washing machines.

Figure 7 shows results of measurements conducted on a single example of a sink grinder waste disposal unit and four makes of dishwasher. Below 250 Hz peaks are evident in both spectra, while above 500 Hz the noise level from the dishwashers, at a measurement distance of 1.5 m, falls off approximately 6.5 dB per octave, approximately twice the rate characteristic of the waste disposal unit. Other peaks are present at higher frequencies.

Due to the relatively long operating cycle of the dishwasher, it is likely to cause greater annoyance than the waste disposal unit.

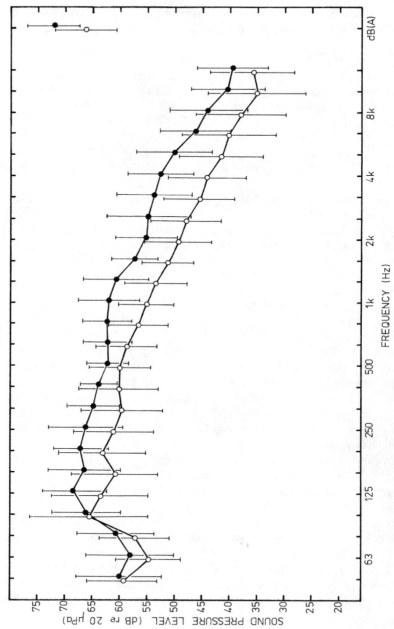

Fig. 5. Noise spectra of washing machines, average and standard deviation, for washing (○) and spinning (●).

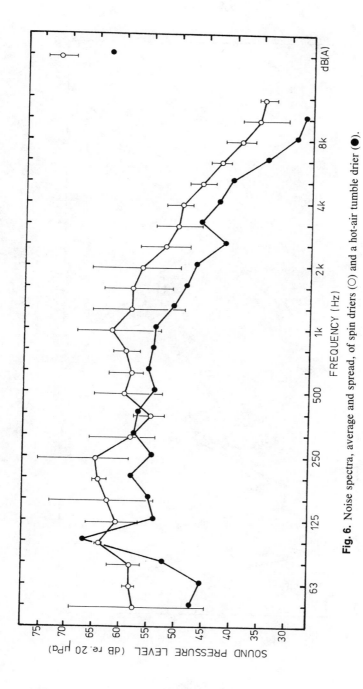

Fig. 6. Noise spectra, average and spread, of spin driers (○) and a hot-air tumble drier (●).

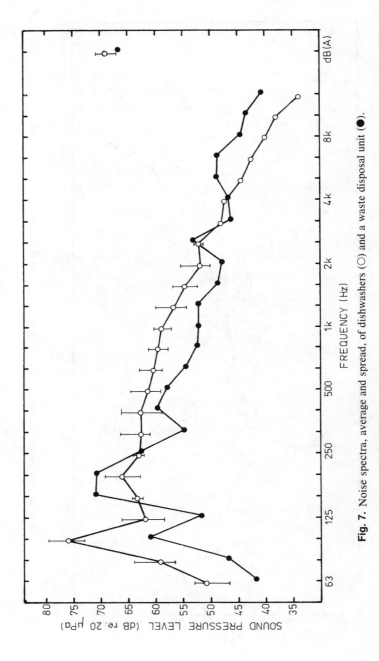

Fig. 7. Noise spectra, average and spread, of dishwashers (○) and a waste disposal unit (●).

Kitchen cleaning devices, with decibels(A) levels ranging from the low 60s to mid-70s are generally quieter than some of the food preparing appliances discussed previously; they are, however, likely to be used for longer periods of time. As many of these appliances are used in the evenings and on weekends, they are likely to cause greater nuisance to neighbours, especially when used in high-density housing units, than are the generally noisier but shorter duration food preparing devices.

3. Miscellaneous Devices

The appliances in this group could equally well be used in domestic rooms other than the kitchen. For instance, the extractor fans (Fig. 8) could also be used in a bathroom or toilet, as could the wall-mounted gas-fired water heater (Fig. 9). Also shown in Fig. 9 are octave band noise levels for gas-fired central heating boilers. These are generally installed in a kitchen, but may be housed in a separate boiler room.

Three extractor fans were measured *in situ* in the windows of kitchens. Third octave band measurements were taken at a distance 1.5 m away at average ear height. Apart from some low-frequency noise peaks, the spectra were fairly broadband, falling at about 7.5 dB per octave above about 500 Hz. With average noise levels of about 58 dB(A) these appliances are unlikely to cause any real noise problem, especially if they are only used for relatively short periods of time. As they communicate with the outside of the house, they may be a source of annoyance to neighbours if the background noise level in the area is extremely low.

Figure 9 shows third octave band averages and spreads in noise level of three Ascot-type gas water heaters measured at 1.5 m. Averages, spreads and standard deviations of octave band noise levels of six gas-fired central heating boilers are also shown. These results include three boilers measured in the simulated kitchen (Jackson 1972), the other three being measured *in situ* at the same distance of 1.5 m.

The unpleasant "roar" or "rumble" of the Ascot-type water heaters is apparent in the 250-Hz region due to the open construction of the flame port and heat exchanger. Above about 500 Hz the noise level decreases at a rate of approximately 4 dB per octave. This rumbling is not so evident with the central heating boilers, except on light up.

Fortunately, the water heaters are generally used for short, intermittent spells so that with decibels(A) noise levels in the mid-60s they are unlikely to cause annoyance to the user. As a class of gas appliance, they are, however, the noisiest, being about 20 dB(A) noisier than gas cookers and central heating boilers and, as will be shown in the next subsection, 30 dB(A) noisier than gas fires.

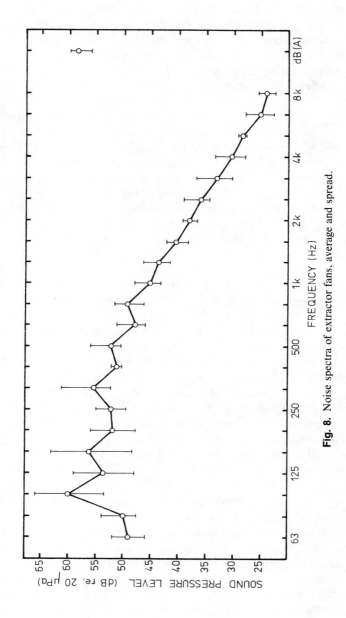

Fig. 8. Noise spectra of extractor fans, average and spread.

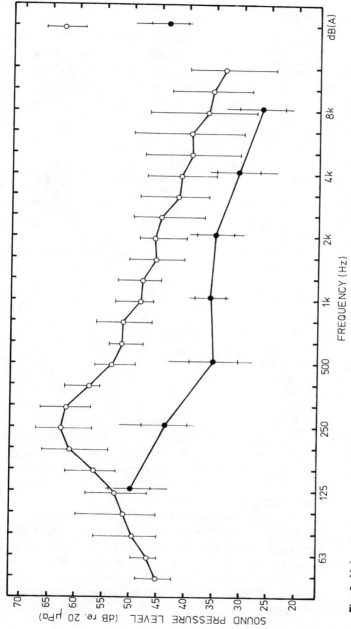

Fig. 9. Noise spectra, average and spread, of wall-mounted gas water heaters (○) and gas central heating boilers (with standard deviation) (●).

B. Living Room/Bedroom Appliances

No attempt has been made to document noise levels associated with household appliances that can be controlled by the user. Television, radio and hi-fi are three such examples. These entertainment aids are generally in operation for longer periods than the labour-saving devices considered in this chapter. This aspect of noise at home is more appropriately the subject of legislation and building regulations on insulation between dwellings. It is sufficient here to note that the noise from neighbours usually figures prominently in social surveys designed to investigate residents' attitudes toward the noise environment of their homes.

A number of appliances have been grouped together under this section, although some, for example, alarm clocks, are seldom used in a living room. As bedrooms and living rooms have similar sound-absorbing qualities, the appliance noise levels reported here are equally applicable to either environment.

Three types of alarm clock were investigated (Fig. 10), 10 ordinary clockwork alarms, 10 travel alarms and 5 electric alarms, each measured at a distance of 1 m. They were all quite noisy, as desired, varying between about 60 and 90 dB(A). Figure 10 shows the average and standard deviations of third octave band levels for each type of clock. Electric alarms relied upon a relatively low-noise-level "drone" or "buzz", whereas the other types used the more impulsive clanging of a bell, as awakening devices. Comparing the travel alarms and ordinary clockwork alarms, we see the effect of the smaller bell in the 400-Hz to 1-kHz region, although at all frequencies the standard deviations overlap. The average third octave band levels of the electric alarms drop off at a rate of about 4 dB per octave above 1 kHz, although individual alarms have peaks, as indicated by the standard deviations. If not switched off, most of these alarms continue buzzing as long as electricity is available. Ringing times for the travel alarms varied between 7 and 40 sec, with an average of 17 sec, compared to an average ringing time of 23 sec, with extremes of 13 and 47 for the ordinary clockwork alarms.

Vacuum cleaners may well be used in different domestic rooms, but are generally employed to clean carpeted areas, for which these measurements apply. Results of octave band measurements of 10 vacuum cleaners are shown in Fig. 11. Seven of these appliances were studied in Jeatt (1972) and measured in the simulated living room (Jackson and Leventhall, 1973). The other three models were measured in actual living rooms. Several of the cleaners had peaks in the 250-Hz region, as shown by the spread and standard deviation for this frequency band.

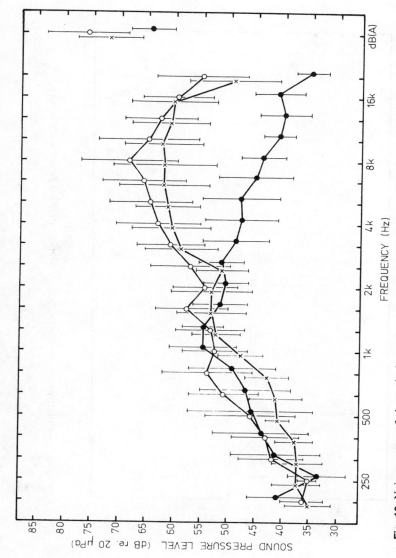

Fig. 10. Noise spectra of alarm clocks, average and standard deviation, of clockwork (○), travel alarms (×) and electric clocks (●).

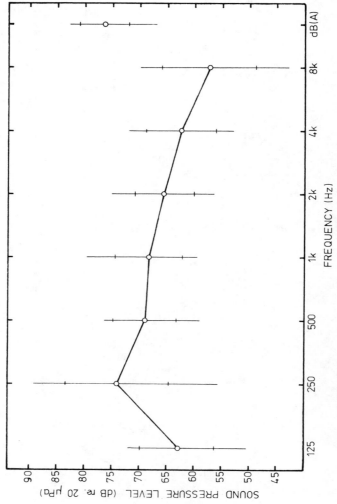

Fig. 11. Noise spectra of vacuum cleaners, average, standard deviation and spread.

The average level at 76 dB(A) at 1.5 m is excessively noisy, although design improvements have been made to reduce the noise of some models now coming onto the market. Greater efforts by manufacturers are certainly required in order to quieten this class of much-used appliances.

Nine different models of electric fan heater were studied, measurements being taken to the front of the appliance, 1.5 m along a line inclined at an angle of approximately 35° to the horizontal. Results of the third octave band analysis are shown in Fig. 12 and include the average levels and standard deviations for different operational modes. Six models had fan-only (i.e., no heat) settings, all nine had 1-kW and 2-kW settings and only two had 3-kW heat settings, so that in this latter mode the standard deviations are the same as the spreads in levels. Some of the models had two-fan speeds; the fastest was used for these measurements.

As can be seen from Fig. 12, the greatest change in noise level when changing from one heat setting to another occurs in the 125- to 250-Hz region. Going from the fan-only to 1-kW heat setting produced, on average, a 5-dB increase in this frequency region, whilst the heat setting increase from 1 kW to 2 kW produced only a 2.5-dB increase. Above about 800 Hz there was generally less than a 2-dB difference between the fan-only and 2-kW settings, the noise levels decreasing at a rate of approximately 7 dB per octave. Even though the 3-kW average noise levels were the result of only two measurements, and hence are none too reliable, it would appear from the standard deviations that this heat setting produces a disproportionate increase in noise in the 1- to 4-kHz frequency bands. These phenomena have been, for air-conditioning equipment, related to the fan design and aerodynamic load (Allen and Potter, 1962).

With noise levels in the region of 40 to 50 dB(A) it is unlikely that electric fan heaters will present any great noise problem. It should be noted that as the fan heater gets older, the fan bearings wear and often produce an irritating rattle.

Unlike electric fan heaters, gas fires are traditionally almost inaudible. This was certainly true of gas fires operated on town's gas, but the advent of natural gas and associated aeration problems in the combustion process introduced additional noise sources. These have been and are being investigated by the gas appliance manufacturers, and certain design modifications are being introduced in an attempt to reduce noise to before-conversion levels. Their varying degrees of success may be seen from the data in Table I, showing the effect of multi-hole instead of single-hole injectors and the utilisation of sound-absorbing shields around the burner heads (Jackson, 1972).

Figure 13 shows averages, standard deviations and spreads in octave band noise levels measured 1.5 m to the front of 10 gas fires. All these

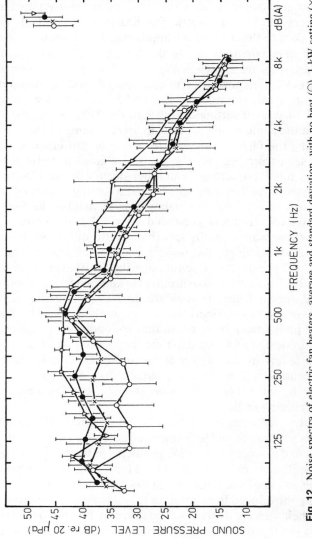

Fig. 12. Noise spectra of electric fan heaters, average and standard deviation, with no heat (○), 1-kW setting (×), 2-kW setting (●) and 3-kW setting (▽).

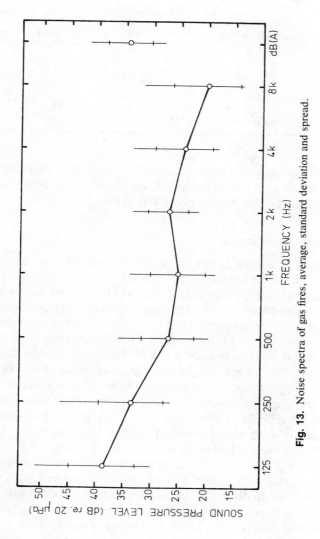

Fig. 13. Noise spectra of gas fires, average, standard deviation and spread.

257

TABLE I. Effect of Design Modifications on Gas Fire Noise

Octave band frequency (Hz)	SPL reduction (dB)		
	Model 1	Model 2	Model 3
250	2.1	1.8	−1.4
500	1.9	−2.8	1.1
1000	4.2	−2.9	5.8
2000	9.0	2.8	8.6
4000	11.8	1.1	14.3
8000	13.3	−1.6	15.5

models were, compared to other commonly used appliances, extremely quiet, with decibels(A) levels ranging from 28 to 42, with an average of 34 [about 14 dB(A) quieter than electric fan heaters].

C. Bathroom Appliances

Appliances considered in this subsection are electric hair driers, razors and toothbrushes and toilets.

Ten hand-held electric hair driers were studied whilst blowing hot air. Nine of them also blew cold air. Averages and standard deviations of third octave band analyses, conducted at a distance of 10 cm from the appliances, are shown in Fig. 14, from which it will be seen that over the whole frequency range there was only a 1.5-dB increase in noise level when changing from cold to hot air. With an average level of 70 dB(A) these appliances are quite noisy and make it difficult, if not impossible, for people to carry on a conversation whilst drying their hair. Due to the relatively short time these appliances are in operation, their bearings are not as prone to wear as are those of electric fan heaters, so that their effective "quiet" life is longer than that of the heaters.

Comparing Fig. 14 with Fig. 12 for electric fan heaters shows that by increasing the frequency axis of Fig. 12 by one octave and also increasing the noise levels by 20 dB, we arrive at spectra quite similar to those of the hair driers. This is as might be expected, considering the similarity of the noise-producing mechanisms in the two types of appliances.

In Fig. 15 we have shown noise levels of 10 electric razors measured in the simulated kitchen at a distance of 7.5 cm (3 in.). At this distance the direct field of the appliance is predominant, so that the noise levels are independent of the measuring environment and hence reverberant noise

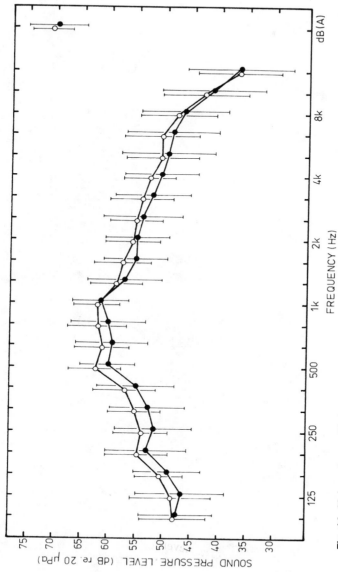

Fig. 14. Noise spectra of hair driers, average and standard deviation, with hot setting (○) and cold setting (●).

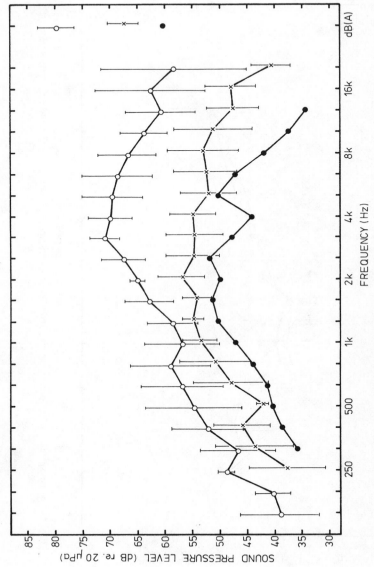

Fig. 15. Noise spectra, average and standard deviation, of rotary razors (○), shuttle razors (×) and battery toothbrush (●).

level. Six of the models tested had rotary heads, whilst the others were of the shuttle type.

Average levels of 80 dB(A) for the rotary razors were about 12 dB(A) above the shuttle type. Even the quietest of the rotary models tested was 3.5 dB(A) noisier than the noisiest shuttle type. Over the frequency range 1 to 8 kHz, the shuttle-type razors had much flatter spectra than the rotary type, which had a broad noise peak in this region.

Also included in Fig. 15 is a single example of a battery-operated toothbrush, also measured at a distance of 7.5 cm. Compared to the razors, it was relatively quiet, with a noise level of about 60 dB(A).

We have only measured the airborne noise of these appliances, but in use they could well appear noisier due to sound and vibration transmission via bone conduction. Manufacturers of rotary-type razors, in particular, should try to reduce the noise levels of their products by at least 10 dB(A).

Results of third octave band measurements of water filling and draining from 10 baths are shown in Fig. 16. Average levels, measured at 1.5 m, and standard deviations of these levels are included. It will be seen that above 250 Hz the baths filling with water are, on average, 9.5 dB noisier than when emptying, although the upper range of noise levels is approximately the same for both situations. The large variations between different baths was due to the loud gurglings produced by some baths when emptying, others being extremely quiet.

Figure 17 shows corresponding measurements conducted on 10 wash basins. Above about 500 Hz the noise levels show much the same variations with frequency as did the baths. The wash basins tended to be noisier when emptying than when filling, the reverse of the situation with baths.

Two types of flush toilet were investigated, those with high-level and low-level cisterns. Ten examples of each of these types were measured; the third octave band average noise level and standard deviations of these levels are shown in Fig. 18. Seven of the low-level and eight of the high-level toilets were in separate water closets, the remainder being in bathrooms. Referring to Fig. 18, we see that below 200 Hz the high-level cisterns produced average noise levels that were 12 dB higher than the low-level cistern noise levels. Above 200 Hz this difference was reduced to an average of 5 dB. This effect is attributed to the water coming down the pipe, often about 2 m long, and rushing into the bowl. The reduction in the intensity of the low-frequency components of noise is particularly important for high-density housing, as these frequencies are generally not attenuated too efficiently by the building structure. As a result of these operational design changes, total noise levels have been reduced by about

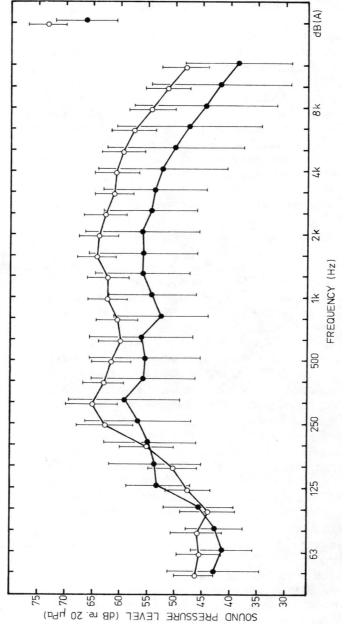

Fig. 16. Noise spectra of bath water, average and standard deviation, when filling (○) and emptying (●).

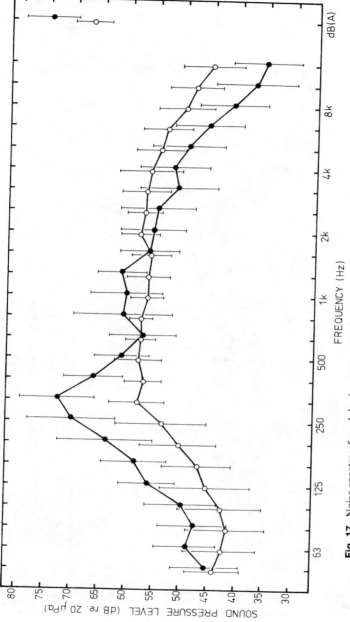

Fig. 17. Noise spectra of wash basins, average and standard deviation, when filling (○) and emptying (●).

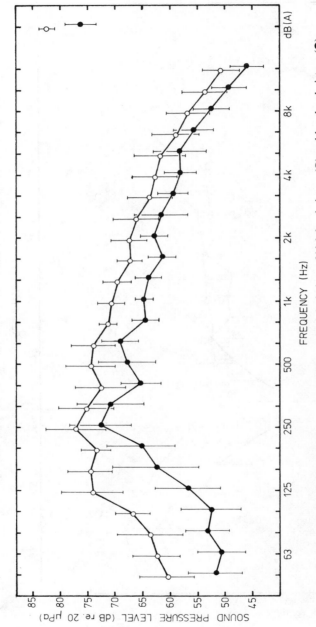

Fig. 18. Noise spectra of flush toilets, average and standard deviation, of high-level cistern (○) and low-level cistern (●).

6 dB(A), from 82 dB(A) for the high-level cisterns down to 76 dB(A) for the low-level cisterns. Even so, decibels(A) noise levels in the mid-70s may still be quite disturbing, especially if the toilet is used late at night.

Noise levels shown in Figs. 16 through 18 were measured in the noise source room. Particular care needs to be exercised in locating water and waste pipes in relation to bedrooms in multi-occupancy dwellings, as the noise of rushing water may be particularly disturbing at night.

D. Miscellaneous Domestic Appliances

This subsection contains results of tests on a number of appliances that do not easily fit into the other categories. Door bells and telephones ringing are shown in Fig. 19, which gives the third octave band averages and standard deviations for four examples of each of these devices. As would be expected for door bells, the standard deviations of the noise levels measured at a 2-m distance are quite high, the maximum being 24.1 dB in the 12.5-kHz band. All these bells were measured in the halls of houses and can be taken as only mildly indicative of expected noise levels, although generally there were peaks in the spectra between 1.25 and 6.3 kHz.

The noise levels of telephones were also measured at a distance of 2 m, two models being in the kitchen, the others in the hall. The variations in noise levels and spectral shapes were much lower than with door bells; the average standard deviation over the frequency range 125 Hz to 20 kHz was 3.5 dB, compared to 12.5 dB for the door bells, over the frequency range 315 Hz to 12.5 kHz. All the telephone spectra measured showed a fairly broad peak in the 315- to 800-Hz region and additional noise peaks in the 1.6-, 4- and 8-kHz third octave bands.

Despite the relatively high noise levels of these devices [about 77 dB(A) at 2 m], they are, due to familiarity, unlikely to cause particular annoyance to neighbours, except when allowed to ring for long periods late at night.

Figure 20 shows third octave band average levels, standard deviations and total spreads in noise levels for nine different models of electric sewing machine, measured at the operator position 0.5 m away. Compared to several other appliances, the spectra were relatively flat and broadband. Decibels(A) levels of about 74 were perhaps a little excessive, but all the measurements were taken with the machines running at maximum speed.

Two examples of hand-operated machines were also measured, with the operator turning the handle as fast as she could. The speed attained was approximately half that of the electric models; the resulting noise

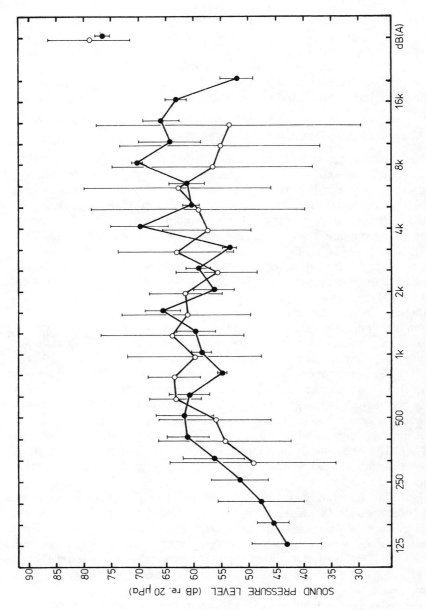

Fig. 19. Noise spectra, average and standard deviation, of door bells (○) and telephones (●).

266

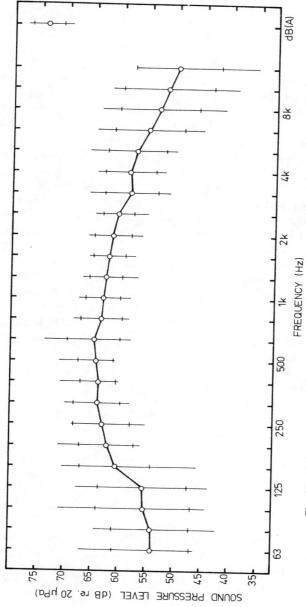

Fig. 20. Noise spectra of electric sewing machines, average, standard deviation and spread.

267

produced had approximately the same spectrum shape as the electric models, but was about 11 dB(A) lower. Due to the excessive expenditure of human energy, a return to the use of hand-operated models is not advocated.

Six electric drills were studied. They all had fast speeds of about 2500 rpm, three also had slow speeds of 900 rpm. Figure 21 shows averages and standard deviations of measurements taken at 0.5 m from the drills working under no-load conditions, and thus allowing direct comparison between models. In practical use the noise level will depend upon the material being drilled and the pressure being applied to the drill bit. The influence of attachments such as sanders, jigsaws, circular saws and hedge trimmers was not studied. With all these attachments the primary noise source is still likely to be the drill motor and gearing, which, from Fig. 21, are seen to produce a level of about 90 dB(A). Over the whole frequency range the average difference in noise levels between the two speed settings was only 1.6 dB, neither speed being consistently noisier than the other. This is reflected in the decibel(A) levels, the averages of which were only 0.1 dB(A) different for the different speeds.

Continuously variable speed drills were not tested; neither were hammer drills, although the Consumers' Association (1973) report levels of about 90 dB(A) at 0.5 m for these drills without the hammer action working. The report describes these drills as "uncomfortably loud". Clearly, all the electric drills tested were objectionably noisy, and while it is admitted that little can be done by way of drill design to reduce vibration transmission when drilling into walls (the hammer drills can only aggravate this problem), it is felt that airborne noise levels must be reduced from the 90 dB(A) at present experienced by the operator. A noise reduction of, say, 10 dB(A) would be a considerable improvement. Another, less desirable solution is that ear defenders should be supplied with each drill purchased.

Figure 22 shows the average and spread of third octave band measurements conducted on two hand mowers. A single electrically operated rotating-blade mower was also studied and shown in the figure. All measurements were taken at a distance of about 2 m from the machine, during a walk past the measuring microphone.

As an outside noise source, lawn mowers are quiet compared to cars or lorries measured at the same distance. Despite the small difference in decibels(A) noise levels between the two types of mowers tested, the general consensus was that the electrically powered machines were more disturbing to neighbours than were the hand-operated ones. The same feeling was shown towards petrol engine mowers, although these were not available for measurement. Let us hope that, for the average-sized

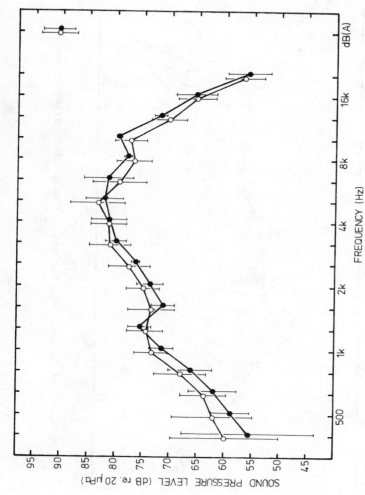

Fig. 21. Noise spectra of electric drills, average and standard deviation, at fast speed (○) and slow speed (●).

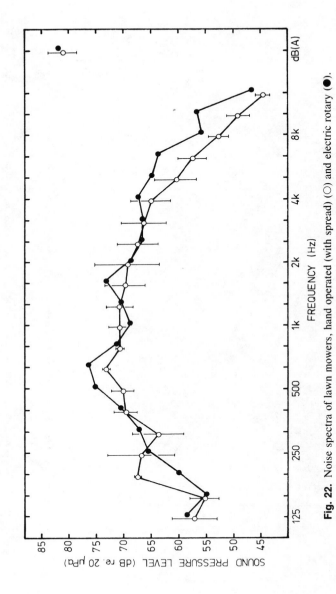

Fig. 22. Noise spectra of lawn mowers, hand operated (with spread) (○) and electric rotary (●).

TABLE II. Decibels(A) Noise Levels Corresponding to Different Degrees of Annoyance

Scale score	Description	Decibels(A) level
1	Not annoying	55
2	Slightly annoying	64
3	Moderately annoying	74
4	Quite annoying	83
5	Extremely annoying	93

garden, people will continue to consider mowing the lawns by hand as a useful form of exercise.

IV. SUBJECTIVE EFFECTS

The degree of annoyance caused by an appliance will depend, in part, upon the benefits derived from its use. Epp and Konz (1975) have shown that for some domestic appliances the degree of annoyance may be related to the decibels(A) noise level by an equation of the form

$$S = 0.1058X - 4.798,$$

where X is the decibels(A) noise level and S the score on a five-point scale defined by various descriptors. The corresponding noise levels are shown in Table II.

On this basis it might be concluded that appliances are generally considered to create no annoyance at noise levels of less than about 60 dB(A). Between 60 and 70 dB(A) slight annoyance will be produced; between 70 and 80 dB(A) they will be moderately annoying; between 80 and 90 dB(A) quite annoying; and over 90 dB(A) extremely annoying. These broad generalisations clearly depend upon the circumstances under which the noise is heard, particularly with respect to time of day, and exclude those appliances such as whistling kettles, door bells, telephones and alarm clocks which depend upon an audible warning signal to fulfill their functions.

V. CONCLUSIONS

Table III summarises the typical noise levels of appliances commonly found in and around the home. It will be seen that while some appliances

TABLE III. Typical Noise Levels of Domestic Appliances

Appliance	Typical decibels(A) level	Appliance	Typical decibels(A) level
Kitchen		*Bathroom*	
Food mixers		Hair driers	
Slow speed	66	Cold	70
Medium speed	72	Hot	71
Fast speed	77	Battery toothbrush	60
Liquidisers		Electric razors	
Purpose-built models		Rotary	80
Food mixer attachments	89	Shuttle	68
Slow speed	62	Bath water	
Medium speed	73	Filling	73
Fast speed	78	Emptying	66
Whistling kettles	81	Hand basin	
Toasters popping up	78	Filling	67
Coffee percolators	54	Emptying	75
Gas cookers	44	Flush toilets	
Washing machines		High-level cistern	82
Washing	66	Low-level cistern	76
Spinning	72		
Spin driers	72	*Miscellaneous*	
Hot-air tumble driers	63		
Dishwashers	69	Door bells	79
Waste disposal units	67	Telephones	77
Extractor fans	59	Electric sewing machines	74
Gas water heaters	63	Electric drills	
Gas central heating boilers	44	Fast speed	91
		Slow speed	91
Living room/bedroom		Lawn mowers	
		Hand operated	81
Alarm clocks		Electric rotary	82
Clockwork	76		
Travel alarms	72		
Electric	64		
Vacuum cleaners	77		
Electric fan heaters			
Fan only	46		
1 kW	46		
2 kW	47		
3 kW	49		
Gas fires (full on)	34		

are in operation the kitchen is more characteristic of an industrial environment than of the home. Fortunately, the noisiest of these appliances, such
as food mixers, are in use for relatively short time periods. However,
automatic washing machines have longer operating cycles and may be
used at unsociable hours. Airborne, coupled with structure-borne noise
from washing machines and vacuum cleaners could give rise to annoyance levels in excess of those indicated in Table II, especially if the
recipient is a neighbour in a multi-occupant building. Cleaning equipment
(vacuum cleaners, washing machines, spin driers and dishwashers) generally seem to be the one group of appliances whose noise levels could and
should be reduced by the manufacturers.

APPENDIX. EFFECT OF ROOM ACOUSTICS ON NOISE LEVELS PRODUCED BY DOMESTIC APPLIANCES

   The noise level produced in a room by a sound source is dependent on
the power output of the appliance, the directivity factor and the distance
from the source. These factors can be taken into account as follows.
   The mean-square pressure in the direct field $\overline{P_d^2}$ is directly proportional
to the directivity factor Q and power output W of the appliance and
inversely proportional to the square of the distance r from the appliance

$$\overline{P_d^2} = QW\rho c/4\pi r^2, \tag{1}$$

where ρ_c is the characteristic impedance of air at room temperature. The
corresponding expression for the mean-square pressure in the reverberant
sound field $\overline{P_r^2}$ is

$$\overline{P_r^2} = 4W\rho c/R \tag{2}$$

and is independent of the distance from the source, being dependent upon
only the power output of the source and the room constant R of the room.
Here R is defined as

$$R = S\bar{\alpha}/(1 - \bar{\alpha})$$

where S is the area of sound-absorbing material in the room and $\bar{\alpha}$ the
average absorption coefficient of these surfaces.
   Hence the total mean-square pressure $\overline{P^2}$ at a point in the room is the
sum of the two components:

$$\overline{P_d^2} + \overline{P_r^2} = \overline{P^2} = W\rho c\left[\frac{Q}{4\pi r^2} + \frac{4}{R} \right]. \tag{3}$$

We are now in a position to answer the question posed earlier in this section, as we can use this equation to derive the difference in sound levels produced by an appliance installed in domestic rooms having different room constants. We shall denote the two rooms by subscripts 1 and 2. The sound level LP_2 in room 2 may be related to that in room 1 by the equation

$$LP_2 = LP_1 + 10 \log_{10} \left[\frac{R_1}{R_2} \right] + 10 \log_{10} \left[\frac{(QR_2/16\pi r_2^2) + 1}{(QR_1/16\pi r_1^2) + 1} \right] \quad \text{dB.} \quad (4)$$

This equation assumes that the power output and directivity factor of the appliance are the same when placed in either room.

To aid our understanding of Eqs. (3) and (4), we may consider the sound field in a room as being analogous to the sea and seashore. Our noise source is on the beach, just above the high-water mark. The sea is relatively flat and corresponds to the reverberant sound field in the room. The beach rises out of the water (the direct sound field) as we come closer to the noise source. Progressively higher tides correspond to the more reverberant rooms with less beach (or direct sound) visible, the water's edge (or dividing line between direct and reverberant sound fields) also progressing up the beach (closer to the noise source). An acoustically average room corresponds to an average tide, so we may conveniently use the distance between noise source and dividing line between direct and reverberant fields as a characteristic distance. In rooms having higher than average room constants, the direct field predominates over the reverberant field, while for lower than average room constants, the reverberant dominates the direct at the same physical distance. The distance r_R at which the direct and reverberant sound levels are equal is known as the 'reverberation radius' and may be evaluated by equating Eqs. (1) and (2), i.e.,

$$r_R = \sqrt{QR/16\pi}.$$

If, in Eq. (4) we let $R_1 = R$ represent the average room, $R_2 = R \pm \Delta R$ represent a non-average room and $r_1 = r_2 = r$, we have, after manipulation,

$$LP_2 = LP_1 - 10 \log_{10} \left[1 \pm \frac{\Delta R}{R} \right]$$

$$+ 10 \log_{10} \left[1 \pm \frac{\Delta R}{R + (16\pi r^2/Q)} \right] \quad \text{dB.} \quad (5)$$

This equation gives the sound level LP_2 in a non-average room in terms of the level LP_1 at the same distance r in an average room. Setting this

TABLE IV. Room Constants and Reverberation Radii of Living Rooms

Octave band	Room constant (m²)		Reverberation radius (m)			
	Average	Standard deviation	$Q = 0.5$	$Q = 2$	$Q = 5$	$Q = 10$
125	11.2	2.4	0.33	0.67	1.06	1.49
250	12.2	2.2	0.35	0.70	1.10	1.56
500	13.8	2.6	0.37	0.74	1.17	1.66
1000	15.2	3.4	0.39	0.78	1.23	1.74
2000	16.7	3.4	0.41	0.82	1.29	1.82
4000	18.1	4.2	0.42	0.85	1.34	1.90
8000	20.1	4.4	0.45	0.89	1.41	2.00

distance equal to the reverberation radius of the average room leads to

$$LP_2 - LP_1 = 10 \log_{10} \left[\frac{1 \pm (\Delta R/2R)}{1 \pm (\Delta R/R)} \right] \quad \text{dB}, \qquad (6)$$

this being the difference in sound level between the non-average and average rooms at a distance equal to the reverberation radius of the average room.

If only the difference in the reverberant levels of two rooms is required, this can be obtained from Eq. (5) with the omission of the term containing the directivity factor; that is,

$$LP_2 - LP_1 = -10 \log_{10} \left[1 \pm \frac{\Delta R}{R} \right] \quad \text{dB}. \qquad (7)$$

In order to apply Eqs. (6) and (7) to our problem of how different rooms affect the noise produced by a particular appliance, we need to know the room constants of typical domestic rooms. For this reason a study of the acoustics of 50 living rooms and 50 kitchens was conducted (Jackson and Leventhall, 1972). From these measurements the room constants and standard deviations of room constants were calculated for different frequencies. These data are presented in Table IV for living rooms and Table V for kitchens.

TABLE V. Room Constants of Kitchens

Room constant (m²)	Octave band mid-frequency (Hz)						
	125	250	500	1000	2000	4000	8000
Average	5.0	5.1	5.4	5.5	5.8	6.0	6.3
Standard deviation	1.8	1.9	1.9	2.0	2.1	2.1	2.2

Table IV also contains values of the reverberation radius for different values of directivity factor and room constant (note that the room constant depends upon the average absorption coefficient and so on frequency).

From Table V we see that the room constant of kitchens is relatively independent of frequency. Taking an average value of 5.6 m² gives reverberation radii of 0.24 m (for $Q = 0.5$), 0.47 m (for $Q = 2$), 0.75 m (for $Q = 5$) and 1.06 m (for $Q = 10$). These results indicate that for both living rooms and kitchens the reverberation radii associated with all but the most directional noise sources are in the range of 0.5 to 1.5 m.

Having determined the average acoustic properties of domestic rooms, we may use the data from Jackson and Leventhall (1972) to calculate the percentage of rooms having a sound level range within a given decibel range of standard rooms. This subject is dealt with in greater detail by Jackson (1972), who shows that the lower values of living-room room constants (noisier than average) were contained within the range $\Delta R = -0.33R$.

Equation (6) gives 1 dB and Eq. (7) 1.7 dB for the differences in level between the average and noisiest living rooms at a distance equal to the reverberation radius of the average room and in the reverberant field, respectively. These results show that up to about 2 m from the noise source, the noise level in any living room is unlikely to be more than 1 dB greater than in the average room, whilst the differences in reverberant field levels (further away than about 2 m) will almost invariably be less than 2 dB. In many rooms, for which ΔR is positive, an appliance will produce less noise than in the average room.

Comparison of Tables IV and V also shows that the standard deviation of the room constants as a percentage of the average values is greater for kitchens than for living rooms. Setting $\Delta R = -0.33R$ shows that about 90% of kitchens will be within 1 dB of the average kitchen. Virtually all kitchens were contained within the range $\Delta R = -0.5R$, leading to sound level differences of 1.8 dB at the reverberation radius of the average kitchen and 3 dB in the reverberant fields. Additional experiments conducted in a further 50 living rooms and 50 kitchens (Jackson and Leventhall, 1973) by using a standard noise source showed that the noisiest living rooms and kitchens were about 4 dB above the average, in contradiction to the 1.7 dB for living rooms and 3 dB for kitchens predicted earlier. These discrepancies were shown to be due to different mounting conditions of the noise source. When placed upon a partially absorbing floor, the power output of the noise source may be reduced, with a resultant decrease in noise level. These results show the importance of the first reflected sound, noted earlier, and illustrate the importance of considering both the appliance and its mounting as the total noise source (Allen and Potter, 1962).

REFERENCES

Allen, C. H., and Potter, A. C. (1962). Design and application of a semi-anechoic sound test chamber, *Sound—Its Uses Control* **1**(1), 34–39.

Bender, E. K. (1973). Noise source impact in: construction/buildings/homes, *Sound Vib.* **7**(5), 33–41.

Consumers' Association (1973). Handyman *Which?* May. 75–79.

Epp, S., and Konz, S. (1975). Home appliance noise: Annoyance and speech interference, *Home Econ. Res. J.* **3**, 205–209.

Jackson, G. M. (1972). A Proposed Method for Assessing the Noise of Domestic Appliances and Its Comparison with Existing Procedures. Ph.D. thesis, Univ. of London.

Jackson, G. M., and Jeatt, K. E. (1974). Domestic appliance noise, *Proc. Int. Congr. Acous., 8th, London.*

Jackson, G. M., and Leventhall, H. G. (1972). The acoustics of domestic rooms, *Appl. Acous.* **5**, 265–277.

Jackson, G. M., and Leventhall, H. G. (1973). A proposed method for measuring the noise of domestic appliances, *Appl. Acous.* **6**, 277–292.

Jeatt, K. E. (1972). Domestic Appliance Noise. M.Sc. thesis, Univ. of London.

PART IV

11

Noise Control

K. A. MULHOLLAND

Department of Construction and Environmental Health
University of Aston in Birmingham
Birmingham
England

THE NOISE HANDBOOK

I. INTRODUCTION

It is perhaps unfortunate that once noise has been established in the air, there is, in all but a few exceptional cases, no way in which the noise can be controlled by means of a power-driven device. We have devices to vary temperature, lighting levels and the condition of the air, and so these aspects of human comfort are well within our control. This is not so with noise; there is no electrical "noise-level controller" that can be bought and installed in homes and work places; worse than this, sound is in fact produced by many of the devices that are used to improve our environment, so that the world abounds with many sources of sound but with no ready means to control noise levels.

The problem will be considered in the following way: noise control at the source of the noise, along the path that the noise takes and at the point of reception. Over all these aspects is the necessity to plan well and to educate people and organisations to consider how avoidable noise can be eliminated and how unavoidable noise can be reduced.

II. GENERAL NOISE CONTROL FACTORS

A. Energy Contained in a Sound Wave

When work is carried out, it can only be produced by means of a heat engine or an electric motor. There is no other source of power, and the ultimate end of all work is waste heat. One step above waste heat comes sound energy: a minute quantity of the work done by most engines is not directly converted into useful work or into waste heat, but is radiated as sound. The smallness of the quantity of energy needed to produce a sound is a major factor in the difficulty of noise reduction. Not only that, but if the noise has a particular characteristic, such as a tone, or an impulsive character or an intermittency, or if it carries information, then the brain rates the annoyance value of the noise at an even higher level than a straightforward measure of the sound level would indicate.

It is therefore difficult at the outset to redesign an existing device so that it is inherently less noisy. On the other hand, if a device has been badly made, it can easily put out a much larger fraction of its power output as noise than it should. In this case, noise control is much easier. However, there are not many things that are built to be unnecessarily loud, unless that was the intention of the designer and the user. It is, of course, pointless to class devices which are intended to produce noise, such as the loudspeaker and the football rattle, as controllable noise sources. Their designers and operators would usually wish them to be

TABLE I. Loudness, Some Sounds

Noise	dB(A)
Large jet airliner	140
Large piston-engined airliner	130
Riveting on steel plate	130
Train on a steel bridge	110
Weaving shed	100
Pneumatic road drill at 30-m distance	90
Heavy road traffic at kerbside	85
Dining room	80
Male speech at 1-m distance	75
Typing office with acoustic ceiling	75
Light road traffic at kerbside	55

even louder if possible. Table I gives the levels, in decibels(A), of some common sounds.

B. Types of Sources

1. Outdoor Sources

First we must consider the types of noise source that must be controlled. Amongst outdoor noise, road traffic noise is the most widespread and frequently the loudest. In recent years, this problem has become so intense that special measures, such as the Land Compensation Act (Anonymous, 1973a), are being taken by the government to limit the growth of this noise and maybe even to reduce it.

A more localised form of noise, but one which produces levels far greater than that produced by road traffic, occurs around airports. Peak levels well over 110 dB(A) can occur miles from the airport, and whole communities can suffer continued exposure to extreme amounts of noise.

The third most important source of noise is that due to construction. A construction site, even though it is a temporary thing, can produce a great deal of annoyance during operation, which can last as long as three or four years. This annoyance is exacerbated by the resentment that persons suffering from the noise feel about the change in their environment caused by the construction work.

Noise from factories is a widespread source of complaint, but reasonable procedures exist for dealing with this phenomenon. Finally, there is domestic noise from one's immediate neighbours, which very often has no general solution, but has to be dealt with as an individual case.

2. Indoor Sources

Indoor noises principally concern the control of noise within a factory or other place of work, and these are receiving increased attention from the government. It is not yet illegal to produce any particular noise level within a factory, but a code of practice (Department of Employment, 1972) exists which lays down guidelines which are likely to become statutory, and trade unions are having increased success in actions seeking compensation for noise-induced hearing loss. The control of indoor noise is therefore important. To a lesser degree, the requirements of privacy and intelligibility for conversations in offices mean that the quieter noises in offices must also be controlled. At the bottom end of the loudness scale, the stringent requirements for quietness that exist for dwelling houses mean that noise control of service systems requires attention.

3. Approach to Noise Control

We shall now continue to consider these various noises and the methods of controlling them. Noise control methods naturally break down into three parts: noise control at the source, noise control along the path that the noise must take to reach the sufferer and noise control at the receiver, which includes his immediate acoustic environment.

III. NOISE CONTROL AT THE SOURCE

A. Containment

The control of noise at the source is usually the cheapest and most reliable method of noise control. Once sound energy has moved away from its immediate source, it becomes an all-pervading problem. The area of the surface of a sphere increases as the square of its radius, and this means that the nearer we move to the source, the smaller the area is which we have to deal with in order to contain the sound. This idea of containment or isolation, then, is the main principle of noise control.

B. Sound Insulation and Sound Absorption

It is important at the outset to understand the difference between sound insulation and sound absorption: These two different but complementary means of noise control are often confused. Sound insulation concerns the reduction of sound as it passes through a wall or barrier, or in the case of vibration through isolators or discontinuities. Here the major contribution to the reduction in sound level is the fact that by far the largest fraction of the sound energy is reflected back from the barrier. For example, if a wall has a transmission loss of 30 dB, this means that only $\frac{1}{1000}$ of the energy falling on the wall is transmitted, but up to $\frac{999}{1000}$ parts may be reflected back. Sound absorption, on the other hand, occurs when a sound is partially reflected from a wall. Sound is absorbed in many ways at a boundary and the reflected sound is reduced because of these absorbing processes. With good sound absorbers 90% of the incident sound energy can be absorbed. Perhaps the most significant difference between insulation and absorption is the fact that absorption involves a degradation of sound energy into heat, whereas insulation can be achieved without any such degradation.

C. Enclosures

1. General Principles

If it is desired to reduce the noise from a particular source, this can often be done by means of an enclosure. However, the mere surrounding of the source by massive partitions is not adequate, since the volume enclosed within the source becomes a reservoir for noise energy. What happens then is that sound levels inside the enclosure increase until the rate of loss of energy within the enclosure is equal to the rate of production of sound energy from the source. If the only way in which sound can escape from the enclosure is through the wall, then it is immediately obvious that the sound energy will build up until the rate of radiation sound from the enclosure is equal to the rate of production of sound from the source and the situation will be as if the enclosure did not exist at all (except for directionality effects). It is therefore necessary to include within the enclosure an alternative means by which the sound energy can be converted into heat. This is done by including sound-absorbing material within the enclosure, and when this is done, most of the sound energy is lost into the sound absorber and the barrier becomes effective.

2. Access Problems

A critical problem with enclosures is that of permitting access to the machinery or noise source that is enclosed. Access can be required for a number of reasons: for monitoring, adjustment of controls, the supply of raw materials and the removal of finished products, as well as for the removal of waste heat. With a little ingenuity most of these problems can be overcome. A small window, preferably double glazed, allows gauges to be read without removal of the enclosure. Controls that only have to be adjusted occasionally can be left within the barrier, with a specially treated doorway allowing occasional access. Controls which have to be continually adjusted require special treatment, and this can often mean re-designing the machine so that the barrier becomes an integral part of the machine. This has to be dealt with at the time of specification of the machine, which we shall discuss later, but in the case of an existing machine, it is often possible to provide extensions to controls or in the case of electrical equipment to remove switches and rheostats to a remote position. For continuous process machinery, the throughput of materials is a more demanding condition. It is often necessary to locate the immediate operator of the machine inside the barrier area, in which case, the operator must either bear a noise greater than that recommended, which involves the use of ear defenders more or less permanently, or if the operator is a member of a team carrying out the work which involves less noisy operations, the team can be cycled through the working day so that no individual operator exceeds the noise dose limit.

The final reason for providing access is the matter of cooling. Unfortunately, all materials which have good sound absorbing properties also tend to have good thermal insulating properties, which means that temperature-sensitive processes cannot be enclosed in sound-insulating barriers. However, it is possible to provide access for cooling air by means of chimney-like constructions. A duct, although it provides a direct air path to the surrounding area, does provide a certain amount of attenuation, particularly if the duct is lined with sound-absorbing material.

Handbooks [e.g., Woods (1972)] are available which quote the value of sound attenuation that can be expected from a given duct cross section and length. These should be used to calculate the length of chimney required to provide adequate sound reduction as well as ventilation.

IV. MACHINERY DESIGN AND SPECIFICATIONS

When one is ordering new machinery to be installed in a factory, it is now prudent to consider the noise that will be produced by the machinery so as to ensure an acceptable working environment for employees. Legislation in this area seems likely. The writing in of noise specifications would appear to be reasonably straightforward, but the main difficulty arises from the interaction of the machine with the environment where it will be installed and the difficulty of measuring the noise produced by the machine in the environment where it is manufactured. It would be convenient to merely specify that the machinery, when installed in a factory, shall produce a noise level not exceeding blank decibels(A). However, this places the supplier of the machine in a difficult situation. He does not know what acoustical environment exists within the factory, but he can measure the noise level from the machine within his own environment. The method employed is usually to specify that the noise level measured at a distance from the machine shall not exceed X dB(A) and to compute the resulting noise field from the simple estimate of the acoustics of the room in which the machine is to be installed. In most cases, the noise level in the room will not exceed the level measured close to the machine. Exceptions occur when the machine is placed by itself in a very small, highly reverberant chamber.

The design of machines for quietness is technical and specialised. A few general rules can be given here, but individual problems would have to be studied on their own merits with the aid of more technical information (Beranek, 1954, 1960; Harris, 1979; Woods, 1972). The general principles are as follows:

(1) Impacting parts of machines should be enclosed.

(2) All enclosures should contain sound-absorbing material suitable for the environment within the enclosure.

(3) Manufacturing tolerances should be kept as small as possible; rotating machinery should be dynamically balanced.

(4) All rotating or impacting machines should be based on anti-vibration mountings.

(5) All rigid connections of the machine in the way of electricity, water, air, gas, etc., should have vibration decouplers around the machine, and air ducts should contain pressure release chambers.

(6) Internal combustion engines should be properly silenced.

(7) Machines should not be sited in small reverberant chambers, but placed within absorbent enclosures sited in large work areas. The overall layout of a building or factory should be designed so that noise critical areas (offices, showrooms, etc.) are kept away from noisy areas by means of buffer non-noise critical areas, such as kitchens, canteens, lavatories, corridors, etc.

V. VIBRATION ISOLATION AND DAMPING

Machines and other sources of impact noise (principally footsteps) can feed acoustic energy into the frame of a building in the form of vibration, which can then run throughout the structure and, by re-radiation from walls, cause noise problems at quite remote parts of the building. Under circumstances where a room has a resonant frequency matching the frequency of some remote vibration source, such as a pump, the noise level in the room can become high due to resonance effects. Adjacent rooms without tuned modes can be unaffected by the noise.

Anomalous cases of high noise level can also be caused by central heating systems where the large-area radiators can act as efficient acoustic radiators and couple the sound within the pipes to the room.

The principal cure for these problems is to provide discontinuities. If a building is constructed in a single steel frame, then the vibration energy can run through the frame virtually unhindered and produce a noise problem wherever high coupling or resonance effects can occur. Discontinuities can be introduced into the system by bolting the framework together with lossy pads between adjacent members, thus attenuating the vibration energy as it passes through the structure. In direct analogy with the airborne case, however, it has been found to be most effective to cut off the vibration energy at the source by placing the vibrating machine on resilient mountings.

Obviously, the weight of the machine must be supported somehow, and this has to be done by providing a strong suspension for the mass associated with the machine. The principal problem here is that the degree of freedom introduced by such suspension allows the machine to vibrate at a resonant frequency determined by its mass and the stiffness of the sup-

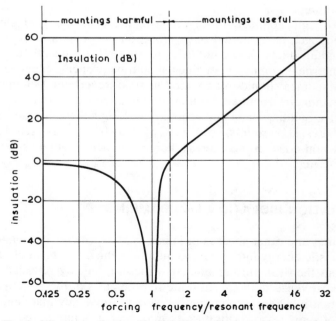

Fig. 1. Vibration reduction (decibels) by resilient machine mountings.

ports. It is necessary to keep this resonant frequency low, thus placing it in the region where little vibrational energy exists. It is also necessary to fit over-ride stops to prevent occasional high-vibration amplitude during, for example, run-up on a machine when high low-frequency amplitudes will exist for a short period of time. Figure 1 shows the isolation that can be obtained from such a suspension. This figure demonstrates that, while correctly designed mounts can be of great value in controlling the transmission of vibration, badly designed mounts can actually make the vibration worse.

VI. OUTDOOR NOISE CONTROL

A. Introduction

As already mentioned, the major sources of outdoor noise are road traffic, aircraft, construction work and factories. Of these, the first is the most widespread, while the second subjects a smaller number of people to higher noise levels. Factory noise subjects a limited number of people to annoyance, and each case tends to be special. In addition, noise from neighbours, in particular loud and prolonged parties, is becoming more common.

B. Traffic Noise

The Land Compensation Act (Anonymous, 1973a) and the Limitation of Vehicle Noise Regulations (Anonymous, 1973b) have resulted in an increased knowledge and awareness of this problem. It is now the duty of local authorities to carry out insulation work or to make grants to persons who are subjected to certain levels of noise from roads if the highway or additional carriageway was first opened to the public after 16 October 1972. Such grants are payable where levels of sound exceed 68 dB(A) for 10% of the time over an 18-hr period. Where a lot of people are likely to be subjected to these noise levels, the provisions of the Land Compensation Act can significantly influence a highway construction project. Therefore, information concerning the method of reducing the 18-hr L_{10} below 68 dB(A) is becoming valuable. The noise from highways can be controlled by the design of the highway, by the erection of barriers between the highway and the buildings nearby and by the placement of speed restrictions on the road or changing traffic-flow conditions.

Free-flowing traffic on a level grade provides the quietest conditions. Roundabouts and traffic lights interrupt free-flowing traffic and increase noise from the road, and the siting of intersections should therefore be considered with care. The alternative scheme is that exemplified by the A38 leading into Birmingham, where through traffic is routed via underpasses below flying roundabouts, which cater for turning traffic and lateral through traffic only. The subject of road construction, however, is a specialised subject and we need not consider it further here. The performance of noise barriers, however, is of more practical use because a noise barrier can be constructed fairly cheaply once the principles are understood.

A noise barrier consists of a wall or fence built to stand between the source of sound and the recipient. It is essential that the barrier be free from holes or gaps, but, in contrast with party wall constructions, it is not necessary for noise barriers to have high sound insulation performance. This is because the principal path of noise is over the top of the barrier, and therefore any construction which provides 20 dB of sound insulation is adequate. To be effective a noise barrier must cut the line of sight between the source and the receiver and in addition oblige the sound ray to bend over the top of the barrier in order to reach the receiver. The effectiveness of a barrier is increased when the effective height of the barrier is increased and the angle into shadow is increased. However, a barrier cannot provide more than 20 dB of attenuation to road traffic noise (see Fig. 2). For further details refer to the document "Calculation of Road Traffic Noise" (Department of the Environment, 1975). While the procedure referred to here applies only to road traffic noise, another

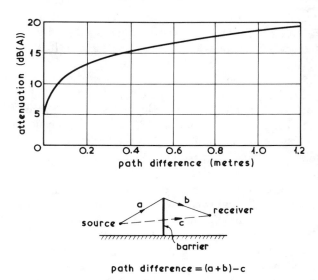

path difference = (a + b) − c

Fig. 2. Reduction of road traffic noise [decibels(A) L_{10}] by a very long barrier.

method of predicting noise reduction which can be applied to other noise sources of differing frequencies is given by Harris (1979), who takes into account the frequency of the noise. This method of predicting barrier performance is therefore of more general application.

C. Aircraft Noise

The duty of local authorities to provide compensation for noise from new installations includes new or altered airports, and, in this case, it is not necessary for property affected to be adjacent to the airport; it can also be merely near to the airport and affected by the noise. The design of quieter aircraft engines is an urgent consideration for aircraft engine manufacturers, and significant improvements have recently been obtained with the use of high by-pass ratio engines which are used on all the large jets of the Boeing-747 type. The noise reduction so achieved, however, has tended to be cancelled out by the increase in air traffic and the tendency of manufacturers who, having discovered a quieter engine for a given thrust, may increase the thrust to the point where the engine is not significantly quieter than its predecessors. The existence of airport noise protest groups has had, in recent years, significant effects, not the least of which is the decision to site the third London airport on Maplin Sands rather than at Wing in Buckinghamshire. Partly because of the activities of these groups, a large number of airport managements everywhere have laid down noise-reducing flight paths and procedures and installed noise-monitoring equipment to ensure that these procedures are obeyed. Although these techniques exist, a knowledgeable pilot can easily close down the power from his aircraft just as it is passing over a known monitoring point.

The domestic sufferer from aircraft noise has a number of options open to him. The most effective of these is to move house. It is known that there is a significant proportion of the population which is not affected by noise and that there is a small proportion (15 to 20%) which is seriously affected. Therefore one need not feel guilty if one is a noise sufferer in passing the problem on to somebody else. The person who buys the house will not be unaware of the noise problem and may have other over-riding reasons for wishing to buy it. Alternatively, a local council may be persuaded to re-site the person affected if he is persistent in his complaints, but one should be genuine in the desire to move since local authorities are very familiar with the person who enjoys making complaints for the sake of it.

Government grants are available for the sound insulation of houses within zones affected by airport noise, and in the United States it has become policy in certain states not to site houses within certain noise level contours around the airport. At Los Angeles, a whole district has been removed because it was found to be within the critical noise contour for exposure.

D. Factory Noise

1. Introduction

Noise coming from a factory can be dealt with by means of common law procedure if the complainant is the occupier of land, or under the provisions of the Public Health Act (Anonymous, 1936), if the local public health officer decides that the noise constitutes a statutory nuisance. The decision that the noise is a statutory nuisance is taken by the public health authority, and if they cannot be convinced, then it is necessary to take action on one's own account. The Control of Pollution Act (Anonymous, 1974) has meant a great increase in the power of local authorities and individuals to take action against noise makers.

2. British Standard 4142

One method of assessing noise from fixed installations is laid down in British Standard 4142 (BSI, 1967), which is meant to be a method of predicting whether or not noise complaints will occur. Briefly, this method starts with a measured value of the noise in question and then applies corrections based on such things as the duration of the noise during a 24-hr period, any specialised characteristics such as tones or impulses and such factors as seasons and intermittency. This corrected measured level is compared either with the measured background noise level or with criteria which are based on the type of area where the noise occurs, the length of time the factory has been in operation, the time of day and the seasonal nature of the sound. If the corrected noise level exceeds the corrected criteria by more than 10 dB(A), then it is probable that the local health inspector will consider the noise to be a statutory nuisance and will take legal action to force the factory to produce less noise within a certain time.

VII. NOISE REDUCTION ALONG SOUND PATHS

A. Introduction

Once sound has left a source and has become established in the surrounding medium, either air or the structure of the building, it must then travel some distance before reaching the point at which the noise nuisance will occur. When this situation arises, we must consider steps that can be taken to reduce the transmission of noise as it travels down the path. Common paths for noise are airborne paths such as ducts and corridors, and it is usual to include in this category walls that break up airborne sound paths, and not to consider these walls as independent vibration paths.

Alternatively, vibration energy can be transmitted through building structures directly and can arise from direct excitation from the source or indirect excitation through a sound field produced in the room containing the source. Where a noise source is directly coupled to conducting paths such as pipes or air ducts, these ducts can carry sound energy.

To control noise along the path, then, we need to consider how sound is transmitted through ducts, corridors, directly through walls and along pipes and through the structure of buildings.

B. Transmission of Sound along Ducts

The manufacturers of fans and ducting systems include some who have paid great attention to the problems of noise. It is possible to contact them directly at the design stage of a building and to have them quote sound levels and sound insulation values together with prices and delivery information for any reasonable noise requirement. Before talking to such people, however, it is advisable to have an understanding of what is a "reasonable noise requirement" and what problems need to be considered. It is not reasonable to expect total silence from a system, but it is reasonable to expect sound levels below ambient noise levels when the building is in use. The noise produced by the system should be graded to the existing noise climate of the building. A second point is the fact that ventilating systems form a link between the inside and the outside of a building, and thus noise can be transmitted through the system and can be produced outside the building by the system itself. It is not usual for such noises to cause trouble except in the case of large industrial plants where the use of giant centrifugal fans can cause discrete frequency noises (at the frequency at which the blades of the fan pass the outlet) or in the case of discotheque noise in built-up areas. Conversely, when a ventilating system is installed in a noise critical area, such as near an airport, the noise transmitted into the building from outside must be considered. In all cases, therefore, the three factors to consider are the noise produced by the system inside the building, the noise produced outside the building and the noise transmitted through the system.

Noise from fans can be found from references to manufacturers' information or by empirical formula derived by Beranek (1960). The coupling of a fan to its associated duct does not greatly affect the noise produced by centrifugal fans, but axial fans are sensitive to obstruction and to bends in the ducting for a length upstream and downstream of twice the fan diameter. It is necessary to point out that (unfortunately) sound radiation from a fan travels equally well upstream and downstream. This is because the speed of sound (343 m/sec) is far in excess of any air speed used in practical systems. It is found that circular ducting is inherently more rigid than rectangular ducting and thus circular ducting transmits sound along the duct to cause a problem at the far end, whereas rectangular ducting allows sound to pass through the walls of the duct, possibly causing problems near to the fan. There are commercially available duct-attenuating systems, but it is necessary to remember that noise is transmitted down a duct in two ways. Firstly, it is transmitted along the air path within the duct. To counteract this the provision of sound-absorbing material within the duct is necessary. Secondly, noise is transmitted along the rigid walls of the duct. To attenuate this noise it is necessary to incorporate a non-rigid bellows section to de-couple the duct from the vibration produced by the fan. Bends within the ducts will attenuate noise travelling along the duct but can cause noise in their own right if turbulence is generated at the bend. The provision of aerodynamic guide vanes within the duct can reduce such noise. At the end of a duct the sound must pass from the duct into the air space of the room. A useful effect occurs here: that of end reflection. This phenomenon occurs at the open end of an organ pipe and helps it to resonate. In the case of general noises travelling along the duct, it is possible that as much as 90% of the sound can be reflected from the end back down the duct. It is to overcome this effect that the mouths of wind instruments are belled to permit the efficient coupling between sound within the tube and the surrounding air. The aim of the designer of ventilating systems is to achieve the opposite effect by providing low coupling to the room while avoiding localised noise sources which are often caused by turbulence produced within constricting grills. Much more detailed information on this extensive subject can be obtained by referring to Woods (1972).

C. Factors Controlling Sound Insulation

1. Transmission Loss or Sound Reduction Index

a. Introduction. The transmission loss of a panel (i.e., of a wall or barrier) refers to the ability of the panel to resist the transmission of energy from one side of the panel to the other. When sound energy falls on a panel, most of it is reflected, but a small fraction is transmitted to form a sound wave beyond the panel. The fraction of energy transmitted is called the transmission coefficient τ and the transmission loss TL (in decibels) is related to τ by

$$TL = 10 \log_{10}(1/\tau).$$

Transmission loss is a property of a panel and is only one of a number of factors that go to make up sound insulation. This is because sound insulation is a property of adjacent rooms and includes the effects of room acoustics and flanking transmission. A number of factors control the transmission loss of a wall: these include the mass of the panel, the number of layers it contains and its structural stiffness.

b. Mass Law. The mass law is the basic rule of thumb for determining the transmission loss of a panel (see Fig. 3).

If the mass per unit area M in kilograms per square metre is known, then the transmission loss at a frequency f will be given by

$$TL = 10 \log[1 + (fM/131)^2].$$

The sound insulation will increase by between 5 and 6 dB per octave (i.e., per doubling of frequency) and by a similar amount for the doubling of the mass of the wall. Subjectively a difference of 5 dB in the sound level of a noise means that only a small difference in the level has been obtained.

Thus it is often difficult to obtain a significant increase in the sound insulation of an existing wall by merely increasing the total weight of the wall. A domestic party wall consists of a double layer of brick which can weigh up to 400 km/m² (90 lb/ft²). It is clearly not practical to add, say, 10 dB to the mass law insulation of such a wall by increasing the weight by a factor of 4! The position is more hopeful with light weight partitions, but

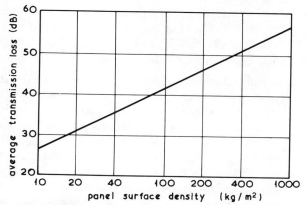

Fig. 3. Average transmission loss as a function of surface density according to the mass law. Averaging is by random incidence over third octave bands.

even so it is better to proceed on a basis of an understanding of the overall problem than merely to increase the weight of the partition.

c. The Coincidence Effect. A panel can support bending waves which run along its face. In certain conditions these bending waves can couple with the sound fields in the rooms on either side of the panel and cause high transmission. This is called the coincidence effect.

This effect is quite common in lightweight panels such as plasterboard, asbestos, glass and wood and is often a limiting factor in the insulation that can be obtained with these materials. Thus control of the coincidence effect can result in a significant improvement in sound insulation. Much research work has been carried out by the manufacturers of these materials (particularly by the manufacturers of glass and plasterboard). These manufacturers will readily advise on the choice of suitable configurations to avoid coincidence troubles.

d. Multiple Layers. One effective method of increasing the sound insulation of a wall is to construct a wall of a number of separate layers. Experimental measurements have shown that the mass law increase of 5 dB for a single panel can be increased to about 10 dB per octave for a double layer panel. There are a number of adverse effects, particularly at low frequencies, where the two layers can resonate with each other against the trapped air between them. This can reduce low-frequency performance of multiple layers.

The beneficial effect of multi-layering can often be obviated by an internal vibration bridge within the panel. For structural reasons it is often impossible to build a double panel without some form of cross bracing between panels either on the edge or over the face of the panel. This produces vibration bridging which can transmit sound through the panel, cancelling out the beneficial effect of air gaps and other discontinuities. In practice the effect of this flanking transmission is to overcome almost completely the beneficial effect of layering, so that a double brick wall has a performance little better than a solid wall of equal mass per unit area.

e. Sound-Absorbing Material. Sound-absorbing material has a role to play in controlling the insulation that can be obtained between two rooms. This will be discussed shortly. The present discussion applies only to sound absorbers which improve the sound transmission loss of a panel. It is thought that by placing sound absorbers inside the cavity, the resonance can be reduced and so the sound insulation of the panel increased. In practice, the results are not as good as might be expected for two reasons: in the first case, the mass resonance occurs at a very low frequency and sound absorbers do not have good sound-absorbing properties in this range. It is then found that sound absorption has little effect on mass resonance. The small amount of sound absorption that may be present already in the cavity is sufficient to drown this resonance very strongly. Hence the addition of further absorbers has little effect. Sound absorbers may also be placed over the face of the panel, and if the panel is lightweight and has small insulation, quite significant improvements in sound insulation have been observed. As much as 15 dB can be obtained at high frequencies, especially where there is a coincidence effect, since the presence of the sound absorber cancels out the increase in transmission obtained by this effect. Sound absorbers by themselves have very little inherent sound insulation. A 4-in. layer of polyurethane foam will have a transmission loss of only 5 to 8 dB.

f. Sound Insulation and Transmission Loss. The quoted transmission loss of a panel is not necessarily the sound insulation that will be obtained between two rooms separated by that panel. Other factors come into play. Principally these are the exposure area S and the effective sound absorption within the room receiving the noise A.

In general the larger the exposure area between the two rooms, the smaller will be the insulation. However, if the absorption in the receiving room is increased, this will increase the insulation. Insulation obtained in practice is thus generally related to the area of sound absorber, but we must note that it is inefficient to try to improve sound insulation by increasing sound absorption since normal rooms have a fair amount of sound absorption anyway, and even if this amount of sound absorption were doubled, the insulation would only increase by 3 dB, which is hardly a significant increase. The quantity S can only be controlled at the planning stage and shows the importance of minimum areas of contact between noisy and quiet areas.

VIII. NOISE CONTROL AT THE RECEIVER

A. Introduction

Noise is received by people and more exceptionally by delicate instrumentation, and it is often necessary to control the level of the noise received. This is normally done by treating the room or area within which the receiver is located, and we therefore have to study the acoustics of this situation.

B. Room Acoustics

1. Introduction

The subject of room acoustics has already been discussed to a certain extent earlier. However, the main points will be mentioned. Consider a

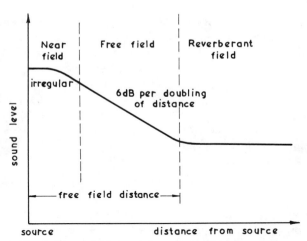

Fig. 4. Sound fields around a source within an enclosure.

source of sound located within a room radiating sound energy. At most frequencies the sound energy travels away from the source and is multiply reflected from the boundaries of the room, and a uniform diffuse sound field (which is called the reverberant field) builds up within the room. The reverberant field is usually uniform within the room except in the region of the source or near highly absorbing areas. Figure 4 shows the normal situation, and we can define three areas of interest. One is the near field within which we are so close to the source that individual characteristics of the source control the high sound level. Farther away from the source there is a free field in which the source can be treated substantially as a point source and the noise levels fall off by 6 dB for each doubling of the distance from the source. This free field can, sometimes, but not always, lead into the reverberant field which fills the rest of the room space.

2. The Near and Free Fields

In the near- and free-field regions the sound level is controlled by the sound being directly radiated from the source and is independent of room acoustics. Any control of the sound level therefore must be by means of direct reduction of the source power or else by means of a barrier between the source and the receiver.

3. Reverberant Field

The reverberant field level is controlled by the power of the source, the size of the room and the amount of sound-absorbing material within the room. Within the reverberant field it is possible to control the sound level by controlling the power from the source, for example, with enclosures, or by means of sound absorbers within the room, but not by means of sound barriers (as opposed to enclosures) near to the source. By increasing the sound absorption within the room the reverberant field level can be reduced, but this is not an efficient process since it is necessary to double the area of sound absorber in order to reduce the reverberant field level by 3 dB. Most rooms already contain a fair amount of sound absorber so that the cost–benefit ratio for this procedure quickly becomes uneconomic and indeed impractical.

IX. NOISE CONTROL IN LARGE ROOMS WITH MANY SOURCES

A. Introduction

In a large room with many sources, the idea of near, free and reverberant field still applies, but there are complications. The level of the reverberant field is controlled by the total sound power that is being emitted into the room and this level can exceed the level in the near field of some of the quieter noise sources. Fortunately, it is quite easy to trace the extent and level of the sound fields by moving a simple sound level meter around the room. However, the problem of reducing the reverberant field level to a specific level [such as 90 dB(A)] to come into line with the code of practice (Department of Employment, 1972) is not straightforward in that if the dominant noise source is treated, the sound field will not necessarily fall by an amount equal to the reduction in the level of the dominant sound source. What happens is that the sound field falls to the point where the next-loudest source becomes the dominant source and not much below this. A process of progressive reduction is thus initiated until enough sources have been tackled to reduce the field to a reasonable level.

B. Pure Tone Effects

All of the preceding theory applies to complex or random sounds which are produced by such sources as moving gas, motion of wheels on floors and the like. The behaviour of pure tones is entirely different. Pure tones are produced by vortices, humming noises from electrical devices and machines, rotating fans and resonating columns of air.

1. Low Frequencies

At low frequencies a typical room will have standing wave resonances which are few in number over quite a wide frequency range. Thus the response of the room to low-frequency sound will be concentrated at the discrete frequencies of these modes, and at low frequencies behaviour tends to be tonal even for random noise sources. The situation is worse when the noise is itself tonal, for example, the noise produced by transformers, and it is possible, depending on the precise geometry of the room, for the source to be in tune with a standing wave within the room. When this occurs, large sound pressure levels can be developed which can be unrelated to sound pressure levels in the adjacent rooms. It is also possible to have high sound levels at some points in a room, while at other points the sound is not audible. Examples of this phenomenon are not frequent, however.

2. High Frequencies

At higher frequencies pure tones can be treated for noise reduction purposes much in the same way as more random sound fields except that there are unusual subjective phenomena normally associated with these fields. Principally, when a pure tone noise fills a room, it can be found that by simply moving the head a small amount, the subjective loudness of the tone will alter considerably. This is because the standing wave nature of the sound is still preserved even though the wavelength may be much shorter than the room size.

X. CONCLUSIONS

Noise control has always been a difficult and frustrating task. First the noise source must be identified and then the noise levels assessed by some method to establish whether the source is causing an unjustifiably high level of noise annoyance to some people. The person who must be made to pay for any quietening has to be identified and actually induced to order the quietening done. Lastly, the action necessary to make the noise quieter has to be found and the work carried out.

In recent years the body of knowledge about noise control has grown very rapidly. There are now in existence many professional organisations specialising in noise control, and both the public and the government have become aware of the necessity to control noise and to treat it as one of the many by-products of technology that can be regarded as pollution. Although there is certainly as yet no reason for complacency, the major noise problems of aircraft, traffic and construction are being tackled. It is within the grasp of present technology to substantially reduce noise from these sources. As so often happens, it is merely a question of whether or not the money, amounting to some billions of pounds, (for example, in the aircraft industry) will be spent.

The body of legislation about noise has also grown, principally during the 1970s so that there is now adequate legislation available to enable people to cause a noise nuisance to stop. Again, though, the question of money causes problems. Local environmental health authorities are forced by cuts to reduce staff, and overtime bans mean that noises "out of hours" are difficult to control. A newer form of noise nuisance (for example, "pay parties") needs further consideration.

In the end the quality of the environment is everyone's responsibility and all should contribute by avoiding noisy activities and by supporting anyone who tries to reduce unnecessary noise.

REFERENCES

Anonymous (1936). Public Health Act. His Majesty's Stationery Office, London.
Anonymous (1973a). Land Compensation Act. Her Majesty's Stationery Office, London.
Anonymous (1973b). Limitation of Vehicle Noise Regulations. Her Majesty's Stationery Office, London.
Anonymous (1974). Control of Pollution Act, Part III, Noise. Her Majesty's Stationery Office, London.
Beranek, L. L. (1954). "Acoustics." McGraw-Hill, New York.
Beranek, L. L. (1960). "Noise Reduction." McGraw-Hill, New York.
BSI (1967). British Standard 4142, "Method of Rating Industrial Noise Affecting Mixed Residential and Industrial Areas." British Standards Institution, London.
Department of Employment (1972). "Code of Practice for Reducing the Exposure of Employed Persons to Noise." Her Majesty's Stationery Office, London. [Re-issued by the Health and Safety Executive, Her Majesty's Stationery Office, London.]
Department of the Environment (1975). "Calculation of Road Traffic Noise." Her Majesty's Stationery Office, London.
Harris, C. M., ed. (1979). "Handbook of Noise Control," 2nd ed. McGraw-Hill, New York.
Woods, I. R., ed. (1972). "Noise Control in Mechanical Services." Sound Attenuators Ltd., Colchester.

12

Noise and the Law
in the United Kingdom

R. GRIME
Faculty of Law
University of Southampton
Southampton
England

I. LEGISLATION, LIABILITY AND NATIONAL INSURANCE

A. Introduction

Traditionally, three different approaches to the promotion of occupational safety by legal means have been adopted in the United Kingdom. The first is not only generally the oldest, but it usually remains the most important. It is the civil law of compensation. Occupational risks produce injuries, and sufferers will use the ordinary civil law of torts to seek compensation in the courts. This creates a law of civil liability for injuries which not only forms the base for insurers, trade unions and their advisers and others in conducting their necessary negotiations, but has a not insubstantial indirect effect upon the way in which employers go about their business. From a legal point of view, this is rarely specialised law at all: it is the application of general principles to particular cases. So it was, and to a very large extent still is, with noise. Those injured by excessive noise may make their cases alleging and proving negligence and claiming damages at common law. Although in the case of noise, this process is barely 20 years old, it creates a set of principles or guidelines of reasonable behaviour which may be applied to the reduction of the risk.

Secondly, Parliament may take a hand by the enactment of public compulsory rules designed to promote the health, safety and welfare of employees at work. We have had such legislation in England since the passage of the Health and Morals of Apprentices Act of 1804, but we are generally more concerned with the modern consolidations: the Factories Act (1961), the Offices, Shops and Railway Premises Act (1963), the Mines and Quarries Act (1954), but above all the administrative consolidation of the Health and Safety at Work Act (1974). The bulk of this legislation is non-specific, too, but there always has been room, particularly through the use of specialised regulations, to deal with individual problems. Recently we have seen a gradual move towards the creation of a code of noise regulations.

Occupational legislation is, in theory at least, quite separate from civil compensation. However, there is overlap. Sometimes the breach of a compulsory statutory provision may be used by a complainant to support a civil claim for compensation against the person who broke the law. This "action for breach of statutory duty" marches alongside claims for compensation based on negligence in employers' liability law. Second, the legislation itself commonly contains provisions which rely upon an assessment of "reasonable" behaviour. Section 2 of the Health and Safety at Work Act (1974), for instance, requires an employer "so far as is reasonably practicable to ensure the safety, health and welfare at work" of all his employees; Section 29 of the Factories Act (1961) requires that every workplace be made and kept safe "so far as reasonably practicable." If such provisions are to be applied to noise, then clearly our understanding of "reasonable" behaviour which we may derive from our knowledge of the civil law of tort will be relevant. Finally, such public legislation spawns highly significant codes of practice and regulations, all of which have the effect of sharpening our understanding of the recognised standards of reasonable behaviour, which is the central concept of the civil law of negligence. Public and private law, therefore, are inextricably mingled together.

The third approach is by way of public or state compensation. Ever since the Workman's Compensation Act of 1897, the United Kingdom has had a system of occupational pensions for those injured through their work. Within that system several industrial diseases are included, and certain types of noise deafness now appear. Legally, the National Insurance (Industrial Injuries) Acts (1965 and 1972) are quite separate from the rest of the relevant law, however much, as a matter of practical politics, the existence of a valid claim under one system may encourage people to utilise the others.

B. Civil Claims for Compensation

The most common form of noise-induced deafness is the gradual attenuation of hearing over a relatively long period of exposure to comparatively low levels of noise. Sudden traumatic injury caused by short-term exposure to very high levels does occur but, from a lawyer's point of view, such cases cause few problems. In any case, the chronic type of injury is far more likely to be encountered.

The lawyer's problems derive from the nature of the injury and, in part, to accepted social attitudes towards the disablement. Compensation depends upon establishing liability. Liability depends upon establishing two matters: the circumstances that caused the injury and that those circumstances are attributable to some person's failure to observe legally required standards. With traumatic injuries—whether they be sudden deafness or broken legs—these tasks are not too daunting. We know when it happened, so we can seek to find out what happened and, applying the appropriate legal rule, we can draw inferences as to responsibility. Moreover, claims about broken legs are likely to be made within a reasonable time of the injury; indeed, the Limitation Acts require that proceedings be commenced within three years of the event.

Typical noise-induced deafness cases are very different. The circumstances giving rise to the injury may well have continued for 10 or 20 years. Employers and physical environments may have ceased to exist, witnesses may be untraceable. Even if the facts be established, or not disputed, there are problems in drawing correct inferences in terms of the law of negligence over such a long period. Finally, the Limitation Acts may present further hurdles for the claimant. Any occupational disease case presents these problems to some degree, but noise cases are often exacerbated by the nature of the injury. Noise-induced deafness typically presents a slow attenuation of the acuity of hearing so gradual that the claimant may be unaware of his condition. In addition, deafness, until recently, was widely regarded as an inevitable concomitant of ageing (as, indeed, in some senses it truly is) or as a necessary side effect of certain types of work, more a nuisance than a disability. Such attitudes may be changing, but their existence has certainly made the tasks of lawyers engaged in this area more difficult.

Over the past 10 years, however, claims at common law for compensation for deafness induced by occupational noise have increased enormously. They are now an established part of the scene and feature in the business repertoire of many insurers and solicitors. As the cases have multiplied, much has been settled, though there are still difficulties which have not yet been finally resolved.

C. Negligence

The most common way in which a claim for damages—compensation— may be made in law is by proof of negligence. In the case of industrial accidents and diseases, such claims are generally made against the employer either in that capacity or in his capacity as occupier of the premises on which the work was carried out, and the conditions of which, it is alleged, gave rise to the injury or disease. The law imposes both upon occupiers and upon employers the duty to take reasonable care of, respectively, persons lawfully on the premises and employees. Failure to take reasonable care will, if it causes loss, provide the basis of a civil claim for damages to compensate the claimant for that loss. This is called the "tort" or civil wrong of *negligence.*

Negligence is generally defined as failure to take reasonable care. This means, inevitably, that it is a variable concept. All depends upon the circumstances, all depends upon the time and place. What is negligence today was not necessarily negligence 10 years ago. What is clearly negligent for a large, well-advised, well-equipped company might not be negligence for a small business with few resources and little expertise. This may be of crucial importance in respect of a claim for damages for industrial deafness, particularly the matter of time. Most noise-induced deafness cases arise as the result of many years of exposure: what may be accounted clear negligence at the end of the period may not be so regarded at the beginning. This could be highly significant.

Although reasonable care is not a precise topic, it does help to make one very important point. Proof of injury brought about by the conditions in which a man is required to work, proof even that the employer's decision brought about the injury or condition in question, will not by itself provide a basis for a claim. Legal liability depends not upon proof that the injury was caused by the employer, but upon proof that his unreasonable behaviour brought about the consequence complained of. In the context of civil claims, the law allows a reasonable risk to be taken. A defendant is only liable if it be shown that the risk he took was unreasonable.

A good way of approaching negligence is to consider *risks* and *remedies*. Commonly, the essence of the complaint against an employer is that he has failed to take *any* action to reduce noise or to protect his employees against the risk of injury. To establish whether or not he was negligent, one may begin by asking whether the risk of noise-induced injury was one which he, as a reasonably careful employer, ought to have known about and done something about. Secondly, if the employer *has* done something about it, one can ask whether the remedy he chose was a reasonable one in all the circumstances.

To take the first question first: will an employer who disregards the risk of noise-induced hearing loss be held to be negligent? In a general sense, this must be answered in the affirmative, at least as regards the present day. It would be absurd to say that the link between noise and hearing loss was not now sufficiently established to be known by the reasonable employer. But general answers do not solve particular questions. Two further questions can be asked. Of what particular sorts of risk ought a reasonable employer to know? Of what ought a reasonable employer of the mid-1960s or late 1950s to have known?

Some cases are easy. In *Berry* v. *Stone Manganese Marine Ltd.* (1971) the plaintiff had been employed from 1957 to the date of the proceedings in the defendants' chipping shop at Charlton, where manganese bronze propellers were shaped with pneumatic hammers. The noise levels reached 120 dB, which was accepted by the judge in the case as "bordering on the threshold of pain". The plaintiff himself gave evidence that, when he first started work there "the noise frightened the life out of him". It could hardly be argued that the defendants in that case, as reasonable employers, need not have known that such noise would cause damage to the hearing of those exposed to it. The defendants did not argue: they made a formal admission that "from 1957 the noise was such that the defendants ought to have known that it would cause damage unless protection was given". That admission was described by the judge as "realistic".

On the other hand, in *Down* v. *Dudley, Coles, Long Ltd.* (1969) the judge held that the employer need not have taken account of the risks of aural injury attendant on the use of cartridge-assisted nail-fixing tools in 1966. The peak sound pressure level generated by the explosions which the tool in question made was about 160 dB, and the plaintiff had, over a 10-day period, been exposed to some 1200 of them. The judge "did not think it safe to conclude that the ordinary reasonable safety officer or employer or foreman in 1966 would have been as alert to the dangers as the expert witness appearing for the plaintiff had been''.

To compare two widely separated examples may be misleading. The fact that different conclusions were reached proves only that each case depends upon its facts. It would be fair to regard *Down* v. *Dudley, Coles, Long Ltd.* (1969) as a difficult case, concerning a rather unusual injury, very close to the line, and *Berry* v. *Stone Manganese Marine Ltd.* (1971) as a much simpler case a good way the other side of the same line. It is fairly clear that in a traditional industry with a well-established, obviously very noisy process, as was the case in *Berry* v. *Stone Manganese Marine Ltd.* (1971), where it may be widely known that operatives commonly go deaf, an employer who takes no notice of the hazard is negligent. But the crucial question is how far is the reasonable employer supposed to know of the risks of much lower levels of noise? For this one may, as a beginning, examine the available public literature.

In 1960 "Noise in Factories" by A. G. Aldersley-Williams was published by the Department of Scientific and Industrial Research. Although this is a "Factory Building Study", primarily concerned with the construction of factory buildings, it in fact covers much more ground. It collects factory noise sources under four heads, impact, reciprocation, friction and air turbulence, and briefly describes processes which involve these sources. It then describes the effects of noise, in the following terms:

The main consequences of excessive noise will obviously be borne by the occupants of the factory, and will influence their welfare, safety and, indirectly, their efficiency. As the level of noise increases, its effect on people will be first annoyance, then interference with communication, and finally damage to hearing.

Damage to hearing is dealt with at considerable length. After pointing out the difficulties, "tentative" criteria are offered:

Thus, in factories where noise levels at the machine operators' positions are above 85 dB in the octave bands between 300 to 4800 c/sec some form of noise control is desirable. When the levels are 95 dB or more, some form of control is necessary if hearing damage to the operator is to be prevented.

It will be noted that the A-weighted scale of sound pressure levels is not used. The octave-band sound pressure level limits referred to do not convert uniquely to the A scale, but the values are approximately equivalent to 88 dB(A) and 98 dB(A), respectively. The paper also contains the recommendation that no one should be exposed to noise levels of, or above, the "pain threshold", set at "130–140 dB at any frequency", without ear protection for even the briefest period. Its general criteria are linked to an 8-hr exposure in any one day, with proportionate increase in permitted levels for reduction in time of exposure.

The legal significance of "Noise in Factories" is perhaps not great. It is a government publication but because of its purpose, perhaps not widely accessible to employers. It was certainly not considered in *Down* v. *Dudley, Coles, Long Ltd.* (1969). There the employer was a building contractor and not the occupier of a factory building.

The Report of the Committee on the Problem of Noise (Wilson, 1963) was intended for wide consumption. But, perhaps unfortunately, it is a good deal less precise in its evaluation of the risk of occupational injury from noise than "Noise in Factories". The report concentrated on noise as "annoyance" and relegated "occupational exposure to high levels of noise" to a separate, short chapter. It is fair to say that that chapter emphasised the difficulty of establishing a criterion. The first sentence of its conclusion runs, "Although it is not possible to state with certainty the characteristics of noise exposure which may cause permanent threshold shift and there are few data on the levels of industrial noise in this country, it is likely that a hazardous noise environment exists in many industries". The report recommends further research, but listed three criteria all of which were roughly equivalent to 90 dB(A).

More significant was "Noise and the Worker" (Ministry of Labour, 1963), a Ministry of Labour safety, health and welfare booklet, first published the same year, 1963. These booklets are intended specifically for industrial employers: they are part of the "armoury" of the factory inspectorate. "Noise and the Worker" in effect adopted the tentative criteria of "Noise in Factories", with a little more detail. It recommended that workers should not, for example, be exposed to noise of 85 dB or more in the octave bands in the range 300 to 1200 c/sec for an 8-hr period. It gave no guidance on exposure for shorter periods or fluctuating noise, but it did suggest a reduction of permitted levels by 10 dB in respect of impulse noise. [The current (1971) edition of the booklet adopts the criteria set out in the Department of Employment (1972) code of practice, which is discussed later in this chapter.]

"Noise and the Worker" is not a technical treatise. Perhaps the most important section is one entitled "Have you a Noise Problem?" This attempts to warn employers of the risk of noise-induced hearing damage in simple everyday terms. It is worth reproducing.

A convenient test of hearing impairment is whether workers can hear and understand everyday speech under everyday (quiet) conditions. If they begin to find this difficult it may be that they are being exposed to excessive noise. This effect may not, however, show itself for some considerable time.

The following points should be considered:

1. Do workers find it difficult to hear each other speak while they are at work in a noisy environment?
2. Have workers complained of head noises or ringing in the ears after working in noise for several hours?
3. Have workers who have been exposed to very high noise levels for short periods experienced temporary deafness, severe enough for them to seek medical advice?
4. Have workers exposed for longer periods complained of a loss of hearing that has had the effect of muffling speech and certain other sounds? Have they been told by their families that they are becoming deaf?
5. Has there been a higher labour turnover in workshops or sections where there is a lot of noise?
6. Has management formed the opinion that noise is affecting production?

If the answer to several of these questions is "yes", there may well be a problem of excessive noise. . . .

This section is retained in its essentials in the present edition except that the reference to "several" is replaced by "If the answer to *any* of these questions is 'yes'. . . ." It is, of course, rather vague and may not be sufficient to put an employer on notice in respect of every noise problem. But it is surely sufficient to make "the reasonable employer" aware of the risks inherent in very noisy workplaces.

The most important public document is the "Code of Practice for Reducing the Exposure of Employed Persons to Noise", first published by the Department of Employment in April 1972. The code of practice is, as its title implies, largely concerned with remedies, but it also establishes a comprehensive "damage risk criterion". This was based upon the results of a joint investigation carried out by the Medical Research Council and the National Physical Laboratory and published (Burns and Robinson, 1970) under the title "Hearing and Noise in Industry". The noise limit is expressed as an A-weighted sound pressure level, and this obviates the need to specify different values for different frequencies. The core of the criterion is 90 dB(A) for an 8-hr exposure. For shorter periods and different sound levels a procedure is described for obtaining a corrected equivalent continuous sound level, the value of which should not exceed 90 dB(A) (normalised to 8 hr). Where it is difficult to measure the exposure, it recommends that there should be no exposure at all to levels above 90 dB(A). Over-riding limits of 135 dB(A) for brief exposures and 600 Pa (150 dB SPL) for instantaneous impulse noise are also specified. The code of practice also suggests that all employers carry out a survey of their premises to identify any areas where noise is received that is above the established limit.

The establishment of a 90 dB(A) damage risk level in 1972 by the code of practice was not the end of the story. In 1975 the Industrial Health

Advisory Committee's Sub-committee on Noise, which had produced the 1972 code, presented a paper to the Health and Safety Executive (1975a) entitled "Framing Noise Legislation". The contents of this paper are described in detail elsewhere. It led, after considerable debate, to two Health and Safety Commission papers published in July 1981. One was a consultative document "Protection of Hearing at Work" (Health and Safety Commission, 1981a) containing relatively precise indications of the contents of proposed regulations, a draft new code of practice and a draft technical guidance note. The other was a technical background paper "Some Aspects of Noise and Hearing Loss" (Health and Safety Commission, 1981b). These papers must be considered, along with the more formal public documents, as having a possible impact upon the state of public knowledge. The main change they contain is, of course, formal rather than substantive. The proposal is that the limiting noise exposure be enacted as regulation. There is, however, no radical shift away from 90 dB(A). There is, on the other hand, a recognition that exposure to lower levels is not to be recommended and a wider obligation is proposed to reduce noise exposure as far below the limiting level as practicable. Further, there is an increased emphasis upon information, instruction and training, including the suggestion that employers be required to produce a "programme of action". It has to be accepted that the publication of these papers must have some effect upon the general level of knowledge and, in particular, upon the fact that noise exposure less than 90 dB(A) for 8 hr may be damaging.

These developments have to be understood against a background of international change and development. In several countries, movements can be seen towards the acceptance of a lower noise limit of 85 dB(A) or maybe even lower. Sweden, for example, works to 85 dB(A) and the United States has been considering a proposal which would use 85 dB(A) as an "action limit" at which, for example, protection might have to be provided and retain 90 dB(A) as a "permissible noise limit" at which penalties might be exacted. But the development of most significance to this country has been the draft directive (EEC, 1982) on the protection of workers from the risk of noise at work published by the Commission of the European Economic Communities in 1982. This proposes a daily noise exposure level L_{eq} of 85 dB(A). It is not intended to be "self-executing" and therefore contains more flexibility than might at first be apparent. Member states would have until 1 January 1985 to comply and may be allowed five years beyond that at 90 dB(A) "whenever it is not reasonably feasible to comply". It will not be long before it becomes difficult for a defending employer to argue that the risks of exposure to 85 dB(A) were not reasonably knowable.

Of what risks, then, is it reasonable to expect an employer to know? When will an employer be held negligent for disregarding a risk of noise-induced hearing loss? To begin with, one thing may be said with some certainty. If the employer runs a noisy process where it is well established and well known that the operatives suffer deafness, as was the case in the chipping shop in *Berry* v. *Stone Manganese Marine Ltd.* (1971), then he must do something about it and is negligent if he does not. He must act when common sense tells him to. In this sort of case, precise limits and technical knowledge are relatively unimportant. In *Carragher* v. *Singer Manufacturing Co.* (*U.K.*) *Ltd.* (1974), the noise in question was produced by a drop hammer. The plaintiff's case was that "the noise level was frequently such that it caused pain to those in the vicinity and was generally at the level of 120 decibels". The defendant's main defence was that, at least until the code of practice in 1972, there was no reason for them to know that the noise was of a dangerous level. In the preliminary hearing, the judge felt that the plaintiffs might well be able to establish their contention that occupational deafness associated with drop hammers had been known for 50 years. This was clearly the crux of the matter. If drop hammers are known to deafen people, it does not matter much how many decibels they produce or whether what they produce is in excess of a limit established in the public documents.

Over the past 15 years, however, knowledge as to the relationship between noise and deafness has grown and become more precise. The documents described here have gradually brought this information to the notice of the public, and of employers. Today, a reasonable employer ought to know that to expose an employee to noise in excess of 90 dB(A) for 8 hr or its equivalent is potentially hazardous. It also seems a fair assumption that the reasonable employer should have known of the criteria set out in "Noise in Factories" and "Noise and the Worker" at least by the late 1960s. This seems to have been the opinion of J. O'Connor in *McIntyre* v. *Doulton and Co. Ltd.* (1978). In that case a china manufacturer, whose production plant contained no noisy processes, maintained a small woodworking shop where display cabinets were made. There noise levels occasionally rose to 100 dB(A) and the judge found that a fair equivalent continuous noise level would be 93 dB(A). In these circumstances, the employers argued that they were not alerted to the risk until the publication of the code of practice in 1972. This the judge rejected: he examined the public documentation, particularly "Noise and the Worker", and concluded that "from the late 1960's the dangers of noise ought to have been appreciated by all major employers".

It is dangerous to be dogmatic: every case depends on its facts. It should perhaps be pointed out that the facts in *McIntyre* were more in favour of the employer than the employee. In a noisier industry, the "date of knowledge" might well be the mid-1960s even for less noisy but still dangerous processes. It was not until 1979 that the courts took the opportunity of an exhaustive examination of the "date of knowledge" issue in an ordinary case—one not involving extreme noise levels like *Berry* nor the relative peace of the china factory in *McIntyre*. In *McGuiness* v. *Kirkstall Forge and Engineering Co. Ltd.* (1978) the plaintiff was employed for some 33 years in noise levels of up to 104 dB(A). The particular facts of the case do not, perhaps, cause any problem. As a matter of fact, it was well known that the drop-forging industry was aware of noise risks for much of the period at issue in the case. However, the judge, J. Hodgson, examined the available literature in detail and concluded that 1963, the date of "Noise and the Worker" was a watershed and that, although some time might be allowed to the reasonable employer for response, such response to knowledge had to be reasonably prompt and such response itself became part of the "continuum of knowledge" which imposed upon employers the continuing duty of keeping abreast.

There is dilemma here. Knowledge of the risk coupled with inaction is almost inevitably negligence. Suspicion that there may be a risk is often not, in the real world, sufficient ground for action. On the other hand, an employer who willfully shuts his eyes to the possibility of a risk is also likely to be held negligent. The code of practice (1972) recommends noise surveys. Even before 1972, an employer who on the basis of "Noise and the Worker" failed to conduct a survey may well be held negligent.

An employer who identifies a risk may yet be held negligent if he fails to adopt reasonable remedies to combat it. This in part requires the examination of "the state of the art" at any relevant time to discover what remedies could be said to be available to the reasonable employer. Thus, again, guidance can be sought from the publicly available documents. That which is officially recommended can hardly be said to be unavailable.

But that is not all. The law requires only reasonable remedies to be adopted. An employer may properly reject remedies that are unreasonably expensive or too difficult to carry out. This involves a difficult balancing process. It depends upon some assessment of the likely severity of the injury, upon the resources which it might be reasonable to expect an employer of the type in question to make available for the purpose, as well as upon the design of the equipment, the structure of the workplace, and the size and organization of the workforce. It is impossible to establish firm rules or principles.

In this context it must be remembered that, as was pointed out firmly in *McGuiness* v. *Kirkstall Forge,* the provision of remedy is also something that changes over time. To provide a remedy this year might avoid a finding of negligence; to be discovered providing only the same remedy in a decade's time might well be evidence of negligence.

There are many obvious remedies, often cheap and easy to adopt. Technical advice and expert assistance are also to hand. "Noise in Factories", "Noise and the Worker" and the code of practice all place considerable emphasis upon planning the environment so as to reduce noise. Of course, the opportunities for large-scale planning of this sort are likely to be limited: new machinery is not bought every week, nor is accommodation planned or re-planned very often. But minor work can often be carried out cheaply and to great effect—the resiting of noisy processes away from other workplaces, the use of sound-absorbing or insulating material, even the use of temporary screens can reduce the transmission of noise. The same is true of noise reduction at source. There may often be a variety of remedies which may be technically feasible and useful. [In this connection, the Department of Employment have published Technical Data Note 12, "Notes for the Guidance of Designers on the Reduction of Machinery Noise", Health and Safety Executive (1975b).]

It is perhaps more immediately obvious to centre the approach on the potential victim of the noise. One remedy which is often easy to carry out is the reduction of exposure by reorganising the work. Personal ear protection is also relatively cheap and apparently easy to supply. But here some caution is needed. An employer does not discharge his duty of care by giving employees something to stuff into their ears and doing no more. That does not protect them: it may even increase the risk by giving them a totally false sense of security. In two cases, *Berry* v. *Stone Manganese Marine Ltd.* (1971) and *Bolton* v. *Hawker Siddeley Aviation Co. Ltd.* (1973), employers who had provided earplugs were held negligent, in the first case because no attempt had been made either to fit the plugs or to instruct or train employees in their use, in the second because, in the circumstances, earmuffs were the only practicable form of protection. So not only must a properly informed choice be made, but the employer bears some responsibility for seeing that the protection is used.

This raises one of the most vexed questions in the whole area. It cannot be said what level of responsibility an employer bears in seeing that the use of ear protection is understood and that it is worn. Obviously, this is a matter of discipline, organisation and industrial relations. What is possible or reasonable will depend upon time and place. However, an employer with no system of training and discipline with respect to the use of earmuffs, etc., is very likely to be negligent. On the other hand, one who has arranged for proper fitting, instruction in the use and the effectiveness of protectors, who has an administrative system for recording the issue and maintenance of protectors and who has made the failure to wear them a matter of discipline is probably not negligent if the man has still failed to protect himself. There is, however, still some doubt on the issue of monitoring audiometry—the regular testing of the hearing of the workforce so as to detect damage. This may have two functions. It may help to identify any individual who is peculiarly susceptible to noise and who should be given special treatment. This point was raised in *Bolton* v. *Hawker Siddeley Aviation Co. Ltd.* and the judge, J. Faulds, was prepared to hold the employers negligent for not having carried out such tests which would have indicated that the plaintiff in that case was specially sensitive to noise. This has particular relevance when new starters are considered. An employer would certainly be wise to carry out audiometric tests on persons he is about to employ. This will not only give him a record of the state of each individual's hearing at that date, which would be of great value in the event of a subsequent claim, but would also enable him to identify any individual who, because of his already damaged hearing, ought not to be exposed to high noise levels at all. In relation to the latter point, failure to take such steps certainly could be accounted negligence.

Monitoring audiometry may also be used on a wider scale to identify among the workforce as a whole those whose hearing has begun to deteriorate, so that they may, if possible, be removed from danger. In this there is, of course, some slight element of "locking the stable door after the horse has bolted", but since the damage is cumulative, there is less in that objection than might appear. However, from a legal point of view, it might be hard to hold an employer negligent for, in effect, failing to go out among his workforce to inform those who might have a claim against him of that fact. This objection was raised in *Berry* v. *Stone Manganese Marine Ltd.,* and the judge, J. Ashworth, seemed prepared to accept it. In neither case was the issue crucial, but the issue cannot be said to be closed.

Monitoring audiometry occupies, it would seem, a slightly ambiguous place in common law claims. Nor has its position as yet been made much clearer in official requirements. "Framing Noise Legislation" proposed the production of a code of practice on monitoring audiometry but did not propose the imposition of any statutory requirement. The consultative document of July 1981, "Protection of Hearing at Work", however, proposes regulations requiring both monitoring audiometry and individual monitoring of exposure of individuals whose exposure, disregarding the effects of personal hearing protection, is likely to exceed 105 dB(A) for 8 hr. At the very least, this must affect the position so as to make it difficult to argue that the reasonably prudent employer has never heard of the use of monitoring audiometry as part of the plan to protect employees peculiarly at risk.

A claimant who successfully makes a claim for damages will be awarded lump-sum compensation from his employer. This in practice means that the employer's insurers will have charge of the case and may well reach a settlement. The calculation of the proper amount of the compensation is a difficult matter and will not be described in detail at this point, but some general comments might be made.

Compensation is intended to compensate for the disability suffered by the successful claimant. This is not a matter of precision, but the amount of disability obviously rises as the condition progresses. But it does not increase in the same way and at the same rate as the measurable hearing loss. A man may suffer quite substantial identifiable hearing loss before he has a disability which would merit compensation. His losses will, typically, appear first in frequencies higher than the normal speech frequencies and the interference with daily life that such losses represent to an ordinary person are slight or negligible. Later, when the losses are clearly affecting ordinary life, the extent may be slight. The claimant may, in everyday language, be a little hard-of-hearing. This is a condition to which he may well have adapted himself over the years so that its effect is further minimised. Later, the hard-of-hearing man will have become seriously deaf and the disability suffered is correspondingly very much greater.

This factor may be relevant in various ways. An employer in receipt of a claim may be able to show that the industrial noise exposure for which he is responsible only accounts for the last few years of relevant noise received by the claimant's ears. He may have had other employment or war service which might have affected him. But the employer is nevertheless faced with a substantial claim.

Again, although the risk of noise-induced deafness has been recognised for a very long time in some form or other, the past 15 years have seen a growth in the scientific knowledge about it and in public consciousness. These are both relevant considerations when a long-term exposure is in question. An employer such as the building contractor in *Down* v. *Dudley, Coles, Long Ltd.* (1969) may be held quite blameless for not recognising a noise risk in 1960, but he would clearly be held negligent for the same inaction in 1970. A great deal, of course, depends on the risk. "Boilermaker's deafness" has been known for years, but the risks inherent in 8 hr of exposure to 90 dB(A) might arguably be said to have been part of public knowledge for less than 10 years.

Similar considerations apply to remedies. Meeting a noise problem by wholesale and indiscriminate issuing of earplugs is no longer considered good practice, if it ever was. There is a greater emphasis on noise reduction at source and reduction of exposure than on personal protection in modern writing and thought. Earmuffs are also now generally regarded as a better form of protection, in many cases, than plugs, despite the greater expense and the greater difficulty in ensuring use. As attitudes change, so must reasonable care.

The consequence is that a claim based upon 10 years of exposure to noise may not receive a simple answer. It may be that the employer may be considered negligent only for, say, the last 4 years of the period, either because the reasonable employer would only have become aware of the risk some 4 years previously or because what he did to meet the risk, while reasonable in the light of current practice and knowledge for the first 6 years, became unreasonable with the advance of knowledge or change of practice which can be said to have taken place in the last 4. In any case, the claimant will be left with a half-successful claim, effective only for the period immediately preceding the action. (The picture may, of course, be further complicated if the employer altered his practice at some stage, so as to render negligent behaviour blameless, but for the sake of argument we will assume a relatively simple case.) The result would be that the claim would succeed only as regards the loss attributable to the period in which the employer had been held to be negligent, which would in practice mean a reduction of the amount of damages recoverable.

D. Awards

Damages for personal injuries in civil claims are a lump sum awarded by the court and intended to compensate the successful plaintiff for the losses caused by his injury. They can broadly be looked at under three heads. Two of these heads are fairly straightforward attempts to rectify financial loss. First, the plaintiff may have suffered immediate loss; he may have been off work as a result of the injury, which both loses him income and costs him money. Second, the injury may have disabled him from carrying on in the same employment, so that he has been obliged to take less-well-paid work and has so suffered a loss of earnings. Neither of these heads are likely to be very important in noise deafness cases.

The third head is much less easy to assess. Damages are awarded for the actual physical damage suffered, for loss of physical amenity. A man must be compensated simply for being deafened or for losing his left leg. But how can this be done? In the end, it cannot really be anything better than guesswork, roughly linked to some sort of scale of values of various faculties and parts of the body together with age and way of life. The presence of tinnitus (ringing in the ears) may also be a relevant factor. After a while, it is possible, with experience, to estimate fairly accurately the level of damages likely to be awarded for different types of injury. Unfortunately, there have as yet been so few cases of compensation of deafness by itself that such experience is fairly sparse. There may have been cases where hearing loss has been compensated as part of more complex and extensive injuries, but such cases cannot be a very reliable guide. In the noise deafness cases that have so far been decided, damages ranging between £600 and £27,000 have been awarded, though the latter was a Northern Ireland jury award and probably included a large element for loss of earnings. At present the best that can be offered is a careful guess: that a maximum of something like £10,000–15,000 might be awarded in an English court for loss of amenity for very serious damage to hearing in both ears and that an average award of some £7000–8000 might be expected. There have been awards of up to £7000 for tinnitus. In severe cases this may represent a considerable handicap as it is present all the time, whereas a hearing loss only affects the voluntary use of the hearing facility. These figures should be treated with caution. In addition, there may be special damages awarded for immediate losses or for loss of earnings, where appropriate. In this context, the so-called "Smith v. Manchester Corporation (1974) element" must be considered; that is to say, the award of damages to compensate not loss of earnings (deafened people do not commonly lose their jobs as a consequence of their deafness), but the effect of the disability upon the plaintiff's chances in the labour market were he, voluntarily or involuntarily, to seek other employment in the future. This is sometimes regarded as a single premium insurance policy to cover the risk of such loss in the labour market, and its amount must vary with the state of the market, the skill and age of the plaintiff and the security and stability of his present employment. "Smith element" damages of around £1000 to £1500 have been awarded in deafness cases. All these figures assume full liability for the whole period of exposure, with no reduction for contributory negligence, etc.

E. Protective Legislation

It is only recently that statutory provisions have been made specifically to deal with the problem of noise. There are, of course, general provisions in the various statutes which could be used to cover noise. One prime candidate for such use is Section 29(1) of the Factories Act of 1961, which requires that every workplace "so far as is reasonably practicable" be "made and kept safe". We have the authority of statements of judges in three cases, *Berry* v. *Stone Manganese Marine Ltd.* (1971), *Carragher* v.

Singer Manufacturing Co. (U.K.) Ltd. (1974) and *McIntyre* v. *Doulton and Co. Ltd.* (1978) that a noisy workplace might be unsafe within this section. This must be taken as a qualification to the view that, in general, Section 29 applies only to risks created by the *place,* not the *process* carried out there. With noise, a clear distinction is not possible. It is, however, fair to say that Section 29(1) has not been widely used by the enforcing authorities.

At present there are only two statutory provisions which deal in specific terms with noise hazards. Regulation 44 of the Woodworking Machines Regulations of 1974 provides that if a person employed in "any factory, or any part thereof, mainly used for work carried out on woodworking machines" is likely to be exposed to 90 dB(A) for 8 hr or its equivalent, then (1) "such measures as are reasonably practicable shall be taken to reduce noise to the greatest extent which is reasonably practicable" and (2) "suitable ear protectors shall be provided and made readily available" to those affected. The first part of this adds little to the common law of negligence, although its importance as introducing a criminal sanction is considerable. The second part, however, which is additional, specifically requires the provision of "suitable" protectors. No matter how unreasonable it might be to expect it, how costly or how difficult, if they are not provided the occupier of the factory (generally the employer) is in breach of the regulations. Furthermore, the regulations go on to require that the protectors shall be properly maintained and "shall be used". Which means that if they are not, for whatever reason, the occupier as well as the operative concerned are in breach again. There is no list of protectors approved as "suitable".

The Agriculture (Tractor Cabs) Regulations (1974) approach the question slightly differently. They require that tractors used in agriculture should be fitted with safety cabs in respect of which a certificate of approval has been issued. Among other things, the cab must not be such as to permit noise levels of more than 90 dB(A) within it for such a certificate to be issued. The use in agriculture of a tractor covered by the regulations without an approved safety cab is a breach of the regulations.

There is little doubt that there will be further close regulation of industrial noise. The power to make regulations on the subject has existed for a long time under various acts of Parliament [e.g., Section 76 of the Factories Act (1961), Section 21 of the Offices, Shops and Railway Premises Act (1965)]. Apart from the woodworking machinery regulations and the other matters already mentioned, these powers have not been used to tackle noise risks. Moreover, the powers are somewhat complicated: regulations made under each act can apply only to those premises that are covered by that act. Total, simple coverage would have been impossible. However, under Section 15 of the Health and Safety at Work Act (1974), the Secretary of State for Employment has a single wide power to make health and safety regulations and paragraph 9 of Schedule 3 of the act specifically mentions noise as a possible subject of regulation. This power could be used to impose conditions on any noisy place of work.

Progress towards general noise regulation has been slow. The Industrial Health Advisory Committee's Sub-committee on Noise, having been responsible for the code of practice in 1972, was asked by the ministry, a year later, to consider the further problems of legislating against noise. The result was "Framing Noise Legislation" (Health and Safety Executive, 1975a), a comprehensive programme for legislative action, which was published in 1975. "Framing Noise Legislation" was passed to the Health and Safety Executive, the body responsible for the day-to-day administration of the Health and Safety at Work Act (1974) (and, indeed, all occupational safety legislation), and the executive adopted the report as a consultative document. Under the Health and Safety at Work Act (1974), there are three occupational safety authorities: the Health and Safety Executive, the body charged with administering the law; the Health and Safety Commission, a representative policy-making body in overall control, with power to approve codes of practice and to propose the enactment of regulations; and the Secretary of State for Employment, who has power to approve and set in motion the enactment of regulations. Thus, 1975 saw "Framing Noise Legislation" inserted into the system for occupational safety legislation.

The report proposed regulations which would impose four obligations. First, all employers would be obliged to carry out noise surveys. If such surveys identified areas where the noise exceeded 90 dB(A), then the other three duties would come into force. Employers would be required to use all practicable means to reduce the level to one below 90 dB(A); if this failed, the remaining noisy areas would have to be appropriately identified (by approved warning notices), and all persons entering such areas would have to be provided with suitable ear protection. In addition, employees would be put under an obligation to use provided equipment and manufacturers of process machinery would be required to take care to ensure that their machines, when used, did not emit dangerous noise.

No action was taken on "Framing Noise Legislation", but in July 1981, the Health and Safety Commission published a further consultative document, "Protection of Hearing at Work", along with a technical background paper "Some Aspects of Noise and Hearing Loss". The consultative document makes proposals which differ in certain respects from those contained in "Framing Noise Legislation". The proposed regulations would impose a general duty on all employers in all circumstances to reduce noise levels and noise exposure to the greatest reasonably practicable extent, even below 90 dB(A). This general duty presumably will include a duty to carry out noise surveys. In any event, where levels (disregarding the effect of personal protection) exceed 90 dB(A), it is proposed that employers carry out surveys, keep publicly available records, provide for instruction and training, issue "suitable and effective" ear protectors, produce a "programme of action", appoint a noise adviser and keep records. If the relevant level exceeds 105 dB(A), then audiometric testing and the monitoring of individual exposure will be required. Employees will be required not only to make use of protective equipment, but also to cooperate with the employer in safety measures. Duties are proposed for manufacturers and designers of "articles for use at work" both for the reduction of sound emission generally and, in respect of likely emissions in excess of 90 dB(A), to provide information in respect of the use of the article. An approval system for ear protectors is also proposed.

The consultative document also contains a draft code of practice to be read in conjunction with (and as an amplification of) the regulations and a draft technical guidance note. The code of practice is intended to be an "approved" code under Section 16(1) of the Health and Safety at Work Act.

These proposals should, perhaps, be seen in the context of the EEC (1982) "Proposal for a Council Directive on the Protection of Workers from the Risks Related to Exposure to Agents at Work: Noise", which proposes a limit of 85 dB(A) to be implemented by the legislation of member states by 31 December 1984 [with a possible "extension" at 90 dB(A) until 1990 "whenever it is not reasonably feasible to comply"].

Also in November 1982, a series of recommendations from the Industrial Injuries Advisory Council (DHSS) was accepted by the Secretary of State.

Finally, the provisions of the Health and Safety at Work Act (1974) must be considered. The overall aim of this act was, by way of implementing the Robens Report on Safety and Health at Work (1972), to set up an administrative and executive structure for occupational health and safety legislation as will enable the existing somewhat confused mass of law to be modernised, generalised and made more flexible and effective. The act itself contains very little by way of substantive law. However, Sections 2 to 9 impose "general duties" upon employers, the self-employed, occupiers of premises, manufacturers of plant and material for use at work and employees. These duties are, in the main, cast in the "reasonably practicable" form. Thus, under Section 2, an employer is obliged "to ensure, so far as is reasonably practicable, the health, safety and welfare at work of his employees". Section 7 imposes on employees the duty "to take reasonable care for the health and safety of himself and of other persons. . .".

The functions of these general duties are to act as a "safety net" to cover matters which may not already be covered by pre-existing legislation such as the Factories Act or codes of regulations and to operate as a generalised extension of protective legislation into occupations hitherto not covered by legislation at all. They are just like any other statutory duty in this field. Breach of them could lead to prosecution and a fine.

Of course, they are not specific, but that does not mean that they cannot be used. "Reasonably practicable" may be vague, but it can be given meaning especially when a code of practice exists. Indeed, this is precisely what is envisaged in the act. Section 16 entitles the Health and Safety Commission to approve codes of practice. Once approved, such codes, by virtue of Section 17, are to be presumptive evidence that what is contained in them is "reasonably practicable". Until a code is approved, it may still be used as evidence—the only difference is that the special rule of Section 17 (which makes it virtually impossible to controvert it) does not apply. It follows that employers could be prosecuted for a breach of Section 2, proof of breach being provided by astute use of the 1972 code of practice. "Framing Noise Legislation" suggested that the 1972 code of practice, with appropriate amendments, be approved; the consultative document (Health and Safety Commission, 1981a) contains a new code designed to be so approved. No action has yet been taken on either proposal.

The Health and Safety at Work Act also contains new powers of enforcement. Under Section 21, an improvement notice may be issued against any person (including a company and an individual, an employee or a manager) who in the opinion of the inspector is or has contravened some part of the law—Factories Act, code of regulations or general duty—and the notice requires the defect to be remedied. It may be appealed to an industrial tribunal, but if it is not, it must be complied with and cannot be questioned later. Failure to comply with an improvement notice is a criminal offence. A prohibition notice, under Section 22, is similar to an improvement notice, save that there is no need for a breach of the law, but instead that activities are being carried on which involve a risk of "serious personal injury". The prohibition notice orders the activities to stop at a specified date.

Clearly both these provisions may be used for noise. Improvement notices, based on the general duties, have been issued on many occasions already.

F. National Insurance for Industrial Injuries

So far we have been concerned entirely with common-law claims, that is, claims made in the ordinary civil courts, the county court or the High Court, for damages for personal injuries caused by the employer's negligence or other breach of duty. If successful, such claims result in an award of damages, a lump sum to be paid by the defendant, usually the employer, to the plaintiff. (In practice, of course, it is usually paid by the employer's insurance company and often by settlement after negotiation with the workman's advisers without going to court.)

This is not the only compensation legally available to those injured at work. The National Insurance (Industrial Injuries) Scheme is the direct descendant of the Workman's Compensation Scheme. It now operates under the National Insurance (Industrial Injuries) Act of 1965. Its aim is to provide basic compensation by way of periodic pension to those who are injured at work. Claims are not made through the ordinary courts, but through a completely separate system of insurance officers, appeal tribunals and the like. There are three benefits: injury benefit, for the time an injured person was away from work through industrial injury; disablement benefit, to compensate for any permanent damage; and death benefit. The benefits (except in the case of certain death benefits) are weekly pensions. The disablement benefit is assessed in accordance with the extent of disablement, expressed as a percentage of total disablement.

Entitlement depends upon two things: that the claimant was an insured person within the ambit of the scheme (and all employed persons are) and that he was *either* suffering from personal injuries caused by an accident arising out of and in the course of his employment *or* suffering from a prescribed industrial disease. Ever since the scheme began in 1897, "accident arising out of and in the course of employment" has caused difficulty. In fact, it has caused several quite separate difficulties. The one which is most relevant to noise-induced injury has to do with the use of the word "accident". The word, in normal use, implies an event, an occurrence. There was an accident on a certain day, which caused the following personal injuries to the claimant. Except in cases of traumatic injury, caused by explosions and the like, most noise deafness cases are difficult to fit into that mould. There is no "accident", rather a long-term "process" which gradually causes deafness.

It was to meet the general problem of diseases and conditions brought about by long-term exposure to adverse working conditions that the notion of prescribed industrial diseases was introduced. Regulations specify a number of diseases and the processes with which they are linked. An applicant, instead of proving that he was injured by an "accident", shows that he worked at the process specified for the specified length of time and is now suffering from the specified disease. He is then entitled to benefit.

The problem with noise deafness is that the number of processes which could give rise to it is very large. It is not the same as, for instance, asbestosis, which is linked only with processes involving asbestos. Noise is everywhere in industry. It is hardly surprising that it took the Industrial Injuries Advisory Council a long time to reach a conclusion or that the conclusion when reached bore the marks of a compromise. Their report was published in November 1973 and recommended that noise deafness should be prescribed as an industrial disease linked to at least 20 years of exposure to pneumatic tools or drop forges or hammers in either the metal manufacturing or the shipbuilding or ship-repairing industries.

The recommendation was implemented in October 1974 when occupational deafness was added to the list of prescribed industrial diseases, with effect from February 1975. The list of prescribed occupations is regularly re-examined and has been added to and amended since 1975. Although there is some doubt on this point, it would appear that, although the terms of the provision would strictly permit any person whose work involves any of the specified processes to any extent at all to qualify, occupations in which the time spent on the process in question are negligible do not count. It should also be noted that there is no requirement at all that the deafness (as defined) should be shown to be related to the occupation in question. A claimant qualifies by being employed in a specified occupation for 10 years and, secondly, by suffering from the specified hearing disability.

The definition of occupational deafness is designed to ensure that only conditions which indicate deafness brought about by industrial noise should qualify. It is defined as "substantial permanent sensorineural hearing loss amounting to at least 50 dB in each ear, being due in at least one ear to occupational noise, and being the average of pure tone losses measured by audiometry over the 1, 2 and 3 kHz frequencies (occupational deafness)." This minimum loss is rated at 20% disablement. More severe hearing loss is rated higher in a series of defined steps up to 100%. Claims must be made within 5 years of the appropriate employment ceasing.

II. THE CONTROL OF ENVIRONMENTAL NOISE

A. Nuisance

English law has a device for the control of pollution which is ancient, flexible and of considerable effect. This is nuisance. Nuisance was known in mediaeval times and its wide application was recognised early. "Nocumenta", wrote Bracton in the thirteenth century, "infinita sunt"—nuisances are infinite. The central idea is and was simple: that the law might be used to suppress activity which interfered with ordinary, decent daily living. Tanneries, prostitutes, church bells, electrical generating stations, printing presses, crowds of people, rats and heaps of rubbish have all, from time to time, been argued or proven to be nuisances. Noise undoubtedly can be a nuisance.

To a lawyer, nuisance is a slightly unusual concept. Generally, legal remedy is confined to reparations for physical damage, the deprivation of a specific property right or identifiable (and quantifiable) pecuniary loss. Nuisance comprehends cases involving neither property damage, personal injury nor financial harm, where all that can be shown is a substantial interference with the complainant's legitimate expectations as to his life-style. The discotheque next door need not lose me money, nor interfere with my health, nor physically damage my house by its vibrations (although it may conceivably do all those things) for it to be a nuisance. It may be enough that it interferes with my sleep.

Such a path may be dangerous if followed too far. The law cannot be used to remedy every interference with our way of life, to support every complaint about our neighbour's behaviour. First, it can only be *substantial* interferences that are nuisances. Further, in expansion of the point, there has to be an objective standard. This is achieved by the so-called "neighbour test" to define what constitutes a nuisance. The basic proposition is that an activity is to be considered a nuisance in accordance with the expectations of the generality of those who may be expected to live at that place at that time. "Specially sensitive" complainants, however real their injury, are disregarded. If you need absolute peace and quiet between the hours of 8 A.M. and 6 P.M., do not choose to live above the shoe-repair shop next to the garage on High Street. If you do, you may not complain that the noise is a nuisance. It follows from this that the amount of noise that constitutes a nuisance cannot be defined with any precision. A noise may be a nuisance in one place and time, but noise of exactly similar volume and characteristics may not be a nuisance at another place or time. "Nor, in my judgement", said J. Veale in *Halsey* v. *Esso Petroleum* (1961), "can Rainville Road, Fulham, properly be compared with the Great North Road". We must all put up with the "wear and tear" of everyday life that is appropriate to the place where we live or happen to be.

This leads at first sight to what appears to be an intractable problem. If a factory or loading bay is producing a lot of noise, the question whether that noise is a nuisance must be decided by reference to the neighbourhood in which the factory or loading bay is situated. But the neighbourhood contains the facility under investigation and its standards are affected by it. How may a proper balance be struck?

Simple answers have to be avoided. It is not enough to show that the noise is like other existing noise and therefore not a nuisance. In *Rushner v. Polsue and Alfieri* (1906) another printing press on Fleet Street was held to be a nuisance. Nor is it sufficient for a noisemaker to argue that he was there first. It might have been a nuisance then, but no one complained. Here we must distinguish "private nuisance"—defined later in this chapter—in which it is sometimes possible to "prescribe for" the right to commit a nuisance. That is to say the activity is held to have ceased to be a nuisance because the potential complainants have all, by lapse of time, lost the legal right to complain. However, there are technical reasons why such circumstances are not very common in practice. Similarly, "You came to the nuisance" is no defence, for if it was a nuisance when one came, then one can complain after arriving. All that the owner of a noise source may do is demonstrate that the neighbourhood has, because of his operations as well as for other reasons, altered its nature in such a way as to lower the expected standards of noise emission.

Behind all this flexibility, or lack of precision, lies a fairly strict legal regime. Once a nuisance is proved, the person responsible for it (normally the occupier of the premises whence it came) is strictly liable. It is irrelevant that all reasonable care was taken to avoid the nuisance, that the activity carried on greatly benefits the public or even that it is not possible to carry on that activity without the nuisance alleged. It remains a nuisance.

One final point of difficulty for noise nuisances might be mentioned. The requirement that a nuisance be a "substantial" interference leads to the conclusion that it be substantial in duration as well. This has two aspects. Noises may not be regarded as substantial nuisances because they are intermittent in duration, that is to say, non-continuous. Second, they may simply be impermanent and may not last for very long. The second aspect is easier to deal with. An activity of limited duration, for example, the construction or demolition of a building, is very unlikely to be held to be a nuisance unless the scale of interference is very large indeed. Such is not likely to be the case where the main element of the alleged nuisance is noise. The first point is more complicated. There is a well-supported requirement that a nuisance be "continuous". However, it is the state of affairs which gives rise to the interference that constitutes the nuisance; to use an old example, it is the obstruction to the watercourse which is the nuisance, the flooding consequent thereon being the effect. So, it is the use of the premises next door as an all-night disco which is the nuisance; the fact that it only opens on Tuesdays and Thursdays is only the effect, and, of course, a measure of its intensity. Thus regular, non-continuous noises may constitute nuisances.

For historical reasons, nuisance has been divided into three species: private nuisance, public nuisance and statutory nuisance.

B. Private Noise Nuisance

Private nuisance is a *tort*, or civil wrong. It is not a matter for legal action by a public authority but a private issue between individuals which can provide the basis of a civil claim. Such a claim may be for damages, for compensation to make good the damage caused by the tort or for an injunction, a negative court order prohibiting further continuation of the tort. Clearly, injunctions are considerably more important in practice than are damages.

Over the years, the tort of private nuisance has become inextricably linked with landholding. It is today a tort committed by one landholder against another. It has two "subspecies". Either it may be constituted by the interference with a specific and identifiable legal right attached to landholding, such as a right of way or a right to light, or it may be proved by showing a general interference with the "use and enjoyment" of the plaintiff's land. Clearly, noise, as a nuisance, always lies in the second subspecies.

The general principles of nuisance apply. It is not necessary that the plaintiff show physical damage or personal injury deriving from the noise (although if it is possible to do so, it will not harm the case), but the interference must be substantial as tested by the "neighbourhood test". In this context, British standards, such as BS 4142 (BSI, 1967), have only peripheral significance.

C. Public Noise Nuisance

Public nuisance is a *crime* for which those responsible may be prosecuted and fined. What is more important is that proceedings may be taken for an injunction to prohibit it. Such proceedings may not, however, be taken by private individuals. Either the Attorney-General or a local authority may take such action. The Attorney-General will act "on the relation of" a private individual or a local authority. The local authority, under Section 222 of the Local Government Act of 1972, may take civil proceedings in its own name when it is "expedient for the promotion or protection of the interests of the inhabitants" of the area for which it is responsible. This provision therefore allows direct action by a local authority in cases of public nuisance. It may also permit actions in *private* nuisance, though this point is not free from doubt.

The simplest way of describing the difference between public and private nuisances is to say that public nuisances are big private nuisances. Like private nuisance, public nuisance may be split into two "subspecies." It is a public nuisance to interfere with a public right, such as by obstructing the public highway; and it is a public nuisance to interfere with the comfort and convenience of the public at large or of a substantial section of them. Thus, in the second category, which is the one of importance in connection with noise, the question is when does a nuisance begin to affect enough people that we can describe the victims as "the public at large" or a "section" of the public? Clearly, no numerical answer can be given. Lord M. R. Denning in *A.-G.* v. *P.Y.A. Quarries* (1957) approached the question by asking whether the nuisance had become so extensive or severe that it was not reasonable to ask that private individuals be left to seek remedies for it. That must be the best practical approach. If it is reasonable for a local authority to act, then it is a public nuisance.

D. Statutory Noise Nuisance

Almost from the beginning of modern local government in the United Kingdom, powers were taken to "suppress nuisances". One of the major contributions of the Public Health Acts was the establishment of a statutory procedure for dealing with circumstances and activities which were "a danger to health or a nuisance" in particular areas. Types of nuisance that were within the act and subject to these procedures were described as "statutory nuisances". In modern times, the procedures were contained in Part II of the Public Health Act of 1936. Under this, local authorities were required to investigate nuisances and were given powers to suppress or "abate" them. By Section 1 of the Noise Abatement Act of 1960, "Every noise that is a nuisance shall be a statutory nuisance", thus including noise in the list of statutory nuisances subject to Part II of the 1936 act. These acts, as amended by the Public Health (Recurring Nuisances) Act of 1969, were repealed as regards noise nuisances and replaced by Sections 57–59 of the Control of Pollution Act of 1974.

At no stage in this long history of the "suppression of nuisances" was the term "nuisance" defined or redefined. A statutory nuisance is the same as a nuisance at common law. It is accepted that common-law *public* nuisances may be statutory nuisances, although there is some question whether a nuisance that cannot be regarded as extensive enough to constitute more than a *private* nuisance may be brought within the terms of the statute. The important point to make is that the statutory provisions proceed from the same base as the ancient common law—the flexible imprecision of nuisance. So, although environmental health officers may, very properly, decide whether or not to use the statutory powers on the basis of scientific evidence (such as sound level readings) and relate such to some accepted or official standard (like a British standard), the law is far less scientific. Whatever the meter says, whatever the British standard lays down, it is at least theoretically possible for the argument to be raised that, in all the circumstances, of time, place and "neighbourhood", the noise was (or was not) a nuisance.

Under Section 57 of the Control of Pollution Act each local authority is required to inspect its area in search of noise nuisances. Once a noise nuisance is discovered, Section 58 obliges the authority to serve an "abatement notice" on the person responsible. An abatement notice is essentially a notice requiring that a nuisance cease. It may, but need not, indicate how this is to be done and require particular building or other works to be carried out. It may be used not only for nuisances which exist at the time of its issue, but also for those "likely to occur or recur". The receiver of the notice has 21 days in which to appeal. The appeal is to the Magistrates' Court and may be upon the ground that the notice was improperly made or served, that there was no nuisance or that its conditions are unreasonable. The Magistrates' Court may uphold, quash or amend a notice.

Subject to the appeal, the abatement notice must be complied with. Moreover, it is a perpetual notice: it will not expire. Any person who without reasonable excuse fails to comply with an abatement notice or with a condition attached thereto, commits an offence for which he may be prosecuted and fined. It follows that once an abatement notice is made, it may be regarded as a particular and continuing set of conditions governing the noise emission of the premises to which it relates, which has to be obeyed at all times. It is, for example, permissible for a noise abatement notice to require that no more than so many decibels of noise, measured at a particular point, be emitted from the relevant premises between the hours of 11 P.M. and 7 A.M. Such a notice, if not successfully appealed against, will govern the operation of those premises for all time.

If the noise was made in the course of a trade or business, the recipient of a notice has a further defence. He may show that he used the best practicable means to comply with the notice. This defence is not available to non-traders.

The Section 58 procedure, in theory, *must* be undertaken whenever a local authority is satisfied that a noise nuisance exists or is likely to occur or recur. However, there is clearly room for disagreements in such matters. Section 59, following a pattern set by the 1960 Noise Abatement Act, provides a supplementary "citizen's" procedure. The procedure may be activated by an occupier of premises who is aggrieved by the nuisance: in essence an affected neighbour–householder. Unlike the Section 58 procedure, here the nuisance must be shown to exist or to have existed. It is not enough that it be "likely". The procedure is by way of complaint to the Magistrates' Court. If the court is satisfied of the existence of the nuisance, it must make an abatement order, an order requiring abatement within a given time, with or without the requirement that "works" be carried out to achieve that end. Failure to obey an order under Section 59 is an offence similar to the offence committed under Section 58. The same defences, including "best practicable means" for traders, are available.

There are provisions, applicable both to Section 58 and Section 59 proceedings, whereby the local authority is empowered to execute necessary works itself, at the expense of the recipient of the notice or order, in order to effect compliance with it.

E. Planning and Noise Abatement Zones

Noise is obviously a matter which a planning authority may legitimately take into account when deciding whether to grant permission for any development. The Scott Report on Neighbourhood Noise (1971) examined the problem, isolated certain matters of high significance, such as aircraft noise and road traffic noise, and supported a generalised noise-zoning approach for the control of environmental noise through planning. A result was a ministerial circular, Circular 10/73 (Department of the Environment, 1973), which set out these policies, together with recommended noise-level standards for different types of neighbourhoods for the use of local planning authorities.

In addition to the recognition of noise as a factor within the ordinary planning system, the Control of Pollution Act of 1974, in Sections 63–67, sets out procedures for establishing noise abatement zones. A local authority may, under these provisions, designate all or any part of its area a noise abatement zone by an order which describes not only the zone but the classes of premises within the zone to which the order shall apply. This order will be effective when confirmed by the Secretary of State for the Environment. Thus, an authority may declare the whole of its area a noise abatement zone and declare that all industrial or light industrial premises shall be subject to the order, the descriptions of premises being taken from the use classification order which governs planning matters.

Once an order has been confirmed, the local authority must then measure the noise emission levels, in accordance with a set procedure, of all the premises within the zone subject to the order. These levels are published, and there is provision for contesting the results. Once the task is complete, it becomes an offence to emit noise from premises governed by the noise abatement zone order greater than that which appears in the published record. Conversely, the emission of such noise is made *not* subject to noise abatement procedures under Sections 58 and 59.

However, the purpose of noise abatement zones is not merely to hold noise levels at the levels attained when the zone was established. The act provides for the issue of noise reduction notices, under Section 66, whereby the published level for any premises within the zone may be required to be reduced, if it seems reasonable to do so, within a specified period of not less than 6 months. There is provision for an appeal. Noise reduction notices can be very flexible. They may require a reduction of noise at particular times of the day, for example. If a noise reduction notice is complied with, that affords a defence to any proceedings under Sections 58 or 59.

The local authority has power to consent to the emission of noise greater than that contained in the original lists of noise levels or in any noise reduction notice. There are provisions dealing with new premises in noise abatement zones.

There are provisions in the Land Compensation Act of 1973 for the compensation of those whose property is affected by public works. In addition, and specifically, Section 20 of that act provides for the sound-proofing, in accordance with certain standards, of certain residential premises affected by newly constructed roads. This provision is similar to earlier provisions made under the Airports Authorities Act of 1965 (now consolidated into the Civil Aviation Act of 1982), allowing the expenditure of public money in soundproofing dwellings affected by airport noise, in accordance with schemes published for particular airports.

F. Miscellaneous Provisions

Particular noise problems have, from time to time, received particular attention. The first which calls for mention arises from the very limitations of the central concept of nuisance. As was already stated, a nuisance must be substantial in point of time. Activities of limited duration will rarely constitute nuisances. A prime example of such an activity, which presents distinct environmental problems, is building and demolition.

Under the Control of Pollution Act of 1974, Sections 60 and 61, construction sites (which term includes demolition and other building works) are specifically controlled. There are two procedures. The first, under Section 60, permits a local authority to serve a notice on a wide variety of persons connected with the works, specifying how the works are to be carried out. The notice may specify times at which work, or work of a particular type, is to be carried out and how such work is to be done, including, where necessary, the type of machinery to be utilised. The second procedure is the consent procedure. Under this, the person intending to carry out works within these provisions may apply for a prior consent. Again, in granting this consent, the local authority has powers that are co-extensive with its powers under the Section 60 procedure. Either way, armed with a consent or an order, provided that the operator of the site complies with it, he has a defence to any proceedings under Sections 58 or 59. Failure to comply with an order or consent is punishable by fine.

Motor vehicles have long been the subject of regulation governing the noise they may emit. The current regulations are the Motor Vehicles (Construction and Use) Regulations of 1978 Regulation 31, 114–117. These apply both to the construction and to the use of motor vehicles and set out particular noise values in respect to vehicles of specified types. They require that vehicles not be manufactured which emit noise above the level required for that type, as measured in accordance with the test procedure set out in the regulations. They also require that motor vehicles not be used on any public road or street so as to emit noise which exceeds another set of required levels, slightly higher than the "construction" levels. Again there is provision setting out the appropriate procedure for measuring the noise. It is the second set of provisions which has the most impact. There is also provision for "type approval" of vehicles from the point of view of noise emission and for the implementation of several EEC directives on motor vehicle noise.

Aircraft noise presents certain particular problems. When civil aviation became a practical possibility, it seemed clear that the ordinary law of trespass and nuisance was likely to provide a serious restriction to the freedom of the air. In strict theory, landholders' rights extend infinitely into the air, so that the possibility of each flight being followed, figuratively, by thousands of writs, drove the legislators to enact an immunity from the ordinary law of trespass and nuisance for aircraft. Such immunity, now within the Civil Aviation Act of 1982, is maintained so long as the aircraft is obeying the regulations which govern flying. The Air Navigation General Regulations of 1981 are the regulations which apply generally. However, they do not impose any specific limit upon the noise that may be emitted by aircraft: they simply permit the emission of an unlimited amount of noise.

A different approach has been adopted. First, originally under the Airports Authorities Act of 1965, now under the Civil Aviation Act of 1982, the Secretary of State may establish the rules for the use of airports within his control—the so-called "designated" airports. Such rules may establish procedures for the use of such airports, including the number of night flights and the way in which aircraft are to take off or land. Secondly, the Civil Aviation Authority, part of whose function is the certification and licensing of civil aircraft, has power to issue noise certificates under the Air Navigation (Noise Certification) Order of 1979 in respect of particular aircraft. Without a valid noise certificate, an aircraft may not take off or land in the United Kingdom.

Bye-laws must not be forgotten. Over the years various local authorities have taken power within their areas to deal with a wide variety of noisy occurrences. Their powers can be quite extensive. Finally, we must not forget the prohibition of the use of unpermitted loudspeakers on the public highway, first found in Section 2 of the Noise Abatement Act of 1960, now in Section 62 of the Control of Pollution Act of 1974.

G. Residual Powers

Under Section 58(8) of the Control of Pollution Act of 1974, the local authority's power to take High Court proceedings is preserved. The specific powers contained in the act do not detract from the general powers to proceed under the common law. In this connection, Section 222 of the Local Government Act of 1972 has already been mentioned. Under that act a local authority may take civil proceedings for the protection or promotion of the interests of the inhabitants of its area. Such action may include proceedings for public or private nuisance.

In *London Borough of Hammersmith* v. *Magnum Automated Forecourts* (1978) a new twist, indeed extension, to these powers was discovered. Proceedings had been taken against a company which operated an all-night facility for taxi cabs. A noise abatement order under Section 58 had been issued. The recipients had promptly appealed. The authority had, as they were entitled to, taken steps to ensure that the operation of the order was not suspended pending the appeal. Thus the defendants remained liable to prosecution for disobeying it (subject to their available defences) despite their appeal. Nevertheless, they took no steps to obey. The local authority thereupon proceeded in the High Court and asked for an injunction simply to restrain the defendants from continuing to break the law, relying upon precedents wherein the court had issued injunctions to restrain flagrant breaches of planning conditions. The Court of Appeal acceded to the local authority's request. An injunction would be granted, notwithstanding the existence of enforcement procedure within the Control of Pollution Act of 1974. The court had an inherent jurisdiction to prevent breaches of the law which would be exercised whenever it appeared that statutory procedures were less than adequate.

Environmental health officers, it would appear, have many weapons at their command for the control of excessive noise.

REFERENCES

A.-G. v. *P.Y.A. Quarries*. (1957). 2 Q.B. 169.
Agriculture (Tractor Cabs) Regulations (1974). Her Majesty's Stationery Office, London.
Airports Authorities Act (1965). Her Majesty's Stationery Office, London.
Air Navigation (Noise Certification) Order (1979). Her Majesty's Stationery Office, London.
Air Navigation General Regulations (1981). Her Majesty's Stationery Office, London.
Aldersley-Williams, A. G. (1960). "Noise in Factories." Department of Scientific and Industrial Research, Her Majesty's Stationery Office, London.
Berry v. *Stone Manganese Marine Ltd.* (1971). *Knight's Industrial Rep.* **12,** 13–35 (April 1972).
Bolton v. *Hawker Siddeley Aviation Co. Ltd.* (1973). Queens Bench Division, Crown Court, Chester. 9th March (unreported).
BSI (1967). British Standard 4142. British Standards Institution, London.
Burns, W., and Robinson, D. W. (1970). "Hearing and Noise in Industry." Her Majesty's Stationery Office, London.
Carragher v. *Singer Manufacturing Co. (U.K.) Ltd.* (1974). Scottish Law Times Notes, p. 28.
Civil Aviation Act (1982). Her Majesty's Stationery Office, London.
Control of Pollution Act (1974). Her Majesty's Stationery Office, London.
Department of Employment (1972). "Code of Practice for Reducing the Exposure of Employed Persons to Noise." Her Majesty's Stationery Office, London.
Department of the Environment (1973). Circular 10/73. "Planning and Noise." Her Majesty's Stationery Office, London.
Down v. *Dudley, Coles, Long Ltd.* (1969). *Mod. Law Rev.* **34,** 195.
EEC (1982). "Draft Directive on the Protection of Workers from the Risk of Noise at Work." *Off. J. EEC* **5.11.82,** P.C. 289, 1–6.
Factories Act (1961). Her Majesty's Stationery Office, London.
Halsey v. *Esso Petroleum* (1961). 2 All E.R. 145 at p. 158.
Health and Morals of Apprentices Act (1804). His Majesty's Stationery Office, London.
Health and Safety at Work Act (1974). Her Majesty's Stationery Office, London.
Health and Safety Commission (1981a). "Protection of Hearing at Work." Her Majesty's Stationery Office, London.
Health and Safety Commission (1981b). "Some Aspects of Noise and Hearing Loss." Her Majesty's Stationery Office, London.
Health and Safety Executive (1975a). "Framing Noise Legislation." Her Majesty's Stationery Office, London.
Health and Safety Executive (1975b). Technical Data Note 12. "Notes for the Guidance of Designers on the Reduction of Machinery Noise." Her Majesty's Stationery Office, London.
Land Compensation Act (1973). Her Majesty's Stationery Office, London.
Local Government Act (1972). Her Majesty's Stationery Office, London.

London Borough of Hammersmith v. *Magnum Automated Forecourts* (1978). 1 All E.R. 401.

McGuinness v. *Kirkstall Forge and Engineering Co. Ltd.* (1978). Leeds, October 16, 1978 (unreported).

McIntyre v. *Doulton and Co. Ltd.* (1978). Queens Bench Division, London, 22nd March (unreported).

Mines and Quarries Act (1954). Her Majesty's Stationery Office, London.

Ministry of Labour (1963). "Noise and the Worker." Her Majesty's Stationery Office, London.

Motor Vehicles (Construction and Use) Regulations (1978). Regs. 31, 114–117.

National Insurance (Industrial Injuries) Act (1965). Her Majesty's Stationery Office, London.

Noise Abatement Act (1960). Her Majesty's Stationery Office, London.

Offices, Shops and Railway Premises Act (1963). Her Majesty's Stationery Office, London.

Public Health Act (1936). His Majesty's Stationery Office, London.

Public Health (Recurring Nuisances) Act (1969). Her Majesty's Stationery Office, London.

Robens, A. (1972). "Safety and Health at Work." Her Majesty's Stationery Office, London.

Rushner v. *Polsue and Alfieri* (1906). 1 Ch. 234.

"Scott Report" (1971). "Report on Neighbourhood Noise." Noise Adisory Council. Her Majesty's Stationery Office, London.

Smith v. *Manchester Corporation* (1974). 118 Solicitors Journal 597 CA.

Wilson, A. (1963). Cmnd. 2056. "Noise—Final Report." Her Majesty's Stationery Office, London.

Woodworking Machinery Regulation (1974). Her Majesty's Stationery Office, London.

Workman's Compensation Act (1897). Her Majesty's Stationery Office, London.

13

Noise and the Law in the United States

P. S. EDELMAN AND A. J. GENNA*

Kreindler and Kreindler
New York, New York

* Present address: Victora & Genna, New York, New York.

I. INTRODUCTION

This chapter aims to give a selective over view of the law applicable to some of the problems created by excess noise. The law regarding noise control is diverse and extensive, including areas such as contract law, lease law, and Constitutional law. Among the areas discussed will be regulation of sound emanating from particular sites (e.g., industrial sites, highways, and airports) as well as remedies and theories of liability—both statutory and nonstatutory—under which industrial workers can be protected and noise abated or under which noise polluters may be prosecuted. Attention will also be given to building codes designed to minimize sound pollution.

II. NONSTATUTORY LIABILITY: NOISE AS A NUISANCE

The word *noise* comes from the same Latin root as the word *nausea*. The distaste of the Romans for excessive sound is not known to us only by etymology; the Romans had laws which specifically regulated noisy chariot traffic passing near residential areas at night. The English common law first took cognizance of the noise problem in the thirteenth century in cases in which noise was termed a "nuisance," and remedies similar to those available to modern plaintiffs (injunction, damages, etc.) originated then. The word *nuisance* has been used so widely in English and American jurisprudence as to be almost impossible to define. The word has been used to describe foreign objects accidentally included in foods, obnoxious advertising, and slanderous statements. The theory of nuisance is still, however, a viable legal force, and therefore some effort is warranted to understand its use as a tool by private citizens against noise polluters who may not for one reason or another be subject to other forms of regulation.

Nuisance itself means little more than some activity which annoys, irritates, or in some other way interferes with a person's quietude or convenience. Nuisances are commonly spoken of as being "public nuisances" or "private nuisances". The terminology would lead one to believe that the two differ only as to the type of disturbance created. This is not so. Historically, public and private nuisance law developed quite separately and served to protect two diverse classes of rights. A private nuisance was, and is, primarily a disturbance—regardless of character—which interfered with the use and enjoyment *of land*. A public nuisance was, and is, a disturbance—regardless of character—that interfered with the rights of the government or the public at large. It is easy to see how the two theories became confused, as there would appear to be little difference between blocking a man's driveway and blocking the public street on which he lives. The creation of a public nuisance is, however, a crime or at least a quasi-crime which may be abated by the government and for which criminal sanctions will be invoked. A private individual disturbed by a public nuisance may not—unless special circumstances are present—recover monetary damages for his inconvenience. The only remedy is abatement, fine, or other criminal type of remedy imposed by the state. Most modern jurisdictions have statutes which either define nuisance very generally or specifically define certain activities as constituting a public nuisance. Noise abuses are well represented here: drag racing, loud record players, and street hawkers are among commonly included items. In the case of *Town of Preble* v. *Song Mountain, Inc.* [1], a "rock festival" was found to be a public nuisance. This case includes a very thorough discussion of the subject, contrasting public and private nuisance, and requiring substantial interference to be demonstrated.

Where there is a private nuisance, essentially affecting a person's enjoyment and use of land, the person adversely affected may seek injunctive relief, self-help, or damages. *Maitland* v. *Twin City Aviation Corp.* [2] is a classic injunction case, won by a mink rancher, who obtained an injunction against low-flying planes, which disturbed the minks during whelping season.

There are also instances where a nuisance is both public and private, allowing redress by both governmental and individual action.

The question of what might constitute a nuisance—public or private—depends largely upon the social utility of the activity sought to be proscribed. Under the common law the rights of habitation were superior to the rights of trade and a residential owner would usually prevail when suing, for example, a tradesman for carrying on a noxious activity such as a tannery. In modern times courts have indulged in a "balancing of rights" approach in which neither habitation nor trade rights are inherently superior and which will turn upon such factors as greatest utility to the community, the cost to the defendant of changing his activity, and the ease with which the plaintiff may be compensated while allowing the defendant to carry on his activity.

The ultimate question in both public and private nuisance suits is whether the defendant's use or maintenance of his premises is reasonable under the circumstances. A factor in this determination will be whether the defendant's use permits the plaintiff to make reasonable use of *his* land.

One salient difference between the law of nuisance and other areas of tort law is that the offensive party does not "take his victim as he finds him". For example, if a man was injured when a negligently driven automobile caused him to fall and hit his head and the man died because he had an unusually thin skull, the driver of the automobile would be liable for the totality of his victim's injury even though a normal person might not have suffered injury to that degree. In the case of nuisance the standard is that of definite offensiveness, inconvenience, or annoyance *to the normal person in the community* [3]. It is not, for example, a nuisance to ring a church bell even though the noise may throw a hypersensitive individual into convulsive fits [4]. The standard in America has long been "the ordinary comfort of human existence as understood by the American people and their present state of enlightenment" [5]. In light of the recent increase in noise pollution one must wonder whether people are becoming less "enlightened" or more hard of hearing. At the turn of the century it was recognized in at least one jurisdiction that under certain circumstances the blowing of a whistle that was not very loud but was persistent was sufficient to constitute a nuisance [6]. Yet a recent attempt to improve the deafening condition of the New York City subway system by judicial action failed. Perhaps the din is now considered part of our "ordinary comfort."

As mentioned earlier, a private nuisance is one which interferes with the use of a person's land. It may be remedied by alleviating the offensive condition or awarding damages to the complainer. A public nuisance affects the rights of the public or at least a large segment of the public. A suit for public nuisance is criminal or quasi-criminal in nature and therefore can be prosecuted only by the government. What then is the remedy available to a private citizen disturbed by a public nuisance if the government fails to prosecute? A private citizen may sue, but only if he has suffered damage qualitatively different from the public affected by the nuisance. This is based upon the theory that the government is charged with redressing wrongs to the public and to allow individual citizens to do so would result in too many law suits clogging our courts. A plaintiff's damage is qualitatively different from his neighbor's when he has a special interest, not shared by others in the community, which is affected by the condition or activity in question. Exactly what constitutes a qualitatively different damage is a matter to be dealt with on a case-by-case basis and causes the courts no end of trouble. Clearly, when a plaintiff suffers physical injury, this is sufficient. Also, if a nuisance interferes with a plaintiff's use or enjoyment of his land, it may then be considered a private as well as a public nuisance and the landowner may sue on either theory.

The increased interest in environmental affairs has seen increased pressure to allow suits by private citizens [7, 8]. Some federal agencies are required, by statute, to permit suit by a private citizen challenging a course of action adopted by the agency. For example, the Rivers and Harbors Act and the Federal Power Act both allow such suits. Based upon statutes such as these, citizens have been able to bring suit to protect the quality of the environment [9, 10]. Absent a statute, the private citizen finds himself in the same position he has occupied in the past. He must prove that he is specially affected by the public nuisance [11].

In addition to suits for damages, or injunctive relief, or both, where a governmental agency produces the noise, to the detriment of a property owner, the courts have held that there is a taking, or confiscation, of property [12–14]. This is discussed in greater detail, *infra,* Subsection IX.E.

Although nonstatutory theories of liability, based on common-law principles, are still of unquestioned importance and vitality, the trend of recent successful noise litigation is based upon the defendant's violation of specific statutory standards rather than the old common law standard of what would be considered obnoxious to the average community member. This is not surprising, for how is noise, measured by the common-law community standard, to be considered "abnormally" loud in the context of modern urban industrial life? The rapid growth of urban centers has rendered the old community standard increasingly meaningless. Reliance solely upon that standard can only ensure our continuance along the path to aural oblivion.

III. WORKMEN'S COMPENSATION AND INDUSTRIAL PROTECTION OF WORKERS

Great Britain was far in advance of the United States in passing workmen's compensation legislation, that is, an assured recovery for injured workmen without proof of an employer's negligence, and regardless of whether or not it was caused by a fellow employee. Great Britain passed such a law in 1897; in the United States each separate state passed its own laws and the federal government passed laws for federal employees and maritime workers not members of a ship's crew (i.e., longshoremen, repairmen, etc.). By 1910 in the United States every state had a law mitigating the harshness of the common law for industrial accidents, but it was not until 1911, in Wisconsin, that an effective workmen's compensation law was passed, and it was not until 1948 that every state had such a law.

I have spoken with insurance industry lawyers concerning noise-caused injury, and at this date it is one of the most difficult problems in the field because now so many claims are filed for injuries to the ears.

Actually, it was in 1948 that the first American decision came down in the highest court of New York [15], allowing for a recovery for hearing damage due to a sustained noisy environment, as opposed to a single "accident" causing injury.

New York has often been hailed as a leader in jurisprudence, and Wisconsin, in 1951, allowed a similar claim for what might be termed an occupational hearing loss [16].

Some courts have considered exposure over time miniature accidents, giving this as a rationale for recovery [17, 18]. Other decisions consider the injury to be from an accident, not as an occupational disease, and allow recoveries for impairment, without provable wage loss (Michigan has such a case). Other states require proof of a wage loss, and some class them as occupational diseases without proof of wage loss [19].

In New York, after the *Slawinski* [15] case, experts were consulted, and a recommendation was made that an employee had to be removed from his noisy environment before evaluation was made of any impairment. A six-month separation was then mandated by statute in 1958.

More recently, a case went up to New York's highest court to determine which employer would be liable to pay compensation on a hearing loss claim. The claimant had been exposed for 22 years to harmful noise, while working for a New York employer, and then took similar work in New Jersey. Under the court ruling and the New York statute, the "last employer" was to pay, and in the case, the court held that the statute meant the last employer over whom the Workmen's Compensation Board has jurisdiction; i.e., the New York employer was made to pay the claim [20]. The dissenting opinion in this case mentions studies dealing with the difficulty of measuring a permanent hearing loss unless the patient is separated from the noise for at least six months.

Following the decision of the Wisconsin court, a statute was passed there, taking effect in 1955, expressly allowing recovery for loss of hearing caused by noise.

As matters stand today, there is a great disparity among the various states as to a recovery for an occupational hearing loss that occurs over a period of time. Some states besides New York require retirement for a period of time before a claim can be filed so as to measure the permanency of hearing loss. Most states do allow a recovery for an occupational hearing loss. Besides Wisconsin and New York, West Virginia, Georgia, and Oklahoma allow recoveries, the latter three if there is an "accident". In Missouri, there is no liability if there was proof of a previous loss. Results in Rhode Island, Louisiana, Arizona, and California are said to have inconclusive decisions. Recovery is allowed under a federal compensation statute, the Longshoremen's and Harbor Workers' Compensation Act.

Present problem areas involve whether the employee has a particular allergy; the majority rule is that, even if there is an allergy, or pre-existing weakness, if particular conditions cause it, a recovery is allowed.

As in many areas of product liability recoveries and other areas of damage law for personal injuries, recovery for industrially caused deafness may be an effective tool for reducing noise in industrial establishments.

Other areas of law for reducing the damage to one's hearing have been relatively backward in most countries, including the United States.

In Canada, there are provincial regulations requiring ear protection for noisy environments, but only a relative handful of American states and Puerto Rico have such laws, and very few of these actually prescribe the noise level allowed in decibels. The federal government has shown some initiative in this area also. Regulations are made by the United States Department of Labor, under the Walsh–Healey Public Contracts Act, which sets certain standards for contracts with the U.S. government of over $10,000. In 1968, a regulation was promulgated, which went into effect in 1972, requiring that an employer provide a working environment of a noise level no higher than 85 dB(A). The Labor Department initiative was the first real national effort in the United States [21].

In 1970, the federal Occupational Safety and Health Act (OSHA) [22] became national law in the United States. This act sets forth certain standards and goals in interstate enterprises. Among its terms is the requirement that no worker be subjected to 115 dB(A) for more than 15 min and to 90 dB(A) for more than 8 hr. A somewhat similar type of statute is law in Arizona, Hawaii, Indiana, North Carolina, and Tennessee. The Department of Labor also has set noise standards in federally supported contracts in construction [23].

The greater national effort in the United States began in 1970 with the passage of the Noise Pollution and Abatement Act of 1970 [24]. The act established an office of Noise Abatement and Control, within the Environmental Protection Agency, and set up a program for investigation and research in noise pollution matters, including public hearings, with the object of coming up with recommendations for legislation. Thereafter, the Noise Control Act of 1972 was passed [25].

In December 1974, the United States Environmental Protection Agency challenged the Labor Department standard referred to above and ordered a review of a standard of 85 dB(A). A compromise may be effected, allowing lengthy time to achieve the goal or to allow temporary variances to certain industries. Present estimates of the cost to industry for such a change range from $13 billion for the 90-dB(A) level to over $30 billion to meet the 85-dB(A) standard. However, a net gain might be expected in reductions in workmen's compensation claims and savings from absenteeism [26, 27].

The New York State Department of Labor regulates the noise in industrial establishments so as to avoid hearing damage. Noise pollution regulations proposed by the New York State Department of Environmental Conservation do not apply within industrial establishments but regulate noise as it either leaves or enters the site of an industrial establishment or sound source.

IV. FEDERAL STATUTES TO CONTROL NOISE

A. The Noise Control Act of 1972

One of the most far-reaching albeit inadequate attempts to control noise pollution was the passage in October 1972 of a federal statute known as the Noise Control Act of 1972 [28], which applies throughout the United States.

Under the act the federal Environmental Protection Agency (EPA) is empowered to set limits on the amount of noise permitted from trucks, buses, and railroad trains operating in interstate commerce. Included among the sources of noise the agency is allowed to regulate are a variety of newly manufactured products such as jackhammers and compressors, automobiles, motorcycles, snowmobiles, motors, and engines. The law also permits the EPA to require labeling of noise emission levels on products such as household appliances. The law calls for a nine-month study of aircraft and airport noise in a way that is consistent with maintaining aircraft safety. Also in the law is a provision for the establishment of a national research and development program for the prevention and control of environmental noise. The act allows the states and municipalities to establish and enforce controls on noise as long as the standards established for carriers covered by the federal law are identical to the federal standards.

Senate Report No. 92-1160 on the bill [28] includes a good general statement of the bill's purposes and coverage and a section-by-section analysis and some of the interesting correspondence giving various senators' views of the statute. Senator Edmund Muskie (D-Maine) felt that the preemption section gave the federal authorities too much power to the detriment of state legislation, which might go further to limit pollution by noise within a state to a higher standard than federal requirements. He felt that the federal standards applying to aircraft noise and noise from manufactured products might be too lenient or nonexistent and that states could not fill the gap. He was particularly critical of the inaction on aircraft noise standards by the Federal Aviation Agency and the lack of strong standards for airport noise.

Among the first regulations proposed by the federal EPA were regulations for limiting noise of interstate trucks and buses. The proposals would cut the noisiest diesel equipment by 5 to 10 dB, an appreciable amount, by requiring mufflers and tire tread changes. Critics wanted the regulations to require relatively inexpensive changes for best-quality mufflers, tire configurations, and engine and transmission shielding, as well as noise limits in residential areas.

Regulations affecting trucks and buses were to be promulgated in October 1973 to take effect in October 1974. The effective date, however, was delayed. Railroad regulations were proposed in July 1974.

In fact, some final regulations were announced on October 22, 1974. The standards applied to trucks traveling interstate and required the replacement of mufflers or tires on an estimated 70,000 vehicles. The new regulations, which were the first to be put into effect under the Noise Control Act of 1972 [28], applied to vehicles weighing more than 10,000 pounds and became effective a year later.

Roger Strelow, an assistant administrator for the EPA, said at a news conference, "These standards will not by themselves reduce urban and freeway traffic noise from larger motor vehicles to a level that will adequately protect the public health and welfare, but they are a significant first step."

At the same time, the agency also announced proposed noise standards for future heavy-duty trucks and for portable air compressors. These standards were to be applied beginning with the 1977 vehicles. The new regulations set three noise level standards and provide for visual inspection of a vehicle's exhaust system and tires.

The performance standards for moving vehicles stipulate that the noise that they produce not exceed 90 dB(A) when the vehicles are traveling over 35 mph or 86 dB(A) at lower speeds. These rules apply to the level of sound 50 ft from the vehicle. A stationary truck must not have a sound level over 88 dB(A) at 50 ft, the new regulations say.

According to the Environmental Protection Agency, the level of most highway traffic noise ranges between 70 and 75 dB(A), while the sound of a passing subway train reaches 95 dB(A).

The U.S. Department of Transportation is responsible for enforcing the regulations. It uses meters to measure noise levels and inspect truck tires and mufflers.

The EPA estimates that the cost of meeting the new standards, mainly for mufflers and exhaust system changes, will average $135 per truck. But a spokesman for nine environmental groups criticized the new standards, saying that they would "legalize noise pollution." "The new regulations will do nothing to alleviate the noise pollution from buses and they do not take advantage of existing noise control technology," Lloyd Hinton said at the EPA news conference.

Prior to the enactment of October 1974 regulations, some environmental groups complained that the regulations allowed too many exemptions of noise sources. Sixteen kinds of sounds are exempted under the present proposed regulations, but many of these could be covered by noise regulations that are now developing in other program areas. These include construction, emergency warning devices, railroads, off-road recreational vehicles, snowmobiles, motorboats, motor vehicle race tracks, air conditioners, and agricultural equipment.

The 1974 regulations followed a conscious policy of the EPA to hold the noise program to a "low level of growth" due to budgetary problems, a policy announced in 1973. Before the October 1974 regulations were announced, a suit was filed against EPA to require issuance of noise regulations. This was done in July 1974 by a drivers' safety council. In addition, in September 1974 environmental groups complained that the EPA had "wilfully" failed to enforce the 1972 act and asked for a congressional investigation. After issuance of the October 1974 regulations, the EPA expected to issue further regulations for noise emission limits and product labeling regulations for major noise sources in addition to the regulations covering interstate trucking, railroads, aircraft and airport noise. The noise emission and product labeling regulations will be discussed in a subsequent section of this chapter.

Another major step was taken in October 1974. Rules governing citizen's suits became effective on October 25, 1974, under the Noise Control Act of 1972. The act creates a cause of action for any person to restrain any other person or governmental agency from violating a federal noise standard or to require enforcement of the standard. Costs of litigation and a reasonable attorney's fee and witness fees can be awarded by the court. This is in addition to any other statutory or common-law rights available.

Under the October 1974 regulations, when a person sues a federal or other governmental agency he must first give 60 days notice of an intention to sue. The notice must cite the nature of the alleged violation on which suit is to be brought and is served by certified mail, return receipt requested [29].

Under the Noise Control Act, the penalties for violation range up to $25,000 per day of violation or imprisonment for one year, or both [28]. It is illegal for a manufacturer or dealer to tamper with or remove the noise abatement equipment on any product.

Standards set by the EPA allow the administrator in his judgment to require a reduction of noise through use of the best available technology, taking into account the cost of compliance.

Insofar as the law affects manufacturers, the manufacturer must warrant that his product will conform to federal noise emission standards for the life of the product's normal use, whose time span is determined by the administrator. The user is responsible for using the product in a normal manner with normal maintenance. Users, but not manufacturers, are subject to more stringent criteria set by local ordinances.

Environmental Protection Agency regulations do not cover noises which originate on residential land, from lawn mowers, for example. These are controlled by local ordinances or by the restrictions on new products being manufactured.

Amendments for noise standards were enacted under the Quiet Communities Act of 1978 [30], at which time further noise emission standards were planned and products, properties, and facilities were to be regulated.

Under the noise control statutes and regulations, there have been cases delimiting the areas of control. Compressors have been regulated [31]. Motor home manufacturers were required to comply with the truck standards [32]. But mobile construction equipment was held to be not covered by truck regulations [33].

One of the 1978 amendments allows a state or political subdivision to ask for more stringent standards, to which the EPA must respond. State noise controls not in conflict with federal noise controls were held not preempted [34].

Pursuant to the 1972 act, the EPA prepared noise exposure standards, which appear in a 1974 publication, "Information of Levels of Environmental Noise Requisite to Protect Public Health and Welfare".

B. Other Areas of Federal Noise Control

Besides the OSHA regulations and the regulations under the Federal Noise Control Act, other bodies in the federal government are involved in noise control. Activities of the Federal Aviation Administration (FAA) are discussed in detail elsewhere, but the FAA is to set noise emission standards for aircraft in conjunction with the EPA.

Interstate motor carriers are subjected to noise control standards, and the Secretary of Transportation is mandated to develop noise standards for highway construction [35].

The Walsh–Healey Act has been mentioned previously [21].

Federal housing efforts have led to Housing and Urban Development Department (HUD) noise standard regulations [36]. Mining regulations are also in force [37].

Noise emission standards for automobiles were required under the Fuel Efficiency Act of 1980 [38].

For manufacturers of regulated equipment, it is apparently the ultimate assembler who is responsible for compliance by the product with EPA noise standards [39].

V. STATE CONTROL OF NOISE

A. State Statutes

At least 44 of the 50 American states have noise laws which prohibit excessive noise in general terms, such as "unnecessary" or "excessive" noise [40]. In 1971, the New York State Legislature voted to amend the Air Pollution Control Act, Article 19 of the Environmental Conservation Law. A one-word amendment added noise to the list of air contaminants regulated by the state. This provided the Department of Environmental Conservation with the authority to establish standards for noise pollution, as for other forms of air pollution, with the same enforcement and fines up to $2500. In order to implement this authority, the department is in the process of formulating Regulations for the Prevention and Control of Environmental Noise Pollution. The Bureau of Noise Control has already written and completed hearings on the first portion of proposed regulations.

The proposed regulations have classified land according to the volume, duration, frequency, and hours of noise that will be permitted. This classification is called Land Use Designation for Noise Control (LUDNC).

Class AA (LUDNC) lands are areas, such as outdoor amphitheaters and public wilderness areas, in which serenity and quiet are very important. In Class AA, the point that sound is considered noise is when the additional sound entering an area exceeds the level that normally exists 90% of the time. In other words, the original quiet is intended to be maintained. Class A lands are places where people sleep, primarily residential areas. Sleep is generally disturbed when the level of sound is greater than 45 dB(A). This then is the night noise limit for Class A. However, due to the expense and inconvenience shown at public hearings, the night time limit was changed, so as not to apply to existing installations initiated prior to July 1, 1974, or modified prior to January 1, 1975. The day limit is 65 dB(A), which is the level which, if exceeded, will interfere with speech.

Such places as restaurants, shopping centers, and other commercial establishments are Class B areas. Here, the main requirement is the need to communicate by speech, so the limit is 65 dB(A). Those places where people are likely to remain for long periods but where speech is not essential are Class C. This includes warehouses, lumber yards, and industrial sites. The main consideration is avoiding hearing damage. Hearing damage can occur at 80 to 90 dB(A) over extended periods. The noise limit for Class C is 80 dB(A).

There is a good deal of noise pollution that does not come under the jurisdiction of the Department of Environmental Conservation. Motor vehicle noise comes under the New York Vehicle and Traffic Law. Two sections of the law deal with equipment specifications requiring cars and motorcycles to have adequate muffler systems. A third section of the Vehicle and Traffic Law, Section 386, passed in 1975, was the first law in any state to set a definite decibel limit. Unfortunately, the 88 dB(A) limit set there was very high, so only the loudest trucks would ever approach such a level.

There was dissatisfaction with the limitations of Section 386, and it was amended in 1976 and 1980. Different noise levels were set forth for different types of vehicles; trucks and trailers are allowed a high of 90 dB(A) over 35 mph and motorcycles only 76 dB(A) at 35 mph or less.

The noise standards for occupational situations are set by the Federal Department of Labor. These standards take into account the decibels(A) levels and exposure time. The baseline is 90 dB(A) for an 8-hr day. As the sound level increases, the exposure allowed is cut accordingly. There is serious question as to whether 90 dB(A) for an 8 hr day is not too high. This is considerably higher than the proposed standard for Class C (LUDNC).

The Proposed Regulations for the Prevention and Control of Environmental Noise Pollution are only a first step in forming noise standards for New York State. The first part of these proposals deals primarily with sound source sites, which are fixed geographic locations of a permanent nature, such as factories and power plants. Also included in the proposals is the procedure for issuing noise permits. Such a permit would be issued if it could be shown that the best practical noise control measures would not stop noise pollution from a particular site. It remains to be seen whether these proposals will be adopted or will require revision and further hearings. More proposals are being formulated to deal with noise pollution from sources other than sound source sites.

Hearings were held in 1973 on proposed regulations. The proposed regulations brought protests from some industry representatives, who claimed they were too restrictive and would "work a hardship on local industry".

In New York, it should be noted that the Department of Environmental Conservation regulations do not apply to noise within the boundaries of industrial establishments. But they do regulate noise as it either enters or leaves an industrial sound source.

In discussing the effect of the state law on local laws, the Department of Environmental Conservation has explained [41]:

Under Section 19-0709 of the Environmental Conservation Law, ordinances and laws of local governments which are not inconsistent with state noise pollution regulations are not superseded. The local laws must comply with at least the minimum requirements of the regulation. The proposed regulation will not preempt local governments from enacting noise ordinances, which may be either identical to the state regulation, or have lower decibel values than the state regulation, or be different in kind; for example a curfew on certain activities during certain hours. This will provide an opportunity for local governments to develop their own noise control programs tailored to local conditions.

The Federal Noise Control Act of 1972 (P.L. 92-574) states that the ". . . primary responsibility for control of noise rests with State and local governments . . .". This Act, administered by the Federal E.P.A., is designed to control the major noise sources in commerce, which require national uniformity of treatment. With the exception of interstate railroad and motor carrier standards, the act states that nothing, ". . . precludes or denies the right of any state or political subdivision thereof to establish and enforce controls on environmental noise (or one or more sources thereof) through the licensing, regulation, or restriction of the use, operation, or movement of any product or combination of products".

In measuring noise, both the A-weighted scale, and octave-band-level limits are specified.

The proposed regulation gives several conditions for which "sound level" becomes "noise". The conditions include the A-weighted sound level, the octave band level, the peak sound pressure level, and prominent discrete tones. If the sound exceeds the limit for any of the conditions, it is defined as noise.

For continuous sounds, both A-weighted and octave band level limits are specified. The A-weighted limit is specified because it is the simplest to measure and therefore simplifies enforcement. The octave band level limit is specified to prevent frequency spectra that might compromise the sound transmission characteristics of normal construction. The octave band limits decrease as the frequency increases because that is the general characteristic of environmental noise, and because high frequency noise is less tolerable.

Impulsive sounds, as might be generated by a drop hammer, produce sharp peaks but contain relatively small amounts of sound energy. Since it is the sharp peak that is annoying, a peak sound pressure level limit has been specified for impulsive sound.

New York's noise regulations are planned to take effect in phases. Those operating a noisy sound source site would be prohibited from doing so in mid-1975. After the 1973 hearings, there were indications that several complex issues remained unresolved. In addition, there seemed little prospect of funding for an effective enforcement force in the field. As of this writing, the only regulations in effect pertain to testing of noise of large vehicles.

The New York State Department of Environmental Conservation provides technical data, consulting services, and administrative guidelines for local governments as well as a handbook on model noise ordinances. The department is now concentrating on reinforcing local regulation. These local measures are discussed hereafter. The prospective penalties involved under the state regulations can be quite severe.

The penalties for violation of the proposed regulation are specified by Article 72 of the Environmental Conservation Law. There is a fine of not less than $250 or more than $2500 for a violation. There is an additional penalty of up to $500 for each day during which the violation continues. Any person who wilfully violates the regulation may be imprisoned for up to one year for each separate violation. Each day on which the violation occurs is a separate violation. In addition, an injunction can be brought against the violator to prevent any further or continued violation of the regulation [41].

California has a statute known as the Environmental Quality Control Act of 1970 [42]. The act contemplates that in activities undertaken by a public agency or supported in some fashion by a public agency through contracts, grants, or other assistance or where a private interest is involved, which requires a permit or lease by a public agency, that environmental data or an environmental impact report be required. In defining "environment", noise is one of the criteria to be considered. Much power is given the Office of Planning and Research to prepare guidelines to be adopted by the Secretary of the Resources Agency. In 1972, California's highest court, in a landmark environmental case, held that the law applied not only to public projects, but also to private developments with possible environmental repercussions.

On December 14, 1974, the California Supreme Court broadened the scope of the law, requiring environmental impact assessments on both public development projects and private projects with a potentially substantial public impact. It was noted in the report that some 21 states had adopted counterparts of the National Environmental Quality Act of 1969 that required impact assessments. California statutes and regulations affecting airport noise are discussed in the San Diego case, *infra* Section IX.C [64, 65].

In Illinois, noise control standards have been adopted for residential, business, commercial, and industrial areas [43]. Maryland has set up standards for health, motor vehicles, and airports [44]. Connecticut has approved vehicle noise regulations for trucks and cars [45]. New Jersey has regulations to control noise from industrial and commercial operations [46]. In New York, an opinion of the Attorney General states that a town may adopt an ordinance to regulate motor vehicle noise emissions, but it must be identical to federal regulations [47]. Other states have environmental agencies that establish noise standards, namely, Florida and North Dakota. Noise legislation has been enacted in the state of Washington, and some form of noise control is prevalent in all states and the District of Columbia, from as little as automobile muffler control to detailed statutes in at least 13 states.

B. Legislative Procedural Changes

In addition to statutory and regulatory limitations on noise pollution, there is a growing movement toward a "consumerist" approach. New Jersey's legislature passed a bill in November 1974 to allow individuals, a corporation, or a governmental agency to bring an action to prohibit "actual pollution, impairment or destruction of the environment in violation of state laws". They may also enjoin "actual" environmental damage even where environmental laws or regulations do not fix standards.

In addition to New Jersey, such so-called "citizen suit" legislation has been enacted in recent years in several other states in the nation, including Michigan, Massachusetts, Minnesota, Indiana, Florida, South Dakota, and California and has served to complement and strengthen local, state, and federal enforcement proceedings. On the federal level, the Clean Air Act Amendments of 1970, the Federal Water Pollution Control Act, and the Noise Control Act all provide for citizen rights of action.

Conceptually, environmental citizen suit legislation grants individual members of the public the right to sue for the protection of natural resources without the need to demonstrate, as a threshold proposition, that they have "standing"; i.e., they have been affected by, or are specially aggrieved by, the activities of a defendant. Traditionally, a would-be plaintiff in order to maintain an action must show that a legally protected interest possessed by him has been violated before a court will hear arguments on the merits of a claim of wrongdoing. Citizen suit legislation eliminates this impediment.

Class-action suits, often confused with citizen suits, are actions brought by a group or by representatives of a group on behalf of a number of individuals—too numerous to sue individually—who have been injured or damaged in a substantially similar manner by the acts of a defendant. For such a suit to be maintained, common questions of law or fact must predominate along with certain other legal prerequisites. New York added provisions for class actions in Article 9 of its Civil Practice Law and Rules in 1975.

Most citizen suit legislation provides that any person, whether or not individually affected, may sue any other person, including municipal or state agencies, to enjoin any activity which results in pollution or impairment of air, water, or other natural resources or which violates a state environmental statute, rule, or regulation.

In New York, citizen suit bills have been introduced in the legislature for several years.

Support for citizen suit legislation in the past has come from a host of individuals and groups, including the Attorney General, the Sierra Club, the Bar Association of the City of New York, the National Audubon Society, the Natural Resources Defense Council, the Environmental Planning Lobby, and New York State Conservation Council, the Adirondack Mountain Club, the Parks Council, and the Council of Environmental Advisors.

The idea of a legislatively created legal right to sue violators of our environmental laws, rules, and regulations is not free of controversy. Opponents of such legislation have voiced fears that the enactment of such a measure would result in a flood of litigation that would further burden our already-clogged court system. This has not been the experience in Michigan, for example, where only 74 suits were brought in the three years following passage of its Environmental Protection Act of 1970, and of these, a scant 13 went to trial on their merits. Money damages are not available under the Michigan act, although the plaintiff may continue to sue for damages under the traditional common-law nuisance doctrine.

Some have argued that the powers vested in the New York Department of Environmental Conservation and the Attorney General are sufficient to protect the environment. Advocates counter that it is sometimes difficult for agencies and officials to move against large industrial polluters for political reasons; if citizens were permitted to sue, state officials could be spared embarrassment. Some groups have predicted a proliferation of frivolous, ill-founded suits motivated by a simple desire to harass legitimate operations.

Although it is generally accepted that courts are capable of dealing with baseless actions, some states have required plaintiffs initiating environmental actions to post monetary bonds or to submit with their pleadings the sworn affidavits of technically competent experts to substantiate claimed wrongdoings. Some commentators have attacked such provisions as unduly restrictive: for example, the posting of a $500 bond would discourage individuals with legitimate claims from using the very mechanism designed to vindicate their rights.

These and other controversial issues have been approached in a variety of ways by states with such legislation. In Massachusetts, environmental suits must be brought by 10 or more persons. Michigan permits a judge to amend existing regulations or standards or to create new ones if he determines it appropriate in a particular case. Some states give the Attorney General or an administrative agency a certain number of days in which to initiate their own proceedings, thereby obtaining an automatic stay of citizen-filed suits.

Most states prohibit plaintiffs from either obtaining money damages from defendants or settling or compromising actions without court consent. A common provision permits the court to appoint special masters or referees to hear evidence and report back to the court. Some state laws allow individuals to intervene in any administrative proceedings arising out of, or relating to, any matter that may result in impacts on the environment.

Litigation is expensive, and for this reason many states allow plaintiffs to recover attorney and expert witness fees while setting an upper limit on costs imposed on unsuccessful plaintiffs. Such provisions encourage citizens to participate in the resolution of issues of concern that arise in environmental decision-making processes.

Suits by individuals for the protection of the environment in the state courts are not new, but the method is expanding and may include actions to abate noise and other types of pollution. Despite some opposition from those desiring to encourage industry and employment, Governor Brendan Byrne, in New Jersey, supported such legislation in 1974 [48]. Other jurisdictions, such as Wisconsin, Michigan, and Florida, have statutes specifically permitting private citizens to sue for the abatement of a public nuisance.

These developments are in harmony with rules promulgated in October 1974 to allow citizen suits under the Noise Control Act of 1972. It is also consistent with the practice in New York City, where any citizen may initiate enforcement proceedings under the New York City Noise Control Code of 1972. Under this code, a citizen may file a "noise complaint affidavit" to begin a prosecution.

VI. NEW YORK CITY NOISE CONTROL CODE

New York City passed Local Law No. 57, the Noise Control Code, effective September 1, 1972. Previously the city had for a long time a law on the books prohibiting "unnecessary noise". With the passage of the code, New York City moved beyond the usual type of statute. Wherever technically feasible, decibel levels were set; there was also a section concerning unnecessary noise in which the term is more clearly defined than anywhere else:

Unnecessary noise means any excessive or unusually loud sound or any sound which either annoys, disturbs, injures or endangers the comfort, repose, health, peace or safety of a person, or which causes injury to plant or animal life, or damage to property or business.

Although some ambiguities exist in decibel enforcement regulations, set forth in a handbook on enforcement, the code contemplates enforcement by both personnel of the Environment Protection Administration and the City Police Department. Specially trained inspectors issue notices of violations, and field inspectors are trained to calibrate noise control apparatuses.

To give an idea of the extent to which the code reaches, it includes control of

(1) air compressors, paving breakers, refuse-compacting vehicles, rapid-transit railroads, passenger stations, tunnels, elevated structures, yards, depots, and garages;

(2) sound reproduction devices, sound signal devices, air horns, steam whistles, and burglar alarms; and

(3) containers and equipment for handling or transporting construction material; and

(4) engine exhausts and motor vehicles.

The code has provisions, which are probably now illegal, concerning aircraft and airports. (See Section IX for details.) Hours of construction activities are limited, but variances may be granted, and there are provisions for emergencies. Areas near schools, hospitals, courts, and "noise sensitive zones" are provided for.

In the first year of the code's existence, 3152 violation notices were issued. More than 46% of the violations were issued for "sound devices". Transportation complaints were surprisingly small.

Enforcement by the Environmental Control Board and an administrative court may have serious consequences. The board has subpoena powers and may order installation of control equipment, issue cease and desist orders, and impose civil penalties up to $1000 per day. It may revoke permits and variances. After a finding of civil liability, operations can be shut down.

Citizens may initiate proceedings by filing a Citizens Noise Complaint Affidavit, and a bounty of 25% of any fine assessed is allowable.

At the end of February 1975, the New York City Council's Committee on Environmental Protection amended the Noise Control Code, extending for two years the deadline concerning a minimum noise level for refuse-compacting trucks. Under the 1972 provisions, refuse-compacting vehicles manufactured after December 31, 1974, were required to have a sound level that did not exceed 70 dB.

In 1975, the New York City Environmental Protection Administration, now the Department of Environmental Protection, also made a report on ambient noise quality. In 1978, a bill was introduced in the City Council to define and set standards for zones, differentiating residential districts, local retail districts, general commercial districts, waterfront districts, and amusement districts, among others. Day time and night time standards were set for the various zones. The proposed bill amended the City's Administrative Code.

San Francisco has a comprehensive noise control ordinance. Noise control ordinances also exist in Chicago, Philadelphia, and Baltimore [49–53].

VII. LOCAL ORDINANCES

In areas smaller than major cities, many of our American town, village, and local governments have enacted ordinances and provisions in zoning codes to control noise. Very often, there are other provisions to control vibration, smoke, dust, and odor.

Perhaps a typical zoning code of this type has been passed in the Village of Ossining, in New York State, a town famous for Sing Sing prison if not for its Hudson River views. In a section of the code devoted to Industrial Park District zoning requirements, certain "performance standards" are required. The Planning Board, in approving a site plan, must approve any plan as conforming to the performance standards. As to already-existing uses and compliance, enforcement is given to the local building inspector. Pertinent "performance standard regulations" include standards on vibrations and noise (with specific measuring techniques) that cover different types at different locations.

In Westchester, New York, the Town of Greenburgh passed a noise ordinance in 1969, which is somewhat similar to Ossining's. Unlike Ossining, however, Greenburgh has equipment for enforcement, including a precision sound level meter and octave filter set. Enforcement is in the hands of the Town Police Department, and the meter is periodically checked against a tester. One of the uses to which the meter has been put is in litigation involving a dog pound, which has had noise levels very upsetting to surrounding residents.

Obviously it is only with such measuring devices that ordinances which depend on measurement of noise levels can be enforced. A more promising method may be to limit the hours of certain noises, particularly in residential areas, such as power mowers, power saws, and even musical instruments.

In my own village of Hastings-on-Hudson, New York, a somewhat similar approach was taken in proposed amendments to the village zoning code. The proposals involve both noise level measurement and time period limitations. Because of an unhappy experience with a discotheque in a basically residential area, an ordinance was passed dealing with music in restaurants after midnight. Violation can bring a fine of from $25 to $250, and each day's violation constitutes a separate offense.

There is available a model noise ordinance produced by the National Institute of Municipal Law Officers (NIMLO). The model code has alternative suggestions, one code being based on types of proscribed noises such as horns, radios, loud speakers, whistles, and animals. It also deals with construction noises, pile drivers, blowers, and fans. In the alternative code suggestion, certain decibel levels are set forth, somewhat like the Ossining and Hastings-on-Hudson provisions already noted. The NIMLO also has a model ordinance dealing with sound trucks and another one dealing with advertising from aircraft. It has published reports on the subject and analyzed much of the legal case law involved with municipal control of noise and air pollution. Specifics can be obtained from the NIMLO in Washington, D.C. A good book on the general subject is Rhyne's "Municipal Law", published in 1957 [54].

VIII. BUILDING AND CONSTRUCTION CODES

A. Introduction

One indication of the problem of controlling noise by building and construction codes is that the State of New York in its proposed noise control regulations in 1973 noted that separate regulations were being developed. This is because "construction noise involves several complex considerations . . . , construction is a temporary activity and . . . the contractor is generally limited in his ability to obtain quieter equipment" [55].

Canada and several European countries have building codes with noise control requirements.

On a national scale in the United States, the Federal Housing Administration has issued construction detail recommendations, as of 1964, that apply to builders who have sought FHA mortgage funds. The recommendations, known as FHA No. 2600, include, for instance, a requirement that sound-absorbent material be installed or the wall thickened where electrical receptacles, heating grilles, and medicine cabinets are to be placed in the same stud space.

B. U.S. Federal Regulations Affecting Construction

One of the major construction hurdles is provided by limitations imposed by the U.S. Department of Housing and Urban Development (HUD) on its grants to housing faced with possible noise problems. In August 1971, HUD came out with Noise Assessment Guidelines which had the effect of blocking subsidized housing projects near highways, airports, and railroads. Unfortunately, the guidelines dealt with projects already long in the planning [56].

Under HUD's regulations, local officials take readings on projects in order to place the project in one of the following categories:

Unacceptable: Noise levels exceed 80 dB for 1 hr or more per 24 hr or 75 dB for 8 hr per 24 hr.

Normally unacceptable (discretionary): Noise exceeds 65 dB for 8 hr per 24 hr or is subject to "loud repetitive noises on site".

Normally acceptable (discretionary):   Noise does not exceed 65 dB for more than 30 min per 24 hr.

Acceptable: Noise does not exceed 45 dB more than 30 min per 24 hr.

The policy of HUD is to have decisions made on the basis of these measurements at the local and regional office. The regulations require that any request for an exception sent to the secretary be accompanied by an environmental impact statement. Most of the requests that come to Washington seek permission to dispense with preparing and circulating an environmental impact statement, a process that normally takes about three months or more.

In general, any project that falls into the discretionary categories may be approved after the sponsor agrees to changes, such as relocating the structures on the site or adding more acoustical insulation. Under HUD's policy, if noise levels at any project are more than 75 dB for 8 hr out of 24, then a full environmental impact statement and the secretary's specific approval is required.

A waiver could be given if mitigating circumstances are found; the housing project is an "in-fill"; that is, housing on a site that might otherwise be used for a less-desirable purpose, such as a factory; or noise-damping changes are made in the structure to reduce interior noise levels for the residents.

So far as the authors know, only the housing project at Yuma, Arizona, has gone through the entire procedure. The units there are near the airport. The Marine air station there is described as "a mainstay of the local economy", and residents are less prone to complain about noise, according to the environmental impact statement. The builder agreed to changes that included adding air conditioning, 0.5-in. drywall inside exterior walls, double-glazed windows, solid-core exterior doors, 30-in. roof overhang, and ceilings with 6 in. of insulation.

For the proposed apartments near Bergstrom Air Force Base, Austin, Texas, the department recommended the addition of double-glazed windows, 1.25-in.-thick insulation board in all wall areas, 0.5-in.-thick insulation board in all ceiling areas, and air conditioning.

In New York City, a 250-unit housing project called Two Bridges on the Lower East Side was held up for more than a year because highway noise exceeded the prescribed decibel limits. The site registered 77 dB during eight consecutive hours, and the HUD Circular 1390.2 made anything over 75 dB(A) unacceptable. An exception was required, necessitating approval by the secretary, an environmental impact statement, and measures for reducing the noise. Local governmental pressure and pressure from civic groups and the public were exerted so that the building was eventually approved. But the delay led to an additional $1 million in construction costs.

In addition to the problem faced and finally surmounted by the Two Bridges project, it appears that due to noise restrictions HUD is shying away from approval of projects in noisy slum areas. Obviously, the residents feel that the areas most in need of subsidized housing are hurt by the regulations. The department seems willing to approve projects with insulated windows, but obviously this affects the cost factor considerably. A *New York Times* article in June 1974 said that the noise regulations affected 3725 units in nine housing projects in Manhattan, Brooklyn, and the Bronx involving federal subsidies of over $8 million a year.

Besides New York City, many other cities faced the effects of the HUD regulations on housing projects.

C. City Construction Regulations

As noted in the section on the New York City Noise Control Code, which began effective September 1, 1972, construction jobs are regulated. Hours of construction are limited, and certain noisy construction equipment is regulated. In recent commentary it was said that in general most complaints about noise were caused by activities on construction sites and that most of the opposition to the code has come from the construction industry.

The code provides that no construction noise is allowed except between the hours of 7 A.M. and 6 P.M. on weekdays. Parenthetically, a similar provision has been in the New York City Building Code since 1936, but enforcement was never very effective until 1972.

The code applies to compressors that require exhaust mufflers and other equipment producing noise. Under the code not only are significant fines possible, but the Environmental Control Board can order violators to cease and desist outside of normal court procedures.

IX. AIRPORT NOISE

A. The Problem

(1)   Airlines in the United States are currently faced with the possibility of having to spend $500 million or more to modify early-model jets such as the 707 and the DC-8 in order to reduce the noise produced by these planes during take-off and landing.

(2)   In San Jose, California, officials decided to close a $700,000 school because the noise of nearby jet traffic rendered the structure all but useless.

(3)   Since 1968 Los Angeles has razed more than 2000 homes at a cost in excess of $200 million because the structures are uninhabitable due to the noise from a nearby jet airport.

The problem of noise generated by air traffic first became acute in the early 1960s with the increased use of jet aircraft in national and international travel. The jets—especially the earlier models, many of which are still in use today—produce a noise which is not only louder but is qualitatively more distressing than the noise of earlier propeller aircraft.

The problem is worldwide. In Osaka, Japan, over 250 people living near a large airport brought suit alleging that the noise from jet aircraft was injuring their health. The local courts agreed and, after protracted litigation lasting five years, banned nighttime flights and awarded each of the plaintiffs monetary damages.

Any effort to deal with the problem of aircraft noise necessarily involves a balancing of the need for fast, convenient, reliable air transportation, provided profitably at a reasonable price, and the right of people living and working in proximity to airports not to have their lives dominated by the deafening whine and roar of jet aircraft. While technological remedies such as better insulation and quieter engines will probably provide us with the ultimate solution to the problem of airport noise, we must be mindful of the legal remedies which will create the impetus for technological developments, as well as provide us with some respite as we advance toward Utopia.

These bases of liability and legal remedies for alleviating airport noise are much the same as for any noise disturbance. However, in the United States the airport situation is complicated by the fact that we live under three governments: federal, state, and local. This creates the problem of determining just who has the authority to deal with a given problem. In the case of airport noise regulation, the conflict is between state and/or local government on one hand and the federal government on the other. The federal government's authority to act in this area derives from our Constitution, chiefly those parts which provide the federal government with the authority to regulate interstate commerce and to promulgate laws toward this end. The states, however, retain the power to regulate activity in their navigable airspace so long as such regulations do not unduly burden interstate commerce [57] and are not inconsistent with federal regulations [58]. As we shall see later, this scheme is not as simple as it appears, and due to federal power to "pre-empt" the field, there is relatively little room for state and local regulation.

B. Federal Administrative Agencies

The Federal Aviation Administration and the Environmental Protection Agency are the chief agencies involved with control of airport noise. Section 1508 of the Federal Aviation Act provides that "[t]he United States of America is declared to possess and exercise complete and exclusive national sovereignty over the airspace of the United States. . .". Section 1348 of that act gives the administrator of the FAA the power to regulate the navigable airspace "in order to ensure the safety of aircraft and the efficient utilization of airspace . . ." and to ensure "the protection of persons and property on the ground. . .". Section 1431 gives the FAA power to regulate aircraft and airport noise.

The Noise Control Act of 1972 directs the administrator of the FAA to consult with appropriate state and local officials regarding the aircraft noise problem and then to undertake a study of the problem and report to Congress. The Noise Control Act also amends Section 611 of the Federal Aviation Act by requiring the FAA to consult with the EPA regarding standards for aircraft noise, including sonic boom. Standards are also to be drawn up for the granting and suspension of the various administrative approvals required. Significantly, these determinations may affect only future approvals and certificates. The FAA is required to consider EPA proposals and to give reasons for any rejection of a proposal. In considering such proposal, the FAA must

(1)   consider relevant available data relating to aircraft noise and sonic boom, including the results of research development and evaluation activities conducted pursuant to this act and the Department of Transportation Act;

(2)   consult with such federal, state, and interstate agencies as deemed appropriate;

(3)   consider whether any proposed standard or regulation is consistent with the highest degree of safety and air commerce or air transportation in the public interest;

(4)   consider whether any proposed standard or regulation is economically reasonable, technologically practicable, and appropriate for the particular type of aircraft, aircraft engine, appliance, or certificate to which it will apply; and

(5)   consider the extent to which such a standard or regulation will contribute to carrying out the purposes of this section [59].

While the EPA was to suggest noise regulations under the 1972 act, an amendment to 49 USC, Section 1439, in 1978 directed the EPA to submit proposed regulations covering aircraft noise and sonic booms to the FAA. Nevertheless, the FAA had the option of not promulgating the proposed regulations so long as a detailed explanation was advanced for not doing so.

To date, it appears that detailed noise regulations have been passed covering large transport planes, turbojets, agricultural, and fire-fighting planes and Concorde supersonic airplanes. Sonic booms by civilian airplanes are also covered. Airport restrictions would be allowed for failure to meet the criteria of the regulations [60], but there are no *federal* airport noise regulations. The regulations apply for certain types of certificate applications or acoustical change applications. Also in 1978, the Secretary of Transportation and the EPA administrator were to study the noise effects of an airport in one state on those in another state.

Thus, the EPA proposes as to aircraft and sonic boom noises, but the FAA disposes. The FAA also consults with appropriate federal, state, and interstate agencies under Section 1431 of the act. Its FAA regulations must rely on standards of the "highest safety", as well as the practicalities of economics and technology.

C. Federal versus State and Local Control

As mentioned earlier, the federal government, by virtue of the commerce clause of the Constitution, is vested with the power to regulate aviation. There is, however, a residuum of power with the individual states. When state and federal power come into play, the question of preemption may arise.

Very simply stated, the Supremacy Clause of the United States Constitution dictates that when the federal government is given power to regulate a certain area, such regulation shall supersede state regulation. This brings up such questions as: Has the federal government exercised its power in a proper area, i.e., constitutionally? Has the federal government controlled the field completely? If not, may the states legislate within the interstices of the federal scheme? Does the mere grant of the power to regulate an area pre-empt state regulation even though the federal government has not actually promulgated rules and regulations?

Detailed answers lie far beyond the scope of this work. However, the relevance of the questions to the control of aircraft noise is demonstrated by even a brief review of the problem. The federal government, while deriving its authority to regulate aviation from the interstate commerce clause of the Constitution, exercises power even over those flights that are wholly intrastate. The rationale for this is that allowing unregulated aircraft to occupy the same airspace as federally regulated interstate aircraft would cause an undue burden on interstate aviation. The necessity for precise regulation and coordination of all air carriage—intrastate and interstate—results in a situation in which activities that appear to be purely local in character are ultimately found to require comprehensive federal regulation.

In *Allegheny Airlines, Inc.* v. *Village of Cedarhurst* [61], the Village of Cedarhurst, on Long Island, New York, attempted to abate noise from aircraft landing and taking off at nearby Kennedy Airport by prohibiting all flights over the village below the altitude of 1000 ft. At the time "navigable airspace" (over which the federal government *concededly* had control) was defined as being above 1000 ft. Hence, Cedarhurst reasoned, its regulation dealt with airspace not within the scope of federal regulation and was therefore not contrary to, or pre-empted by, the federal regulation. The court reasoned otherwise, finding that power to regulate above 1000 ft necessarily entailed the power to regulate the space through which an aircraft must progress in order to reach that altitude. The federal regulation of navigable airspace was found to be so compelling and extensive as to pre-empt state regulation of airspace not at that time within the strict definition of navigable airspace.

In *American Airlines* v. *Town of Hempstead* [57], the Town of Hempstead, also on Long Island, sought to achieve the same end as had the Village of Cedarhurst. Unfortunately for its residents, the decision brought the same result. Rather than prohibiting flights within a given space, they promulgated regulations prohibiting noise in excess of certain limits within town boundaries. Again, jets leaving and approaching Kennedy Airport regularly exceeded these limits. The court was unimpressed with this tack, reasoning that the noise and the aircraft were inseparable and that to exclude one was to exclude the other. The ordinance was struck down.

Doubtlessly, the best-articulated statement of the pre-emption problem as it relates to modern noise law is to be found in a case decided by the Supreme Court of the United States. In the case of *City of Burbank* v. *Lockheed Air Terminal* [62], the Supreme Court ruled on the validity of a Burbank city ordinance which prohibited jet flights from the Hollywood–Burbank Airport outside Los Angeles between 11 P.M. and 7 A.M. The case turned upon the question of whether or not the federal government had by previous legislation pre-empted the field of aircraft regulation to such a degree as to prevent a municipality from passing an ordinance regulating flights from an airport located within its municipal boundaries. The Supreme Court decided by a five–four majority that such regulation was prohibited. The majority's decision was based upon a finding that the federal government (specifically the FAA and EPA) had pre-empted air-craft control, including regulation of the time and frequency of flights. In reaching this conclusion, the Supreme Court looked to the legislative intent behind the Federal Aviation Act of 1958 and the Noise Control Act of 1972. Although neither of these acts contains an *express* provision pre-empting airport traffic management, the Federal Aviation Act does pre-empt "airspace management" [63]. From this, the majority reasoned that, since control of take-off and landing at airports is an integral factor in determining what goes on in the "airspace", management of permissible flight times was also pre-empted. The majority found that while the Noise Control Act of 1972 did not purport to change any pre-existing balance of powers between federal and local governments the prior existing condition was one of complete federal pre-emption, and that the Noise Control Act of 1972 simply expanded federal control within a field that had already been pre-empted.

The City of Burbank case asserted federal pre-eminence in restricting aircraft activity and noise, as against state and local political bodies. However, the decision "singled out airport *proprietors* and gave them special, although undefined, leeway in controlling the sources of aircraft noise directly" (emphasis supplied), according to *San Diego Unified Port District* v. *Gianturco* [64]. Some rights of restriction were reserved to airport owners and operators. "The rationale for this exception is clear. Since airport proprietors bear monetary liability for excessive aircraft noise under *Griggs* v. *Allegheny County*, 369 U.S. 84 . . . (1962) [discussed *infra*], fairness dictates that they must also have power to insulate themselves from that liability" [65].

The extent to which an airport proprietor or local government can still control noise, either by restricting flights or asserting noise level controls, is an issue which, therefore, is still with us, despite federal pre-emption.

In 1978, the Quiet Communities Act [66] was passed, providing for a nationwide program funded by the federal government but involving localities. Under the Quiet Communities Act, the EPA can disburse funds to local governments to plan for abatement of noise in airports and other stationary sources or facilities.

The San Diego case, *supra,* indicated that local governments or proprietors may adopt abatement plans that do not impinge on aircraft operations. This may involve compensating those adversely affected, using zoning powers, baffling existing noise, or resettling homeowners affected by the noise. The distinction between allowable local restrictions and impermissible restrictions was noted in *Air Transp. Assn. v. Crotti* [67], where some restrictions were allowed and some invalidated.

However, it is the proprietor who may impose reasonable standards, taking into account the liability it may face for those dispossessed by excess noise. A state or government entity, which is not a proprietor, could not enforce a curfew or impose a curfew more stringent than the proprietor could [64]. In this case the state had a law directing an owner to impose a stricter curfew or noise-level restriction or to obtain a variance; the state here sought to impose a stricter curfew; this was enjoined.

Not only has the proprietor the obligation to provide for noise control within the confines of national policy, but the Federal Aviation Administration has not promulgated general airport noise abatement regulations. It has the power to do so under the Federal Aviation Act but has, apparently, used the power in a very limited way. Noise abatement design criteria for new supersonic aircraft are, however, regulated [68]. A phaseout of certain noisy subsonic planes is also mandated for 1985.

In *DiPerri v. Federal Aviation Administration* [69], it was reiterated that most courts have given the proprietor primary responsibility for reducing airport noise. The varied decisions, on the extent of proprietor restrictions that have been upheld, are categorized in this case.

The 1976 Noise Abatement Policy Statement of the FAA and the Department of Transportation looks to sharing the responsibility for airfield noise abatement. Local proprietors are to concentrate on the best site location, airport design, ground procedures, land acquisition, and other restrictions that are fair and nondiscriminatory. The importance of the policy statement was referred to in several cases. *Greater Westchester Homeowners Assn. v. City of Los Angeles* referred to the statement to hold that a proprietor is allowed to impose curfews, ban certain aircraft, require acceptable FAA flight patterns, and limit airport expansion within federal guidelines [70].

A suit to require the FAA to regulate a specific airport to prescribe regulations as it is authorized to do has been held improper. If cast in the nature of dangerous flight patterns, a lack of regulation, or a regulation might be held to be arbitrary and an action might compel the FAA to act for safety's sake. But the FAA is not suable for discretionary acts. Landing pattern regulations, however, are subject to review under the Administrative Procedure Act [69].

A city that is the proprietor of an airport has some rights to set a noise curfew. Particularly in a little used airport and where a proprietor faces liabilities for noise, some leeway is allowed [71]. But, as we shall see, the extent of the allowable restrictions is circumscribed. A proprietor's regulations must not be overbroad. One case held that a curfew was unenforceable since it extended to all aircraft during certain nighttime hours without regard to the noise of particular aircraft or the actual number of nighttime flights. This was especially so since federal funds were available to the airport, but on condition that it would be available for use at all hours. The curfew was enjoined in both a state-owned airport and a county-owned airport in another, somewhat similar case [72], where the curfew on the state-operated airport on Long Island was too broad and was enjoined.

Another suit arising out of the same situation was dealt with in *State of New York* v. *Federal Aviation Administration* [73]. New York sought to require the FAA to transfer the operating certificate and invalidate the agreement with the Metropolitan Transit Authority requiring that the facilities be available at all times. The state was held bound by an estoppel effect to comply with the original agreement. The Airport and Airway Improvement Act of 1982 [74] superseded the 1970 Airport and Airway Improvement Act. The FAA's orders were to be reviewed only in the circuit courts.

In August 1983, a federal judge banned a similar curfew at Westchester County Airport, New York. It, too, was held overbroad and discriminatory and an undue burden on interstate commerce. It interfered with all air traffic patterns in the area. Noise studies did not substantiate the county's curfew, and the airport had received federal funds and had not justified the curfew by noise studies [75].

A proprietor, whether a public or private authority, is vested with a limited role imposed by Congress to issue regulations and requirements regarding permissible noise levels at airports [76]. Whether there is a violation of an explicit federal regulation or a conflict with an implicit area of regulation, the federal policy must prevail. To be valid, the requirements of an airport owner must be reasonable and nondiscriminatory and must fit within federal restrictions.

The cited British Airways cases dealt with the issue of allowing the Concorde to fly into Kennedy Airport in New York City. A resolution of the matter was arrived at when the final appeal upheld rulings that an airport proprietor could issue reasonable, nondiscriminatory regulations. However, the Port Authority had delayed its inquiries for so long as to be unreasonable and unfair. An outright ban would not be allowable in view of the federal interest. The Port Authority could adopt new standards, so long as all aircraft, including the Concorde, are given a fair opportunity to meet the requirements.

In 1977, rules were proposed for Concorde flights by the Transportation Department Secretary:

(1)   Concordes could operate if their noise levels were not increased. They would have to operate over land below supersonic speed.

(2)   Local proprietors could apply nondiscriminatory rules.

(3)   New Concordes would have to meet 1969 U.S. subsonic noise standards.

(4)   A national curfew between 10 P.M. and 7 A.M. would be set.

Because Dulles Airport outside Washington, D.C., was a federally operated airport, the issues involving the Concorde landings were somewhat different. Authorization for landings by the Secretary of Transportation were upheld. Since the standard of review was whether it was arbitrary or capricious, his order allowing a trial period for landings was affirmed [77].

In 1982, legislation was passed so that any runway construction or extension that involves two counties requires local approval.

The Quiet Communities Act of 1978 provided for studies involving noise from an airport in one state affecting another state. Grants are available for noise control by state and local government and planning agencies.

The Senate and House Reports on the Aviation Safety and Noise Abatement Act of 1979 include interesting observations on the past and future of federal aviation noise policy [78, 79].

Objectives of the statute are to aid in measuring noise, planning noise abatement with airport operators, and providing funds for enforcement. Operators are encouraged to use preferential runways, barriers, acoustical shielding, and flight procedures and restrictions depending on aircraft type to minimize noise.

Noise impact maps are encouraged, but their use in private suits is prohibited. Recovery for damages is limited by those who buy property with knowledge of a noise impact map unless there is a subsequent change.

The Secretary of Transportation is to promulgate noise regulations for foreign carriers, comparable to U.S. carriers.

The failure of operators to comply with the limits for noise, set by FAR36, can be waived if there is a good-faith reason for inability to comply. Retrofitting has been found to be inadequate and expensive in most cases, and replacement of engines and aircraft is favored in reducing noise. The Civil Aeronautics Board (CAB) has the authority to authorize or require an operator to impose a surcharge as a noise abatement charge for carriers to impose on passengers in order to bring a fleet into compliance with noise regulations. This covers domestic flights and some international flights.

One of the most recent results of the FAA's plans was permission granted in August 1983 to an airline to land and take off a new DC-9-80 at Washington National Airport at night. The night curfew was based on amount of noise, set at 85 dB, but the new model produced less noise [80].

All in all, it appears that the FAA has statutory as well as financial control of airport curfews under the Airport and Airway Improvement Act of 1982 [81].

Airport proprietors are under pressure to try to revise departure and landing routes, to raise flight altitudes, to deal with flight patterns, to restrict touch-and-go operations, to reduce run-ups on runways, and to implement other controls.

One of the more important activities of the FAA affects both American and foreign airlines. It regulates the types of aircraft allowed to be flown in the United States.

In 1969, the FAA promulgated its "Part 36" standards, Noise Standards for Aircraft Type and Airworthiness Certification [82]. Uniform procedures were set out to measure airplane noise, and noise standards were set out. Three noise categories were established for large jet aircraft. Stage 1 aircraft were designated as such if they were designed and built before the effective date of the noise standard and were not made subject to regulation. Stage 2 aircraft were those which met the standards of Part 36 regulations. Stage 3 aircraft were those established for certification after 1975 and so were the least noisy.

As national policy, Part 36 announced the eventual extension of its standards to all passenger jets operating in the United States and the elimination of Stage 1 aircraft.

In 1976, a new program was announced by the FAA [83]. Under this Fleet Compliance Program, certain quotas were set up for compliance in reducing the noise of aircraft in 1981 and 1983 and all aircraft were to be in compliance by January 1, 1985. Elimination or retrofitting were required for noncomplying aircraft models.

Certain exemptions and waivers were granted. Aircraft registered outside the United States were not required to comply if the International Civil Aviation Organization arrived at compatible regulations. The policy statement of 1976 warned that if an international agreement were not reached by January 1, 1980, the FAA would act unilaterally to require international compliance by the January 1, 1985, date. To keep U.S. international carriers on a par, they were also exempted until 1985. In 1979, the U.S. Congress affirmed the FAA's position and the airplane elimination–retrofitting program of Part 36 [84]. Since no international agreement was entered into, in 1980 foreign aircraft were required to be in compliance by January 1, 1985 [85], and American international carriers were kept on a par. No phase-out was required prior to that date [86].

The phase-out requirements are discussed in a most interesting case, *Global International Airways Corp.* v. *The Port Authority of New York and New Jersey* [87]. The district court was to decide whether, in fact, the federal policies and Port Authority rules were in conflict, and to decide other questions raised (such as an unreasonable burden on commerce, discrimination against interstate and foreign commerce) and whether the Port Authority rule was arbitrary and discriminatory. It was also decided that carriers had no federally protected right to increase the interim cumulative noise up until the cut-off dates.

At issue in the case was the conflict between the federal statute and regulations determining the composition of older jet planes in the various air fleets and a regulation of the Port Authority of New York and New Jersey regulating the major airports in the New York metropolitan area. The Airport Authority sought to require a phase-out of noisier planes before the federal regulations in Part 36. In its discretion, the lower court found that the federal regulations pre-empted the proprietor's regulations.

On appeal, the appellate court held that the proprietor's rules regulating take-offs and landings were not necessarily in conflict with the federal Fleet Compliance Program if the local regulations did not directly regulate the composition of the air carrier's fleet. A proprietor could validly limit noise exposure and regulate the number of noisier planes used by carriers. However, a local rule could not force a carrier to change the composition of its fleet earlier than the federal requirements. The case was sent back to the lower court to determine, as a question of fact, the precise effect of the regulation on the composition of the air carrier's fleet.

Recent news on the effect of the Part 36 regulations indicates that DC-8s are being phased out, since they do not meet the international standards now set for January 1988 [88]. The regulations will also require replacement of certain British-manufactured Tridents [89].

As to military aircraft, the FAA has been held to have no control over noise from military planes. Under the Noise Control Act of 1972, the FAA's jurisdiction extends only to air traffic control [90].

D. Environmental Impact Statement Requirements

Changes in an airfield must comply with the National Environmental Policy Act of 1969 when federal funds are involved and are subject to state law which may require similar types of environmental impact analyses. One case dealt with a decision to construct an additional runway at a county airport which accepted federal funds. An environmental impact statement prepared by the FAA must discuss and analyze the reason and need for the project and take into account the effect of noise. Where the sound impact was improperly evaluated, federal participation was enjoined until requirements of the act were met [91].

In another case, an airport was to be converted from military to civilian use. The Air Force prepared the environmental impact statement, but although it dealt adequately with the effect of noise on nearby schoolrooms and general noise problems, it was held deficient as to the effect on people's sleep in the vicinity. During the time the statement was to be reanalyzed, nighttime operations were enjoined as were major structural alterations [92].

An environmental impact statement must contain sufficient information provided by the promoter to enable a reasonable and independent choice among available alternatives. Noise studies are required and direction of traffic patterns and configurations to mitigate noise must be studied [93].

E. Aircraft Noise as a Taking of Property

Where aircraft noise adjacent to an airport interferes with the use and enjoyment of private property and where there is a measurable decrease in property value, there may be a "taking" of property. In California, an intermediate appellate court allowed recovery for plaintiffs in an action in inverse condemnation against the City of Los Angeles as owner and operator of the airport there [94]. Plaintiffs alleged a diminution of market value of their land due to the noise of the airplanes overhead and consequent fall of soot, oil, and fuel, which damaged painted surfaces, automobiles, and clothing kept outside. Appraisers testified to the decrease in the home values.

The City of Los Angeles defended, alleging that [94] only those parcels directly under the glide path of the airplanes had suffered a "taking" and

[95] that federal regulation of flight pre-empted the field. In its decision, the court rejected a "direct overflight rule" [95] and the narrow interpretation of this "rule" in a federal appellate case, holding that there was no taking unless an overhead flight makes the property uninhabitable [96].

California's appellate court, in the Aaron case, followed the broader view of other courts and rejected restrictive and technical views about trespass directly over one's land [97]. In dictum, the California court noted that under its civil procedure code a compensated taking is allowed when "necessary to provide an area in which excessive noise, vibration, discomfort, inconvenience or interference with the use and enjoyment of real property located adjacent to or in the vicinity of an airport and any reduction in the market value of real property by reason thereof will occur through the operation of aircraft to and from the airport".

As already noted, the justification for putting the initial burden on airport proprietors to arrange for noise abatement is because they bear the liability for excessive aircraft noise, for which they must pay those affected [95].

If the United States is involved as owner–operator of an airfield, suit must be brought in the Court of Claims within six years of the government aircraft's operating regularly and frequently over a person's property [98].

The taking occurs if the flights are "so low and so frequent as to be a direct and immediate interference with the enjoyment and use of the land [95]. There can even be a second taking if noisier aircraft are brought in [98]. As a general rule, flights below 500 ft constitute a taking if they are regular and frequent [98]. Sonic booms do not come within the purview of a navigation easement.

In a recent case in California, owners of homes near John Wayne Orange County Airport filed inverse condemnation actions against the county and asserted suits for nuisance, trespass, negligence, and damages for personal injury and emotional distress caused by noise and vibrations of jet airplanes using the airport. A jury trial resulted in a verdict for the defense, but a new trial was granted after misconduct by some of the jurors.

Damages were sought for a diminution of the market value of the homes, and other damages are allowable in California for unreasonable delay in actually condemning the land [99]. Also, an action for personal injuries and emotional distress caused by aircraft noise will lie in a proper case [70].

The Andrews case [99] also notes that evidence of actions which the proprietor could take to mitigate noise should have been given to the jury. Such activities involve extension of the curfew, permit more sleep on Sunday mornings, ban certain types of jets, ban jet sales and flight instruction, and restrict jet aircraft to those conforming to Part 36 of the federal air regulations, flight rules, and enforcement of noise limits.

APPENDIX. U.S. GOVERNMENTAL REGULATIONS COVERING NOISE CONTROL*

Aircraft
General aircraft operating and flight rules, 14 CFR 91
Type and airworthiness certification, noise standards, 14 CFR 36
Airport noise compatibility planning, 14 CFR 150
Army Department, environmental protection and enhancement, 32 CFR 650
Citizen suits, prior notice, noise abatement programs, 40 CFR 210
Construction equipment noise emission standards, including portable air compressors, 40 CFR 204
Construction industry, safety, and health, 29 CFR 1926
Defense Department, air installations compatible use zones, 32 CFR 256
Health standards, mandatory
Surface coal mines and surface work areas of underground coal mines, 30 CFR 71
Underground coal mines, 30 CFR 70
Hearing protectors (earmuffs), 40 CFR 211(B)
Highway traffic noise and construction noise, abatement procedures, 23 CFR 772
Housing construction, noise abatement and control, HUD environmental criteria and standards, 24 CFR 51

Labor provisions, federal supply contracts, safety and health standards, 41 CFR 50–204
Low noise emission products, 40 CFR 203
Metal and nonmetal open pit mines, safety and health, 30 CFR 55
Metal and nonmetal underground mines, safety and health standards, 30 CFR 57
Motor carrier noise emission standards, interstate, compliance, 49 CFR 325
Motor carriers engaged in interstate commerce, noise abatement programs, 40 CFR 202
Noise Control Act of 1972, practice rules governing proceedings, noise abatement programs, 40 CFR 209
Occupational safety and health, 29 CFR 1910
Product noise labeling, 40 CFR 211
Railroad noise emission compliance regulations, 49 CFR 210
Railroad noise emission standards, 40 CFR 201
Sand, gravel, and crushed stone operations, health and safety, 30 CFR
Transportation equipment noise emission controls, 40 CFR 205
Transportation equipment noise emission standards, interstate rail carriers, 40 CFR 201

REFERENCES

1. 62 Misc. 2d 353, 308 N.Y. Supp. 2d 1001 (1970).
2. *Maitland* v. *Twin City Aviation Corp.*, 241 Wisc. 541, 37 N.W. 2d 74 (Sup. Ct. 1949).
3. Restatement of Torts, Sec. 882, comment g.
4. *Rogers* v. *Eliot,* 146 Mass. 349, 15 N.E. 768 (1888).
5. H. C. Joyce, "Treatise on the Law Governing Nuisances," Matthew Bender, Albany, New York 1906, §20.

* Reference is to Code of Federal Regulations.

6. *Meeks* v. *Wood,* 66 Ind. App. 594. 118 N.E. 591 (1918).
7. K. C. Davis, The liberalized law of standing, *37 U. Chi. Law Rev.* **450** (1970).
8. E. Hanks and J. L. Hanks, An environmental bill of rights: The citizen suit and the National Environmental Policy Act of 1969, *24 Rutgers L. Rev.* **230** (1970).
9. *Scenic Hudson* v. *Federal Power Commission,* 354 F. 2d 608 (1965).
10. *Citizens Committee for the Hudson Valley* v. *Volpe,* 425 F. 2d 97 (1970).
11. *National Audubon Society* v. *Johnson,* 317 F. Supp. 1330 (D.C. Tex. 1970).
12. *U.S.* v. *Causby,* 328 U.S. 256 (1956).
13. *Griggs* v. *Allegheny Co.,* 264 U.S. 84 (1961).
14. *Aaron* v. *City of Los Angeles* (Cal. App. 1974), 115 Cal. Rptr. 162, 40 Cal. App. 3d 471.
15. *Slawinski* v. *J. H. Williams & Co.* (Appellate Div. 3d Dept. 1948), 273 App. Div. 826, 76 N.Y.S. 2d 826, aff'd. (1948), 298 N.Y. 546, 81 N.E.2d 93 (tinnitus; recovery not limited to literal wording of the act).
16. *Wojcik* v. *Green Bay Forge Co.,* 265 Wis. 38, 60 N.W. 2d 409.
17. *Hinkle* v. *H. J. Heinz Co.,* 462 Pa. 111, 337 A. 2d 907 (1973).
18. *Romero* v. *Otis Int'l.,* 343 So. 2d 405 (La. App. 1977).
19. The cases are collected in A. Larson, "Workmen's Compensation," Matthew Bender, New York, 1983.
20. *Matter of Russell* v. *Union Forging Co.* (1969) 24 N.Y. 2d 763, 300 N.Y.S. 2d 33.
21. Regulations are set forth in 41 CFR, §50-204.10.
22. 29 USC, Sec. 651, P.L. 91–596, 84 Stat. 1590. Regulations were set out at 29 CFR, Sec. 1910.95. Amendments have since been proposed.
23. 29 CFR Sec. 1926. 52.
24. 42 USC, Scs. 1858, 1858a; P.L. 91–604, Sec. 14, 84 Stat. 1709.
25. 42 USC, Sec. 4901 *et seq.,* 49 USC Sec. 1431 (dealing with standards for control and abatement of aircraft noise); P.L. 92–574, 86 Stat. 1234.
26. Council on Environmental Quality, 7th Annual Report 55 (1976) and EPA report of February 1976 quoted therein.
27. Hearings before the Senate Subcommittee on Government Regulation, 94th Cong. 1st Sess. (1975).
28. Senate Report No. 92–1160; P.L. 92–574, 86 Stat. 1234, Sec. 11.
29. 40 CFR, Sec. 210; Fed. Reg. 36011 (October 7, 1974). The notice of rule promulgation was dated September 27, 1974, to enter into force 30 days afterwards.
30. P.L. 95–609, 92 Stat. 3079 amended certain sections of Title 42 and Title 49.
31. *Atlas Copco, Inc.* v. *EPA,* 642 F. 2d 458, 206 U.S. App. D.C. 53 (1979). Regulations of portable air compressors were amended in 1977. 42 Fed. Reg. 61453, effective 1978, 40 CFR 204.52.
32. *Recreation Vehicle Industry Assn.* v. *EPA,* 653 F. 2d 526, 209 U.S. App. D.C. 307 (1981).
33. *Harnischfeger Corp.* v. *USEPA,* 515 F. Supp. 1310 (D.C. Wis. 1981). ·
34. *Rockford Drop Forge Co.* v. *Pollution Control Board,* 389 N.E. 2d 212, 27 Ill. Dec. 400, 71 Ill. App. 3d 290, aff'd., 402 N.E. 2d 602, 37 Ill. Dec. 600, 79 Ill. 2d 271 [forging facility regulation not pre-empted by 42 USC, Sec. 4905(e)].
35. P.L. 91–605, Federal Aid Highway Act of 1970, 23 USC, Sec. 101, *et seq.* The Federal Highway Administration has set forth noise standards for the planning and design of federally aided highway construction. 38 Fed. Reg. 15953 (June 19, 1973). Regulations for railroad engines and cars and medium and heavy trucks were promulgated in 1976. 41 Fed. Reg. 2161-95 and 15337-15558. Truck regulations have since been added in 1977. 42 Fed. Reg. 61457. Solid waste compacters mounted on trucks had regulations proposed in 1977. Newly manufactured garbage trucks were subjected to standards in 1979.

40 CFR, Part 205, 44 Fed. Reg. 56524. Proposals have been made for motorcycles, mopeds, and buses.

36. 24 CFR Part 51, 44 Fed. Reg. 40860 (1979). Also see Noise Assessment Guidelines, 1972. These are discussed *infra*, Section VIII.

37. Coal mines, 30 CFR, Secs. 70.500–70.510. Metallic and nonmetallic mining, under the Bureau of Mines, 39 Fed. Reg. 28433 (1974) and 39 Fed. Reg. 35999 (1974).

38. 15 USC, Sec. 2002; 40 CFR 205.1, *et seq.* Construction equipment controls are at 40 CFR 204.1, *et seq.*

39. *Chrysler Corp.* v. *EPA,* 600 F. 2d 904 (D.C. Cir. 1979).

40. State statutes are listed in "Third Annual Report of the Council on Environmental Quality" (1972).

41. "Prevention and Control of Environmental Noise Pollution—Explanation of Noise Control Regulations," August 1973, pp. 1, 2. (Regulations in 6 New York Code R. R., pp. 450–454.

42. Public Resources Code., Sec. 2100, *et seq.,* Stats. 1970, Ch. 1433, Sec. 1; Stats. 1972, Ch. 1154, Sec. 1.

43. Regulations of the Illinois Pollution Control Board were upheld in *Ill. Coal Operators Assn.* v. *P.C.B., 59* Ill. 2d 305, 319 N.E. 2d 782 (1974); *Shell Oil Co.* v. *Ill. PCB,* 37 Ill. App. 3d 264, 364 N.E. 2d 212 (1976) and *Union Oil Co.* v. *Ill. PCB,* 43 Ill. App. 3d 927, 357 N.E. 2d 715 (1976).

44. *Environment Reporter* **5,** 119 (May 1974).

45. *Environment Reporter* **3,** 949 (December 1972).

46. *Environment Reporter* **4,** 1660 (February 1974).

47. 1976 Op. Atty. Gen. (Inf.) 194.

48. *New York Times,* November 22, 1974.

49. *Environment Reporter* **3,** 631 (October 1972).

50. C. R. Bragdon, Municipal noise ordinances, *Sound Vib.* **7**(12), 16 (1973).

51. Third Annual Report of the Council on Environmental Quality, (1972), 206–219.

52. Second Annual Report of the Council on Environmental Quality (1971), pp. 47–49.

53. Sixth Annual Report of the Council on Environmental Quality (1975), pp. 89–90.

54. C. S. Rhyne, "Municipal Law," National Institute of Municipal Law Offices, Washington, D.C., 1957.

55. Explanation of Noise Control Regulations, New York State Department of Environmental Conservation, August 1973, p. 10.

56. Regulations are at 24 CFR, Part 51, 44 Fed. Reg. 40860 (July 12, 1979). Noise Assessment Guidelines were published in 1972 (HUD NTIS, PB-210 590). U.S. Government Printing Office, Washington, D.C.

57. *American Airlines* v. *Town of Hempstead,* 272 F. Supp. 226 (E.D.N.Y. 1967) *aff'd,* 398 F. 2d 369 (2nd Cir. 1968); *Huron Portland Cement Co.* v. *City of Detroit,* 362 U.S. 440 (1960).

58. *Southeastern Aviation, Inc.* v. *Hurd,* 209 Tenn. 639, 355 S.W. 2d 436 (1962), *appeal dismissed,* 371 U.S. 21 (1962).

59. Sec. 611(d), Federal Aviation Act.

60. 14 CFR, Sec. 91.309(b)(2) and Part 36 noise limits.

61. 132 F. Supp. 871 (E.D.N.Y. 1955) *aff'd* 238 F. 2d 812 (2nd Cir. 1956).

62. 411 U.S. 624 (1973).

63. §1508, Federal Aeronautics Act of 1958.

64. 411 U.S. at 635-636, n. 14, cited, *San Diego Unified Port District* v. *Gianturco,* 651 F. 2d 1306 (9th Cir. 1981) at p. 1316.

65. *San Diego Unified Port District* v. *Gianturco, supra,* citing authority from the Second and Ninth Circuits, Noise Abatement Policy report of 1976, a senate report, and the Federal Register.
66. P.L. 95–609, 92 Stat. 3079 (1978), codified in 42 USC, Sec. 4905 and other sections, and 49 USC, §1431.
67. *Air Transp. Assn.* v. *Crotti,* 389 F. Supp. 58 (N.D. Cal. 1975).
68. See 14 CFR, Part 36 and 91.301-311. Certain limitations as to nighttime flights are prescribed where design criteria are not met. General authority over aircraft noise is contained in 49 USC, Sec. 1431.
69. 671 F. 2d 54 (1st Cir. 1982).
70. *Greater Westchester Homeowners Assn.* v. *City of Los Angeles,* 26 Cal. 3d 86, 160 Cal. Rptr. 733, 743, 603 P. 2d 1329, 1340 (1979), cert. den. (1980), 449 U.S. 820. Statement referred to in Di Perri case, *supra* [69].
71. *National Aviation* v. *City of Hayward,* 418 F. Supp. 417 (D.C. Cal. 1976). See also *Santa Monica Airport Assn.* v. *City of Santa Monica,* 9 E.L.R. 20763 (D.C. Cal. 1979), where a night curfew and weekend and helicopter limitations were upheld, but a complete ban on jets was not allowed.
72. *U.S.* v. *State of New York,* 552 F. Supp. 255 (N.D.N.Y. 1982), aff'd., 708 F. 2d 92 (2d Cir. 1983).
73. *State of New York* v. *Federal Aviation Administration,* 712 F. 2d 806 (2d cir. 1983).
74. Airport and Airway Improvement Act of 1982, 49 USC, Sec. 2201 *et seq.*
75. *U.S.* v. *County of Westchester,* 83 Civ. 3499 (RJW) decided August 24, 1983, 571 F. Supp. 786.
76. *British Airways* v. *Port Authority of N.Y.,* [I] 558 F. 2d 75 (2d Cir. 1977) and [II] 564 F. 2d 1002 (2d Cir. 1977).
77. *Environmental Defense Fund, Inc.* v. *U.S. Dept. of Transportation,* 14 Avi. 17,140 (D.C. Cir. 1976), cert. den. (1976).
78. 2 U.S. Code Congress and Admin. News, 96th Cong., 2d Sess. 1980, p. 89, *et seq.*
79. "Treatise on Environmental Law, Grad," Vol. 1A, Chapter 5, Matthew Bender, New York (1977).
80. *New York Times,* August 25, 1983, p. B2.
81. 49 USC, Sec. 2201, superseding the 1970 Airport and Airway Development Act.
82. 14 CFR, Part 36 (1983), 34 Fed. Reg. 18, 355 (1969).
83. "Fleet Compliance Program," Operating Noise Limits, 14 CFR, Secs. 91.301–311, as amended (1983), 41 Fed. Reg. 56,046 (1976). The policy statement was set forth in a release of the Department of Transportation and FAA Aviation Noise Abatement Policy (1976).
84. Aviation Safety and Noise Abatement Act of 1979, 49 USC, Secs. 2101–2125 considered the FAA Fleet Compliance Program.
85. 14 CFR, Sec. 91.305(c).
86. FAA Operating Noise Limits, 45 Fed. Reg. 79.302, 79.310 (1980).
87. 727 F. 2d 246 (2d Cir. 1984), Docket No. 83-6167, Slip. Op. 7531 (January 31, 1984); clarification opinion, p. 2345, *et seq.,* March 14, 1984.
88. *Herald Tribune* (International Edition), September 30, 1983, p. 1. Japan Air Lines will be replacing the DC-8s with Boeing 767s. In the *New York Times* of November 25, 1983, Delta Airlines was said to be the first airline to use muffled engines for its DC-8s to comply with FAA noise requirements by January 1985.
89. British Airways is to lease new Boeing 737-200s to replace the Tridents. MD-9 Super 80s and Airbus Industries' A320s are also being offered for the future.
90. *Westside Property Owners* v. *Schlesinger,* 597 F. 2d 1214 (9th Cir. 1979).

91. *City of Romulus* v. *County of Wayne,* 392 F. Supp. 578 (D.C. Mich. 1975).
92. *Davidson* v. *Department of Defense,* 560 F. Supp. 1019 (D.C. Ohio 1982).
93. T. E. Shea, The judicial standard for review of environmental impact statement, *Boston College Environmental Affairs Law Rev.* **9,** 63 (1980–1981).
94. *Aaron* v. *City of Los Angeles* (Cal. App. 1974) 115 Cal. Rptr. 162, 40 Cal. App. 3d 471.
95. *U.S.* v. *Causby,* 328 U.S. 256 (1946); *Griggs* v. *Allegheny County,* 369 U.S. 84 (1962).
96. *Batten* v. *U.S.,* 306 F. 2d 580 (10 Cir. 1962), cert. den., 371 U.S. 955, reh. den., 372 U.S. 925.
97. *Thornburg* v. *Port of Portland,* 233 Or. 178 (1962), 376 P. 2d 100; *Martin* v. *Port of Seattle,* 391 P. 2d 540 (1964). See also Recent developments in inverse condemnation of air space, *J. Air L. Commerce* **81,** 39 (1973). Regulations on portable air compressors were amended in 1977, 42 Fed. Reg. 61453.
98. *Powell* v. *U.S.,* 17 Avi. 17,988 (Claims Ct. 1983). Regulations for railroad engines and cars and medium and heavy trucks were promulgated in 1976, 41 Fed. Reg. 2161-95 and 15337-15558. Truck regulations have since been added in 1977, 42 Fed. Reg. 61457. Solid-waste compacters mounted on trucks were affected by regulations proposed in 1977. Newly manufactured garbage trucks were subjected to standards in 1979, 40 CFR, Part 205, 44 Fed. Reg. 56524. Proposals have been made for motorcycles, mopeds, and buses.
99. *Andrews* v. *County of Orange,* 130 Cal. App. 3d 944 (Fourth Dist. 1982), citing *Aaron* v. *City of Los Angeles,* (Cal. App. 1974), 115 Cal. Rptr. 162, 40 Cal, App. 3d 471; and *Smart* v. *City of Los Angeles* (Cal. App. 1980), 169 Cal. Rptr. 174, 112 Cal. App. 3d 232.

14

EEC Directives on Noise in the Environment

B. HAY

Department of Civil Engineering and Building
Coventry (Lanchester) Polytechnic
Coventry
England

I. EEC DIRECTIVES ON NOISE IN THE ENVIRONMENT

A. Legal Status of EEC Directives

The European Economic Community (EEC) was established by the Treaty of Rome [1], which therefore forms a primary source of community law. The Council of Ministers is the supreme organ of the community. It is composed of 10 Ministers, one from each member state. The council acts mainly on proposals from the commission, a body whose 14 members are under oath to act independently in the interests of the community as a whole.

Directives are regarded as a secondary source of community law, because their authority is derived from the provisions of the founding treaty. Article 100 of the Treaty of Rome states that "the Council, acting by means of a unanimous vote on a proposal of the Commission, shall issue directives for the approximation of such legislative and administrative provisions for the Member States, as have a direct incidence on the establishment or functioning of the Common Market". Article 189 states that "directives shall bind any Member State to which they are addressed, as to the result to be achieved, while leaving to domestic agencies a competence as to form and means". The Court of Justice has powers under Articles 169 and 170 to impose its jurisdiction upon a member state for failure to implement a directive.

B. Directives on Environmental Noise

1. Introduction

Of the areas covered by directives and proposed directives, by far the most important from the viewpoint both of noise policy and of trade barriers is vehicle noise. Airplane noise is a major environmental problem, but the EEC's involvement is secondary to that of the International Civil Aviation Organisation (IACO) [2, 3]. The family of five construction plant directives should prove to be of environmental importance, as the permissible sound power levels have been adopted by the European Council, (see Subsection 3 below). The proposed directive on household appliances is by comparison of much less environmental importance. It is primarily a measure to eliminate trade barriers.

All the directives and proposed directives are based on Article 100 of the Treaty of Rome [1] with the exception of those concerned with subsonic airplanes and helicopters, which are based on Article 84(2). Six subjects are covered by adopted directives: airplanes (subsonic), construction plants, motorcycles, motor vehicles which have four wheels (excluding tractors), tractors, and lawn mowers. Two other subjects are covered by proposed directives: helicopters and household appliances.

2. Where to Find Directives

The texts of directives which have been adopted by the council are published in the L (legislation) series of the *Official Journal of the Euro-*

pean Communities (see Table I). The member states are usually notified of an adopted directive a few days afterwards. Then the member states bring into force the laws, regulations and administrative provisions necessary to comply with the specific directive within the stated time limit after its notification, (e.g., 12 or 18 months). The texts of commission proposals

TABLE I. Adopted and Proposed Directives: Environmental Noise

Text reference number	Subject	Council directive number	Date of adoption
[2]	Aeroplanes (subsonic)	80/51/EEC	20.12.79
	Amendment	83/206/EEC	21.04.83
[4]	Construction plant and equipment	79/113/EEC	19.07.78
	Amendment	81/1051/EEC	07.12.81
	Construction plant and equipment	84/532/EEC	17.09.84
[5]	Air compressors	84/533/EEC	17.09.84
[6]	Power generators	84/536/EEC	17.09.84
[7]	Welding generators	84/535/EEC	17.09.84
[9]	Powered hand-held concrete breakers and picks	84/537/EEC	17.09.84
[8]	Tower cranes	84/534/EEC	17.09.84
[31]	Lawn mowers	84/538/EEC	17.09.84
[11]	Motor cycles	78/1015/EEC	23.11.78
[14]	Motor vehicles	70/157/EEC	27.07.70
[18]	Amendment	77/212/EEC	08.03.77
[15]	Amendment	81/334/EEC	13.04.81
[16]	Amendment	84/372/EEC	03.07.84
[20]	Amendment	84/424/EEC	03.09.84
[21]	Tractors (agricultural or forestry)	74/151/EEC	04.03.74
[22]	Driver-perceived noise level in the cab	77/311/EEC	29.03.77

Text reference number	Subject	Proposal for a council directive	Date of proposal
[10]	Excavators, dozers, loaders and excavator-loaders	COM(80)468 COM(81)541 6061/84	28.11.80 09.10.81 04.04.84
[25]	Helicopters	COM(81)554	27.10.81
[28]	Household appliances	COM(83)694	12.11.83

for directives are published in the C (information and notice) series of the official journal (see Table I). Preliminary drafts and council revisions of commission proposals are not published.

3. Discussion on Adopted and Proposed Directives

a. Airplanes (Subsonic). On 21 April 1983, the council adopted the directive 83/206/EEC, which amended the directive 80/51/EEC of 20 December 1979 on the limitation of noise emissions from subsonic airplanes [2]. This directive was notified to the member states on 26 April 1983.

Member states were to bring into force the provisions necessary to comply with this directive by 26 April 1984. Each member state shall ensure that any civil subsonic jet or propellor-driven airplane which is registered in its territory may not be used in the territory of member states unless it has been granted a noise certificate based on the relevant ICAO [3] standards. Noise certification will now be granted on the basis of satisfactory evidence that the airplane complies with requirements which are at least equal to the applicable standards specified in Part II, Chapters 2, 3, 5 or 6 of Volume I of Annex 16/5 to the Convention on International Aviation, as applicable from 26 November 1981 in accordance with Amendment 5. As from 1 January 1988, member states will not permit the operation in their territory of civil subsonic jet airplanes which are not registered in a member state and which do not comply with the ICAO standards.

b. Construction Plant and Equipment. The Council of the European Communities adopted the directive [4] 79/113/EEC on 19 December 1978, which applies to the determination of the airborne noise emission of construction plant and equipment, which are used outdoors. Under the directive, the acoustic criterion for the environment of a sound source shall be expressed either by the A-weighted sound power level of the sound source L_{WA} or by L_{WA} supplemented by the directivity index. Then, on 7 December 1981, the council adopted a directive 81/1051/EEC, which amended certain articles in the directive 79/113/EEC. The acoustic criterion for the noise emitted to the operator's position(s) of the machines is expressed by the equivalent continuous A-weighted sound pressure level $L_{Aeq}(t_1, t_2)$, where $t_1 - t_2$ is the measurement interval.

A major step forward was made on 17 September 1984, when the Council adopted a "framework directive" 84/532/EEC [4a] laying down procedures for EEC type-approval, EEC type-examination, EEC verification and EEC self-certification for construction plant and equipment. Five separate directives [5–9] for particular classes of construction equipment were adopted by the Council on 17 September 1984 and notified to the

TABLE II. Adopted Directives for the Permissible Sound Power Construction Plant and Equipment

Class of construction plant and equipment [reference number]	Permissible sound power level in dB(A)/1 pW		Technical details[a]
	From 26.03.1986	From 26.06.1989	
Air compressors [5]	101	100	$Q \leq 5$
	102	100	$5 < Q \leq 10$
	104	102	$10 < Q \leq 30$
	106	104	$Q > 30$
Power generators [6]	104	102	$P \leq 2$
	104	100	$2 < P \leq 8$
	103	100	$8 < P \leq 240$
	105	100	$P > 240$
Welding generators [7]	104	101	$I \leq 200$
	101	100	$I > 200$
Hand-held concrete breakers and picks [9]	110	108	$m < 20$
	113	111	$20 \leq m \leq 35$
	116	114	$m > 35$ and appliances with an internal combustion engine incorporated
Tower cranes [8]	102	100	Lifting mechanism[b]
	Levels laid down in [6] apply		Energy generator[b]
	Highest values of the two components		Assembly comprising lifting mechanism and energy generator

Proposal for a council directive	Permissible sound power level in dB(A)/1 pW		Technical details[a]
	From the date of the entry into force of the directive	From 5 years from the entry into force of the directive	
Excavators, dozers, loaders, and excavator-loaders [10]	107	104	$P_1 \leq 150$
	110	107	$150 < P_1 \leq 250$
	112	110	$250 < P_1 \leq 375$
	115	113	$P_1 > 375$

[a] Q = standardized nominal air flow in cubic meters per minute; P = electric power in kilovolt amperes; I = nominal maximum welding current in amperes; m = mass of appliance in kilograms; P_1 = net installed power in kilowatts.

[b] Components of system.

Member States on 26 September 1984. The date for implementing these five directives is March 26, 1986. By contrast, there is still only a proposal for a Council directive for the permissible sound power levels of excavators, dozers, loaders, and excavator-loaders [10]. The reason is that the issue of the use of a mobile test, as distinct from a static test, is still being discussed in Brussels. On 4 April 1984, the European Commission issued an internal paper for discussion. This was followed by a meeting in September 1984 in Brussels of the Committee for European Construction Equipment. The next step would be for the European Commission to publish a Commission document with a COM number, in the Official Journal of the European Communities.

  c. *Motorcycles*. The new noise limits and test procedure for motorcycles introduced by the directive 78/1015/EEC [11] apply in the United Kingdom [12], for motorcycles first manufactured on or after 1 October 1982 and first used on or after 1 April 1983 (see Table IIIa). Furthermore, Article 8 of the directive 78/1015/EEC, specifically provided that the

TABLE III. (a) Permitted Construction Limits for Motorcycles and Tractors Coming into Use in 1983

Class of vehicle	Noise limit [dB(A)]	EEC directives	U.K. statutory instrument
Motorcycles		78/1015	1980 No. 1166
Less than or equal to 80 cm³	78		
Less than or equal to 125 cm³	80		
Less than or equal to 350 cm³	83		
Less than or equal to 500 cm³	85		
Greater than 500 cm³	86		
Tractors (agricultural or forestry)		74/151	1980 No. 1166
Engine power not exceeding 90 hp DIN	89		
Engine power exceeding 90 hp DIN	92		

(b) Proposal EEC Motorcycle Noise Limits from 1 October 1986 (from Proposed Directive COM(84)438)

Categories by cubic capacity [cm³]	Maximum permissible sound level [dB(A)]	
	With effect from 1 October 1986	With effect from 1 October 1995
≤80	77	75
>80, ≤175	80	78
>175	82	80

council should decide on a further reduction in the maximum noise limits of motorcycles by 31 December 1984.

Accordingly, the Commission has submitted to the Council a proposal COM(84) 438 [13] for a Council directive amending the directive 78/1015/ EEC. The purpose is to make the prescribed method of measurement more representative of the conditions of actual use of motor cycles in urban traffic, and to reduce the number of categories of motor cycle. The sound level of motor cycles, when measured under the prescribed conditions, shall not exceed the limits stated in Table IIIb. The implementation date for the proposed directive COM(84)438 [13] is 1 October 1986.

d. Motor Vehicles (Four-Wheeled). Measurement Procedure: The council's first directive on motor vehicle noise, 70/157/EEC [14], laid down a method of measuring the noise they emit during type-approval tests. This test procedure remained valid until 1 October 1984.

Directive 70/157/EEC has subsequently been amended by directive 81/ 334/EEC [15] and further amended by directive 84/372/EEC [16]. (See Table IV and Fig. 1.) The amendments apply to vehicles in category M_1 and N_1. Category M_1 means vehicles used for the carriage of passengers and comprising no more than eight seats in addition to the driver's seat. Category N_1 means vehicles used for the carriage of goods and having a maximum weight not exceeding 3.5 metric tonnes.

With effect from 1 October 1984, member states shall not, on grounds relating to the permissible sound level and the exhaust system,

  (i) refuse to grant EEC type approval or to grant national type approval in respect of a type of motor vehicle or
  (ii) prohibit the entry into service of vehicles if the sound level and the exhaust system of this type of vehicle complies with the provisions of the directive 70/157/EEC [14] as amended by the directive 84/ 372/EEC [16]. The relevant U.K. national type-approval regulations are the Statutory Instruments 1984 Nos. 1401 and 1402 (51) [17].

Noise Limits: Directive 70/157/EEC [14] laid down a noise limit of 82 dB(A) for private cars. A subsequent directive, 77/212/EEC [18], lowered the limit to 80 dB(A). This noise limit of 80 dB(A) was made mandatory in the United Kingdom [12, 19] for private cars manufactured on or after 1 April 1983 and first used on or after 1 October 1983. However, on 3 September 1984, a new noise limit of 77 dB(A) for private cars was adopted by the Council of the European Communities in directive 84/424/ EEC [20] (see Table V). This states that "Member States shall bring into force the provisions necessary to comply with directive 84/424/EEC be-

TABLE IV. Vehicle Noise Measurements (from Directive 84/372/EEC Amending Annex I to Directive 70/157/EEC)

Point 5.2.2.4.3.2 is hereby replaced by the following:

5.2.2.4.3.2. *Approach speed*

The vehicle must approach the line AA′ at a steady speed corresponding to the lower of the following two speeds:
—the speed corresponding to an engine speed equal to three-quarters of the speed, S, at which the engine develops its rated maximum power,
—50 km/h.

However, if in the case of vehicles equipped with an *automatic transmission* having more than two discrete ratios there is a change-down to first gear during the test, the manufacturer may select either of the following test procedures:
—the speed, V, of the vehicle shall be increased to a maximum of 60 km/h in order to avoid such a change-down, or
—the speed, V, shall remain at 50 km/h and the fuel supply to the engine shall be limited to 95% of the supply necessary for full load. This condition is considered to be satisfied:
—in the case of a spark-ignition engine, when the angle of the throttle opening is 90%, and
—in the case of a compression-ignition engine, when the movement of the central rack of the injection pump is limited to 90% of its travel.

If the vehicle is equipped with an *automatic transmission which has no manual override,* it must be tested at different approach speeds, namely 30, 40 and 50 km/h, or at three-quarters of maximum road speed if this value is lower. The test result shall be that obtained at the speed which produces the maximum sound level.

Point 5.2.2.4.3.3.1.1 is hereby replaced by the following:

5.2.2.4.3.3.1.1. *Vehicles in categories M_1 and N_1 equipped with a manually operated gearbox having not more than four forward gear ratios must be tested in second gear.*

Vehicles in these categories equipped with a manually operated gearbox having more than four forward gear ratios must be tested in second and third gears successively. Only overall gear ratios intended for normal road use shall be considered. The arithmetic mean of the sound levels recorded for each of these two conditions shall be calculated.

However, the vehicles in category M_1 having more than four forward gears and equipped with an engine developing a maximum power greater than 140 kW, and whose permissible maximum-power/maximum-mass ratio exceeds 75 kW/t, may be tested in third gear only, provided that the speed at which the rear of the vehicle passes the line BB′ in third gear is greater than 61 km/h.

Point 5.2.2.4.3.3.2 is hereby replaced by the following:

5.2.2.4.3.3.2. *Automatic transmission equipped with manual override.*

The test shall be conducted with the selector in a position recommended by the manufacturer for "normal" driving.

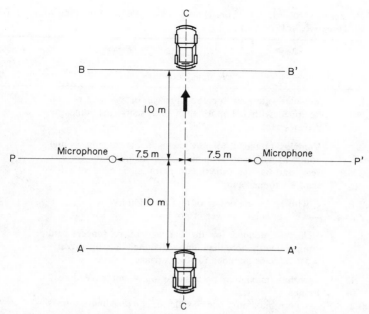

Fig. 1. Measuring positions for vehicles in motion as given by directive 84/372/EEC (Table IV).

fore 1 January 1985''. With effect from 1 October 1988, member states may refuse to grant national type approval in respect of a type of motor vehicle of which the sound level and the exhaust system do not comply with the requirements of directive 70/157/EEC [14] as amended by directive 84/424/EEC [20].

e. Tractors. By virtue of the directives 74/151/EEC [21] and 77/311/EEC [22], no member state may refuse to allow the use of tractors which are type approved to the EEC's requirements, provided that they satisfy the limits for noise emitted by tractors and for the driver-perceived noise level. The directive 74/151/EEC [21] set a limit of 85 dB(A) for tractors weighing not more than 1.5 tonnes (unladen), and 89 dB(A) for tractors weighing more than 1.5 tonnes (unladen).

Notwithstanding the directive 74/151/EEC [21], the noise limit applicable in the United Kingdom [12] for tractors first manufactured on or after 1 October 1982 and first used on or after 1 April 1983 are 89 dB(A) for tractors fitted with engines of less than 90 hp and 92 dB(A) for tractors of over 90 hp. With respect to the driver-perceived noise level in the cab, the directive 77/311/EEC [22] set a limit of 90 dB(A). This limit had already

TABLE V. Vehicle Noise Limits (from Directive 84/424/EEC Amending Annex I to Directive 70/157/EEC)

Vehicle categories		Noise limit expressed in dB(A)
5.2.2.1.1.	Vehicles intended for the carriage of passengers and equipped with not more than nine seats, including the driver's seat	77
5.2.2.1.2.	Vehicles intended for the carriage of passengers and equipped with more than nine seats, including the driver's seat, and having a maximum permissible mass of more than 3.5 tonnes and:	
5.2.2.1.2.1.	—with an engine power of less than 150 kW	80
5.2.2.1.2.2.	—with an engine power of not less than 150 kW	83
5.2.2.1.3.	Vehicles intended for the carriage of passengers and equipped with more than nine seats, including the driver's seat; vehicles intended for the carriage of goods:	
5.2.2.1.3.1.	—with a maximum permissible mass not exceeding 2 tonnes	78
5.2.2.1.3.2.	—with a maximum permissible mass exceeding 2 tonnes but not exceeding 3.5 tonnes	79
5.2.2.1.4.	Vehicles intended for the carriage of goods and having a maximum permissible mass exceeding 3.5 tonnes:	
5.2.2.1.4.1.	—with an engine power of less than 75 kW	81
5.2.2.1.4.2.	—with an engine power of not less than 75 kW but less than 150 kW	83
5.2.2.1.4.3.	—with an engine power of not less than 150 kW	84

However,
—for vehicles of categories 5.2.2.1.1 and 5.2.2.1.3, the limit values shall be increased by 1 dB(A) if they are equipped with a direct injection diesel engine,
—for vehicles with a maximum permissible mass of over two tonnes designed for off-road use, the limit values are increased by 1 dB(A) if their engine power is less than 150 kW and 2 dB(A) if their engine power is equal to or greater than 150 kW.

been anticipated in the United Kingdom by the 1974 Agricultural (Tractor Cabs) Regulations [23].

f. Helicopters. In May 1979, the Sixth Meeting of ICAO's Committee on Aircraft Noise (CAN 6) recommended noise certification standards for helicopters [24]. The CAN 6 standards are contained in Chapter 8 of Part II of Annex 16/5 to the Convention on International Civil Aviation [3] as applicable from 26 November 1981 in accordance with Amendment 5. The European Commission, on 6 October 1981, proposed a directive [25] [COM(81)554] to make the introduction of standards mandatory into the laws of the member states of the European Community. However, a number of countries with major helicopter-manufacturing industries felt that the CAN 6 standards were too stringent, and revised standards were considered at the Seventh Meeting of ICAO's Committee on Aircraft Noise (CAN 7) held in Montreal, (2–13 May 1983 [26]. As a result, the CAN 7 committee recommended to the Council of the European Communities that the CAN 6 noise certification standards for helicopters should be relaxed. However, it is stressed that the power to make this decision rests with the council.

g. Household Appliances. In 1982, the commission proposed a directive [27] concerning airborne noise emitted by household appliances. In 1983, the commission submitted to the council proposed amendments [COM(83)694] [28] which would entail replacing Annex I by a precise reference to an international standard. The noise levels of each family of household appliances shall be determined in accordance with the methods defined in IEC Publication 704-1, 1982 [29], first edition, of the International Electrotechnical Commission, which has been endorsed by the European Committee for Electrotechnical Standardization (CENELEC) in their Harmonization Document HD 423-1 [30]. The values obtained in this way constitute the basic data for checking the accuracy of their noise emission with the declared values. Unless otherwise indicated, these values are inclusive of tolerances. This test code is applicable unless separate directives lay down different or supplementary provisions, taking into account the special characteristics of certain types of household appliances.

h. Lawn Mowers. The purpose of the adopted directive 84/538/EEC [31] is to restrict noise emissions from motorized lawn mowers by specifying upper limits and the methods for measuring such emissions. Table VI sets out the adopted noise limits. The implementation date for the adopted directive 84/538/EEC [31] is 1 July 1987.

TABLE VI. Permissible Sound Levels for Lawn
Mowers from 1 July 1987

Levels in dB(A)/1 pW	Cutting width of lawn mower L
96	$L < 50$ cm
100	$50 < L \leq 120$ cm
105	$L > 120$ cm

II. NATIONAL LAWS ON EXTERNAL INDUSTRIAL NOISE

Protection of the external environment has been secured to a certain extent through national legislation in 7 out of 10 member states of the EEC, who have all issued general laws for the control of industrial noise emanating from industrial installations.

A. Belgium

The Noise Abatement Act [32] of 18 July 1973 is still the relevant law for controlling environmental noise in Belgium. The power to create protection zones corresponding to residential areas, industrial zones, recreational areas and those areas where quiet is particularly required was given to the King in Article 1. The ability to impose particular conditions to ensure the application of the 1973 law, and to make special orders, was conferred in Article 3. However, no implementing orders have so far been passed to establish noise limits inside zones around industrial estates.

B. Denmark

The Environmental Protection Act No. 663 of 16 December 1982 [33] is now the relevant statute for controlling noise emitted from industrial premises in Denmark. In 1984, the Ministry of the Environment revised their 1974 guideline [34, 34a], which gave recommended maximum limits for external industrial noise, which is stated as the equivalent, steady, corrected noise level in decibels(A) outdoors. The main differences occur in districts type 4 and 5, where the residential area is now defined by its background noise when it is high and low, respectively (see Table VII). Noise measurements are carried out according to the first part of the International Standard ISO 1996/1–1982 [35], which has been adopted as a Danish standard. This international standard [35] defines the basic quantities to be used for the description of noise in community environments and describes basic procedures for the determination of these quantities.

C. France

The main relevant French law [36] is Law No. 76-663 of 19 July 1976 relating to establishments classified for environmental protection and which imposes the obligation of a study of the impact of the premises on the environment by virtue of Law No. 76-629 of the 10 July 1976. The Minister of the Environment issued Circular No. 3055 [36] to the prefets

TABLE VII. Recommended Maximum Limits in Decibels(A) for the External Industrial Noise Level L_r in Denmark[a]

Type of district	Monday–Friday daytime 7 A.M. to 6 P.M.	Monday–Friday evening 6 P.M. to 10 P.M.	Every day nighttime 10 P.M. to 7 A.M.
	Saturday 7 A.M. to 2 P.M.	Sundays and holidays 7 A.M. to 10 P.M.	
1. Predominantly industrial area	70	70	70
2. Industrial areas with prohibition against annoying activities	60	60	60
3. Urban residential with mixed residential and industrial settlements	55	45	40
4. Residential area with high background noise level	50	45	40
5. Residential area with low background noise level	45	40	35
6. Summer residential area and recreational area	40	35	35
7. Summer cottages in town areas	Levels depend on paragraph 2.2.3	Levels depend on paragraph 2.2.3	Levels depend on paragraph 2.2.3
8. The open land including villages	Levels depend on paragraph 2.2.3	Levels depend on paragraph 2.2.3	Levels depend on paragraph 2.2.3

[a] L_r is the equivalent continuous A-weighted sound pressure level during a specified time interval, plus specified adjustments for tonal character and impulsiveness of the sound.

of the regions on the 21 June 1976. This circular applies the methodology of the existing French Standard NF S 31-010 [37] to the establishments classified for environmental protection. A criterion of discomfort C_r is calculated, according to the circular, as

$$C_r = 45 \quad \text{dB(A)} + C_T + C_Z,$$

where the constants C_T and C_Z depend on the time of the day and the type of zone, respectively.

Thus the circular permits the prefets to set up external noise limits above a base of 45 dB(A) as a function of the local situation. Hence, in the same town the noise limits can vary from zone to zone. For the moment, there is no change in the French legislation [36]. However, it is expected that Circular No. 3055 will be modified soon by the Minister of the Environment. Also, the French Standard NF S 31-010 [37] will be modified, taking the first part of the new International Standard ISO 1966/1–1982 [35] into consideration, as well as the second and third parts, which exist only in draft form at present.

D. Federal Republic of Germany

The relevant legal documents are the 1974 Federal Emission Control Law [38] and the 1968 Technical Instruction [39] on Protection against Noise. This technical instruction [39] applies to plants which require official approval in accordance with Clause 16 of the Industrial Order of 4 August 1960 [40]. The enforcement of the technical instruction [39] is the responsibility of the individual Lander (i.e., the federal states). When the Lander are examining applications for permission to erect new plants, to change the location of operation of a plant and to make major changes in the operation of a plant, they can lay down noise emission limits. The technical instruction specifies in Subclause 2.321 the standard emission values according to the type of zone and whether it is nighttime or daytime (see Table VIII). The reference time intervals are 16 hr for the day and 8 hr for the night, starting at 10 P.M. and ending at 6 A.M. However, the nighttime period can be shifted forwards or backwards by 1 hr if this is necessary because of special local conditions, provided that the neighbours are guaranteed an 8-hr night's rest.

Allocation of a plant's sphere of influence to the areas listed in Subclause 2.321 should be based on the following principles: (a) If the development plan shows built-up areas conforming to the areas listed in Subclause 2.321 (see the Utilisation for Building Purposes Order of 26 June 1962 [41] and the subsequent Order of September 1977 [42]), then the development plan should be used as a basis. (b) If the actual utilisation of land within the plant's sphere of influence differs considerably from the "utilisation for building purposes" set down in the development plan, then the actual utilisation should be used as a basis, taking into account the intended development of the area. (c) However, if there is no development plan, then the actual use must be established; any predictable changes in use should be taken into account.

TABLE VIII. Noise Limits in the Federal Republic of Germany

Standard immission values in dB(A) in various locations	Day, 6 A.M. to 10 P.M.	Night, 10 P.M. to 6 A.M.
1. Areas in which only business or industrial establishments exist (and dwellings for owners and directors of the factories and the supervisory personnel)	70	70
2. Areas in which primarily business establishments occur	65	50
3. Areas which are neither predominantly business nor residential but which contain both	60	45
4. Areas which are primarily residential	55	40
5. Areas which are entirely residential	50	35
6. Rest areas, hospitals and nursing homes	45	35
7. Dwellings which are attached to the industrial installation	40	30

In the case of special circumstances where the technical instruction [39] is supposed not to provide sufficient protection of the environment from industrial installations, the individual Lander have adopted the Association of German Engineers' Guideline VDI 2058, Part 1 [43]. This guideline does not have the status of a law or regulation, but has the character of a generally agreed-upon statement of experts, which may be used to find a solution in difficult cases. For this purpose, the most important aspects of Guideline VDI 2058, Part 1 [43], are the following:

(i) The reference time interval at night is 1 hr instead of 8 hr as in the technical instruction [39];

(ii) during the day, the A-weighted sound pressure level of noises of short duration shall not exceed the limit appropriate for the situation under consideration by more than 30 dB(A);

(iii) inside buildings, and for structure-borne noise, the limits for the rating level are 35 dB(A) by day and 25 dB(A) at night. The level of noises of short duration should not exceed these limits by more than 10 dB(A); and

(iv) a penalty of 6 dB(A) is added to the levels during the time intervals from 6 A.M. to 7 A.M. and from 7 P.M. to 10 P.M.

The results of measurements, performed for the characteristic operation conditions of the noise source [44] under the weather conditions prevailing at the location under consideration, must comply with the noise limits in the technical instruction [39]. In a paper presented at the Inter-Noise

Conference in July 1983, Martin [45] discusses the problem of what weather conditions can be considered as prevailing: "Downwind conditions are preferred for the noise measurements, because the results show a good reproducibility and the level of the noise from the source under consideration is the highest of all the wind directions. A widespread practice by the Lander requires that the arithmetical average of three measurements under downwind conditions, minus 3 dB(A), has to comply with the limit. The subtraction of 3 dB(A) is nowadays sometimes interpreted as a correction to determine the long term (i.e., the yearly) average of the rating level, L_r, over all weather conditions, from the downwind data". Furthermore, it was expected that a revision of the Association of German Engineers draft guideline on outdoor sound propagation [46] could be ready for publication in 1984.

E. Greece

At present, no legislation exists specifically on external industrial noise limits in Greece.

F. Republic of Ireland

At present, no legislation exists specifically on external industrial noise limits in the Republic of Ireland. However, the Planning Acts [47] enable the planning authorities to specify conditions relating to noise control in the granting of planning permission. Section 39(c) of the Local Government (Planning and Development) Act of 1976 amended Section 26(2) of the 1963 Principal Act by the insertion of an extra paragraph (bb) after paragraph (b) in the 1963 Principal Act:

(bb)   conditions for requiring the taking of measures to reduce or prevent

(i)   the emission of any noise or vibration from any structure comprised in the development authorised by the permission which might give reasonable cause for annoyance either to persons in any premises in the neighbourhood of the development or to persons lawfully using any public place in that neighbourhood, or

(ii)   the intrusion of any noise or vibration which might give reasonable cause for annoyance to any person lawfully occupying any such structure.

G. Italy

At present, no Italian legislation exists specifically on limiting values for noise emanating from industrial installations.

H. Luxembourg

At present, no legislation exists specifically on limiting values for noise emanating from industrial installations in Luxembourg.

I. The Netherlands

Chapters IV and V of the Noise Abatement Act [48] came into force on 1 September 1982. Chapter IV, Section 16, states that "categories of establishments which may be a major source of noise nuisance must first be designated by General Administrative Order". One such order was made on 15 October 1982 under Section 16(1).

By virtue of Chapter V, Section 46, the maximum permissible "noise load" from an industrial estate on the walls of dwellings within the specified zone shall be 50 dB(A). The "noise load" is defined as the 24-hr value of the equivalent noise level in decibels(A), occurring at a certain place and caused by the combined installations and appliances on the estate (not including the noise of motor vehicles on the estate, which do not form part of the installations). However, by virtue of Section 47, the provincial executive (subject to rules laid down under the order) may set a limit for the maximum permissible "noise load" which is higher than 50 dB(A), provided that the limit does not exceed 55 dB(A) or 60 dB(A) in the case of existing dwellings or those under construction.

J. United Kingdom

The Department of the Environment in the United Kingdom has issued Circular No. 10/73 [49] which gives guidance to local planning authorities as to the conditions they may properly impose when granting planning permission under the Planning Act [50]. Limiting values may be specified for the maximum noise level in decibels(A) at the site boundaries within specified hours from Mondays to Fridays (or Saturdays) and a maximum noise level at any other time. In practice, however, local planning authorities often prefer to lay down conditions requiring a factory to be insulated against noise in accordance with a scheme approved by the authority. The use of British Standard BS 4142 [51] as a basis for deciding on the acceptability of new industrial development has been criticised by Christie [52] and Hay [53]. Its most serious shortcoming lies in relation to the prediction of the background level. The notional background level (as calculated by BS 4142) will overestimate the actual background level in most situations. It is suggested [53] that BS 4142 could be withdrawn and regulations made by the Secretary of State for the Environment for limiting the level of noise which may be caused outside a factory.

The secretary has the power to do this under Section 68 of the Control of Pollution Act 1974 [54], but so far has not exercised it. Such regulations may apply standards laid down in documents not forming part of the regulations. Suitable standards could be based on the first part of the new International Standard ISO 1996/1-1982 [35] as well as the second and third parts, which are due to be finalised in April 1985 and published in 1986. By contrast, regulations [55] have been made under Section 64(8) of the Control of Pollution Act 1974 [54] with respect to the methods to be used by local authorities when measuring levels of noise from premises of a specified class, in a "noise abatement zone". When the noise source is more than 50 m from the measuring position, a positive wind component of up to 2 m/sec towards the measuring position is desirable.

III. PROPOSED DIRECTIVE ON NOISE EXPOSURE IN THE WORKPLACE

On 27 November 1980, the European council adopted the "framework" directive 80/1107/EEC [56] on the protection of workers from the risks related to exposure to chemical, physical and biological agents at work. Articles 3 and 4 of the directive 80/1107/EEC provide for the possibility of laying down limit values and other special measures in respect of the agents being considered. Noise is an agent to which the provisions of the directive 80/1107/EEC apply. Accordingly, on 18 October 1982, the European commission submitted to the council a proposal for a directive COM(82)(646) [57] on the protection of workers related to exposure to noise at work. This proposal was amended in 1984 [57a], and the main amendments are outlined below.

Article 4(1) states that the peak sound pressure to which the ear of the worker is subjected at work must not exceed the limit value of 200 Pascals (i.e., 140 dB in relation to 20 micro-Pascals). Article 4(2) now states that the daily sound exposure level $L_{EX,d}$ to which the ear of the worker is subjected at work must not exceed the limit value of 90 dB(A). This is expressed by the formula

$$L_{EX,d} = 10 \log_{10} \int_0^T P_A^2(t) \, dt/E_0,$$

where $E_0 = 1.152 \times 10^{-5}$ Pa^2 sec, T is the actual daily period spent at work in seconds, and $P_A(t)$ is the instantaneous level of A-weighted sound pressure in Pascals.

Article 6 states that where the worker's exposure to noise is likely to exceed a peak sound pressure level of 200 Pascals and/or a daily sound exposure level of 85 dB(A), then their exposure must always be reduced as far as reasonably practicable by means of technical or organisational measures. But where their application does not achieve compliance with the limit values stated in Article 4, and exposure cannot reasonably be reduced by other means, hearing protectors must be used to achieve compliance.

Article 9(1) now states that health surveillance shall be carried out on those workers exposed to a daily sound exposure level in excess of 85 dB(A), no account being taken of hearing protectors. This surveillance shall include an audiometric examination.

Article 12(2) now states that where any article (too, machine, apparatus, etc.) intended for use at work causes, when properly used, exposure which is likely to exceed the limit values laid down in Article 4, the designer, manufacturer, importer or supplier of the article must take the necessary steps to make available adequate information on the maximum noise likely to be produced in standard conditions when so used. No implementation date for this Council directive has been fixed.

A. Implications for U.K. Industry

The Health and Safety Commission (HSC) is collecting comments on the proposed European community directive COM(84)426 [57a] on noise in the workplace, from the Confederation of British Industries, the Trade Union Congress and other interested parties. The U.K. government will then determine their attitude to the proposed directive in the light of the HSC's advice.

IV. COMPARISON OF NATIONAL LAWS ON NOISE EXPOSURE IN THE WORKPLACE

The maximum permissible noise level for continuous exposure to noise for workers has been reduced by law in Germany to 85 dB(A) L_{eq} for an 8-hr day. However, it remains at 90 dB(A) L_{eq} by law in Belgium, Denmark and France for exposure over a 40-hr week. (See Table IX.) The Irish Republic has a limit value of 90 dB(A) by law, but no mention is made of duration of exposure in the regulations [62]. The Netherlands [63] have laws forbidding a person younger than 18 years from carrying out work whereby the sound level in the auditory canal (without hearing protectors) exceeds 90 dB(A). Five out of the 10 EEC member states still do not specify the maximum noise exposure at the workplace by law for adult

TABLE IX. Laws Specifying the Maximum Noise Level at the Workplace in EEC
Member States

EEC member state	Continuous noise maximum permissible level [dB(A)]	Maximum exposure [hr]	Trade-off per halving exposure [dB(A)]	Impulse peak SPL [dB]	Maximum number of impulses per day	Law [reference number]
Belgium	90	40	5	140	100	[58]
Denmark	90	40	3	—	—	[59]
France	90	40	3	—	—	[60]
Germany	85	8	3	—	—	[61]
Greece	—	—	—	—	—	None
Irish Republic	90	—	—	—	—	[62]
Italy	—	—	—	—	—	None
Luxembourg	—	—	—	—	—	None
The Netherlands	—	—	—	—	—	None
UK	—	—	—	—	—	None

workers. The nations lagging behind are Greece, Italy, Luxembourg, the
Netherlands and the United Kingdom. However, if the European council
adopts the proposed limit of 90 dB(A) for the daily sound exposure level
[57], all the member states will have to produce legal regulations to eventually comply with this limit.

REFERENCES

1. European Economic Community (1957). "Treaty Establishing the European Economic
   Community." Her Majesty's Stationery Office, London.
2. Council directive 80/51/EEC (1979). *Off. J. Europ. Commun.* **L18,** 26–28. Council directive 83/206/EEC (1983). *Off. J. Europ. Commun.* **L117,** 15–17.
3. International Civil Aviation Organisation (ICAO) (1981). "International Standards and
   Recommended Practices—Environmental Protection: Annex 16/5 to the Convention on
   International Civil Aviation", Vol. I, "Aircraft Noise", 1st ed. ICAO, Montreal.
4. Council directive 79/113/EEC (1978). *Off. J. Europ. Commun.* **L33,** 15–30; Council
   directive 81/1051/EEC (1981). *Off. J. Europ. Commun.* **L376,** 49–55.
4a. Council directive 84/532/EEC (1984). *Off. J. Europ. Commun.* **L300,** 111–122.
5. Council directive 84/533/EEC (1984). *Off. J. Europ. Commun.* **L300,** 123–129.
6. Council directive 84/536/EEC (1984). *Off. J. Europ. Commun.* **L300,** 149–155.
7. Council directive 84/535/EEC (1984). *Off. J. Europ. Commun.* **L300,** 142–148.
8. Council directive 84/534/EEC (1984). *Off. J. Europ. Commun.* **L300,** 130–141.
9. Council directive 84/537/EEC (1984). *Off. J. Europ. Commun.* **L300,** 156–170.
10. Proposal for a council directive COM(80)468 (1980). *Off. J. Europ. Commun.* **C356,** 3–
    10; Amendment COM(81)541 (1981). *Off. J. Europ. Commun.* **C302,** 7–10.
11. Council directive 78/1015/EEC (1978). *Off. J. Europ. Commun.* **L349,** 21–31.

12. Statutory Instrument No. 1166 (1980). "The Motor Vehicles (Construction and Use) (Amendment) (No. 6) Regulations." Her Majesty's Stationery Office, London.
13. Proposal for a council directive COM(84)438 (1984). *Off. J. Europ. Commun.* **C263,** 5–7.
14. Council directive 70/157/EEC (1970). *Off. J. Europ. Commun.* **L42,** 16–21.
15. Council directive 81/334/EEC (1981). *Off. J. Europ. Commun.* **L131,** 6–7.
16. Council directive 84/372/EEC (1984). *Off. J. Europ. Commun.* **L196,** 47–49.
17. Statutory Instruments Nos. 1401 and 1402 (1984). "The Motor Vehicles (Type Approval) (Great Britain) Regulations." Her Majesty's Stationery Office, London.
18. Council directive 77/212/EEC (1977). *Off. J. Europ. Commun.* **L66,** 33–34.
19. Statutory Instrument No. 1422 (1982). "The Motor Vehicles (Construction and Use) (Amendment) (No. 5) Regulations." Her Majesty's Stationery Office, London.
20. Council directive 84/424/EEC (1984). *Off. J. Europ. Commun.* **L238,** 31–33.
21. Council directive 74/151/EEC (1974). *Off. J. Europ. Commun.* **L84,** 25–32.
22. Council directive 77/311/EEC (1977). *Off. J. Europ. Commun.* **L105,** 1–9.
23. Statutory Instrument No. 2034 (1974). "The Agricultural (Tractor Cabs) Regulations." Her Majesty's Stationery Office, London.
24. International Civil Aviation Organisation (ICAO) (1979). "Committee on Aircraft Noise, Sixth Meeting. Can 6." Doc. 9286, 23 May–7 June, Montreal. ICAO, Montreal.
25. Proposed Council directive COM(81)554 (1981). *Off. J. Europ. Commun.* **C275,** 2–4.
26. International Civil Aviation Organisation (ICAO) (1983). "Committee on Aircraft Noise, Seventh Meeting. CAN 7", 2–13 May, Montreal. ICAO, Montreal.
27. Proposal for a council directive (1982). *Off. J. Europ. Commun.* **C181,** 1–29.
28. Amendments to the proposal for a council directive COM(83)694 (1983). Commission of the European Communities Documents, 18 November.
29. International Electrotechnical Commission, IEC Standard 704-1, (1982). Test Code for the Determination of Airborne Acoustical Noise Emitted by Household and Similar Electrical Appliances. Part 1: General Requirements, 1st ed. Bureau Central de la Commission Electrotechnique Internationale, Geneva.
30. European Committee for Electrotechnical Standardization, Harmonization Document HD 423-1 (1983). European Committee for Electrotechnical Standardization, Brussels.
31. Council directive 84/538/EEC (1984). *Off. J. Europ. Commun.* **L300,** 171–178.
32. Belgium (1973). Loi relative a la lutte contre le bruit.
33. Denmark (1982). Environmental Protection Act No. 663, 16 December.
34. Denmark (1974). Ministry of the Environment Guideline No. 3.
34a. Denmark (1984). Ministry of the Environment Guideline No. 5.
35. International Standards Organisation (1982). ISO 1996/1: "Acoustics—Description and Measurement of Environmental Noise. Part 1: Basic Quantities and Procedures"; ISO/DIS 1996/2: "Acquisition of Data Pertinent to Land Use"; ISO/DIS 1996/3: "Application to Noise Limits." To be published in 1986.
36. *Journal Officiel de la Republique Francaise* (1976). Loi No. 76-663 du 19 juillet relative aux installations classees pour la protection de l'environment; Loi No. 76-629 du 10 juillet relative a la protection de la nature; Circulaire No. 3055 du 21 juin.
37. L'Association francaise du Normalisaion (1974). NF S 31-010, "Mesure du Bruit dans une Zone Habitee en Vue de l'Evaluation de la Gene de la Population."
38. Federal Republic of Germany (1974). Bundes-Immissionsschutzgesetz vom 15 Marz, Bundesgesetzblatt IS.721,1193.
39. Federal Republic of Germany (1968). Technische Anleitung zum Schutz gegen Larm, vom 16 Juli, Beilage zum Bundesenzeiger No. 137 vom 26 Juli.
40. Federal Republic of Germany (1960). Gewerbeordnung vom 4 August, Bundesgesetzblatt IS.690.

41. Federal Republic of Germany (1962). Baunutzungsverordnung vom 26 Juni, Bundesgesetzblatt IS.429.
42. Federal Republic of Germany (1977). Baunutzungsverordnung vom 20 September, Bundesgesetzblatt IS.1764.
43. Verein Deutscher Ingenieure (1973). VDI 2058 Blatt 1: Beurteilung von Arbeitslarm in der Nachbarschaft. Kommission Larmminderung, Graf-Reche-Strasse, 84-4000 Dusseldorf.
44. Deutsche Normen (1977). DIN 45 645 Teil 1. Einheitliche Ermittlung des Beurteilungspegels fur Gerauschimmissionen. Alleinverkauf der Normen durch Beuth Verlag GmbH, Berlin 30.
45. Martin, R. (1983). *In* "Proc. Inter-Noise Conference, 13–15 July, Edinburgh." Institute of Acoustics, Edinburgh.
46. Verein Deutscher Ingenieure (1976). VDI 2714 Schallausbreitung im Frein.
47. Republic of Ireland (1976). Local Government (Planning and Development Act Section 39(c); (1963) Local Government Planning and Development) Act Section 26(2). The Stationary Office, Dublin.
48. Netherlands (1979). Noise Abatement Act No. 99, 16 February.
49. United Kingdom (1973). Department of the Environment Circular 10/73, "Planning and noise", 19 January. Her Majesty's Stationery Office, London.
50. United Kingdom (1971). Town and Country Planning Act Section 29(1).
51. British Standard 4142 (1967). Inc. Amendment No. S.1661, January 1975; 2956, May 1980; 4036, September 1982. Method of rating industrial noise affecting mixed residential and industrial areas. British Standards Institution, London.
52. Christie, D. (1982). *In* "Institute of Acoustics Conference Proceedings: Noise Nuisance Assessment", B.S. 4142 reviewed and criticised, 8 February.
53. Hay, B. (1982). *Appl. Acous.* **15,** 459–468.
54. United Kingdom (1974). Control of Pollution Act. Her Majesty's Stationery Office, London.
55. United Kingdom (1976). Statutory Instrument No. 37. "The Control of Noise (Measurement and Registers) Regulations." Her Majesty's Stationery Office, London.
56. Council directive 80/1107/EEC (1980). *Off. J. Europ. Commun.* **L327,** 8–13.
57. Proposed council directive COM(82)(646) (1982). *Off. J. Europ. Commun.* **C289,** 1–6.
57a. Amended proposal council directive COM(84)426 (1984). *Off. J. Europ. Commun.* **C214,** 11–16.
58. Belgium (1974). Arrete royal etendant la surveillance medicale des travailleurs confiee aux services medicaux du travail et modifant les tires II et V du reglement general pour la protection du travail, 10 Avril; (1972). Arrete royal relatif a la lutte contre les nuisances, 23 Mai.
59. Denmark (1975). The Working Environment Act, Law No. 681.
60. *Journal Officiel de la Republique Francaise* (1969). Decret No. 69-348, 12 April; (1971). Circulaire relative a la protection des travailleurs contre les effets nuisibles du bruit, 26 November; (1972). Arrete fixant les conditions et les modalities d'agrement des organismes habilites a proceder a des mesures d'intensite globale et des analyses spectrales des bruits en milieu de travail, 16 March; (1975). Arrete fixant une methode de mesure des niveaux sonores en milieu de travail en vue de la protection de l'audition, 12 August.
61. Federal Republic of Germany (1974). Unfallverhutungs vorschrift larm der berufsgenossenschaften, 1 December; (1975). Verordnung uber Arbeitstatten, 20 March.
62. Republic of Ireland (1975). Statutory Instrument No. 235; Factories Act No. 10.
63. Netherlands (1919). The Labour Act Section 10; (1972). The Labour Decree for Young Persons, Section 5.

Index